For Alberta!

FROM MIRACLE TO MENACE

 FriesenPress

Suite 300 - 990 Fort St
Victoria, BC, V8V 3K2
Canada

www.friesenpress.com

ISBN
978-1-5255-4516-0 (Hardcover)
978-1-5255-4517-7 (Paperback)
978-1-5255-4518-4 (eBook)

1. HISTORY, CANADA

Distributed to the trade by The Ingram Book Company

FROM MIRACLE TO MENACE

Alberta, A Carbon Story

DAVID YAGER

Table of Contents

Preface

"That's what is supposed to happen when the world decarbonizes by switching away from fossil fuels. Massive economic dislocation and disruption. Did someone forget to tell you that?"

The current debate about climate change and what mankind should do about it has deteriorated to the point of absurdity.

We're told repeatedly the planet's future is in peril and therefore we must change the composition of the atmosphere. This can only be accomplished if everyone on the planet uses less and ultimately no fossil fuels; coal, oil and natural gas.

The problem is that despite decades of warnings, there is no evidence carbon resource consumption is declining on a global basis, and there exists no workable plan to make this happen on an international scale or in the short timeline put forth by the so-called experts.

Worse yet, all data and forecasts indicate worldwide demand for fossil fuels will continue to increase for the foreseeable future, driven by economic and population growth. Because no matter what climate crusaders claim, there are currently no practical substitutes for the majority of carbon's myriad of products and energy sources, things most people regard as essential.

The divide between what we're told must be done and what we're doing is staggering. It's as if people live in parallel universes where we all occupy the same physical space, but our brains are on separate planets where emotions and feelings have replaced facts.

The climate-concerned have successfully convinced hundreds of millions of people that the world is on a one-way path to disaster unless massive changes occur immediately. Those who warn and worry about the earth's future believe changing this is just a matter of stronger collective political will.

But when you and I and everyone else fully understand how dependent we have become on fossil fuels and what life looks like without them, nothing happens. Why? Because faced with the reality of life without cheap energy and the thousands of products derived from carbon resources, the vast majority conclude the cure is worse than the disease.

The challenge is fundamentally economic. It is simply not possible to focus on the future when, out of necessity, so many remain focused on surviving today and tomorrow. This applies to billions of people worldwide, and not necessarily your friends and neighbors.

This is not to say climate change isn't a problem, or that carbon resources are not a major cause. I'm not a climate change denier. In my lifetime I've seen the giant glacier at the headwaters of the Athabasca, North Saskatchewan and Columbia rivers recede significantly. If and when it is gone, life in western Canada will be much different.

But if this problem is as serious as so many believe and we want to get something useful accomplished, we must change the channel and have a more intelligent debate than the one currently underway. Because right now this is going nowhere, even backwards.

Being in the middle of the carbon/climate debate all for over 30 years, I decided to write this book.

In Canada, the direction of climate policy and the country has become a virtue-driven preoccupation about the actions of one-half of one per cent of the world's population representing 1.6% of its emissions. The Canadian debate ignores the needs, aspirations and sheer mass and might of the world's other 7.36 billion people.

Those concerned about climate change believe their local governments can and should do something; that taxes, policies, programs and even lawsuits will somehow make a difference. But we share a world where billions of other people want cheap energy and all the wonderful technological and lifestyle advances that come with it.

Millions of Canadians are convinced Canada can help change the global climate by undertaking actions that ultimately inflict economic misery on millions of their fellow citizens. Unfortunately, the main result is growing hostility and polarized positions within the country.

Meanwhile, the greenhouse gas content of the global atmosphere continues to rise, oblivious to platitudes, speeches, policies, litigation and outright anger growing across the country.

My home province of Alberta has become ground zero for the Canadian version of the war on climate change. The third largest deposit of crude in the world—Alberta's oil sands—has been singled out and vilified as a major contributor to the problem. Therefore, blocking expansion and reducing production have been identified as solutions.

The so called "tar sands campaign" has been very successful. By obstructing export pipelines required for market access and branding oil sands as public enemy number one for climate destruction, Alberta has gone from boom to bust in only five years.

Which is fine. That's what is supposed to happen when the world decarbonizes by switching away from fossil fuels. Massive economic dislocation and disruption. Did someone forget to tell you that? Surely nobody is surprised that millions of people will lose their jobs, homes and companies in the quest to save the planet.

Except this is only happening in one place; Alberta. World fossil fuel demand grows unchecked and not one other major oil and gas producing jurisdiction in the world is enduring Alberta and Canada's economic misery.

In targeting Alberta and its oil sands, the anti-carbon movement has won a battle but lost the war. Oil not supplied from Alberta will come from somewhere else unless demand declines. Global events are proving that forcing people to change won't work either.

What follows is a multitude of facts and statistics from a multitude of sources organized in a manner that I hope will inspire the thoughtful and open minded to think differently about the climate change challenge, and ask all stakeholders to do something materially different than what we're doing today.

I'm an Alberta carbon creation. Besides being one of trillions of the Earth's "carbon-based life forms," my entire working life has been dedicated to finding and producing carbon energy. As the old bumper sticker goes, "Oil Feeds My Family And Pays My Taxes."

My great grandparents on both sides emigrated to Canada in the late 1800s. They were "sodbusters," leaving an ethnically German settlement in Russia and Sweden with one-way tickets to the New World. They took the government offer of a quarter section of (almost) free land in Manitoba and a chance to build a new life and a new country.

After being raised on the family homesteads, my parents met in Winnipeg 1942 and married in 1943. Mom worked at an explosives factory during WWII doing her bit as a Canadian "Rosie the riveter." My father trained as an accountant. Limited wartime job opportunities and health issues convinced Dad to move to a warmer climate. In 1944 they relocated to Vancouver.

No jobs there either so Dad went to Calgary to find work. In late 1944 my twin older brothers were born in Vancouver in Dad's absence. He was hired in inventory administration, which paid, as he wrote in his family memoirs, "a huge sum of $26 per week."

The war created a housing shortage. But in 1945 he found a three-bedroom house with a wood-burning furnace and stove, spartan furnishings, an outhouse, and no running water. There was a standpipe about 100 yards away on the next street. I'm here, so they obviously got by.

The discovery of oil at Leduc in 1947 changed the province. Dad wrote, "… and the oil boom began. Oil companies from the United States began to move in and brought with them their own senior employees. Also, a great many local oil development companies were formed and, of course, support industries such as drilling, service and supply companies followed, and some being locally owned. The oil industry soon became a major employer in Alberta, and the wages were generally higher than those paid by other companies."

In 1949 he went to work for a Houston-based oil service company as an accountant. He retired as president of that company's Canadian division in 1982. He got first "oilpatch" bonus in 1953. Dad wrote, "Imagine, age 36, and finally got a car, a 1950 Pontiac. $800 was a big bonus in those years. An excellent salary then was anything over $750 per month," seven times what he earned nine years earlier. I arrived in 1953 and my sister in 1954.

Because of carbon and the good fortune of being born in Alberta, from my youngest memories we had a nice house, a new car in the driveway and took regular family vacations. Having come from nothing my parents never took their economic success for granted. Thanks to carbon energy the Yager family was, as I would learn later, firmly planted in Alberta's growing middle class. Lucky me.

Shortly after high school I dropped out of university and worked in the oilfields for years. Like hundreds of thousands of others, I've toiled in the middle of night in the middle of winter in the middle of nowhere. In my corporate career I employed hundreds of oilfield workers and because of my personal experience, I relate to them all. This has left me with an indelible soft spot for those who, to paraphrase Robert Service, "moil for oil" to pay their bills and feed their families.

I've tried for years to put a face on this business. I'm still trying today.

———————————

The book relies extensively on the writing and thoughts of others for two reasons.

First, of all the quotes and references from dozens of writers and authors, only a handful have worked in the oil industry. Because nowadays, virtually all voices from the fossil fuel business that question the accepted climate change narrative are immediately branded as conflicted. Therefore, most of the sources are academics, scientists, politicians, journalists and authors.

Second, thanks primarily to Alberta, Canada is the fifth largest oil and gas producing jurisdiction in the world. It is a huge business and financial contributor relative to the total size of the Canadian economy. Canada's carbon resources have been of massive benefit to the creation of the country as we know it today. There's a lot of numbers. To achieve something useful, the climate debate must include economics along with the endless supply of warnings, platitudes and pontification.

No country of Canada's size, geography and climate would exist in its current form without cheap and plentiful coal, oil and natural gas. The west's vast quantities of coal greatly reduced the cost of building and operating the transcontinental railroad. Without coal, I wonder if Canada would exist it all. If so, what would it look like?

There are hundreds of people who have helped me understand complex energy policy issues and shaped my views on how and why things work. For this book and in no particular order I thank Jim Dinning, Rod Sykes, Dunnery Best, Trevor Tombe, Jack Mintz, Bob MacLean, David Collyer, Hossein Hosseini, Michael Kelly and Dennis McConaghy for their time, thoughts and support. I have quoted several and consulted the rest. There are many more too numerous to list.

My brother Barry Yager, who grew up in Calgary but now lives in Montreal, provided continuous perspective on how to write about this subject in a way which will hopefully engage people who strongly believe in the climate change challenge but whose only exposure to fossil fuels is as a consumer of the products.

The person who has done the most to shape my thinking about business and economics throughout my writing career is author Peter Foster. We met in Calgary when he was writing his first book, *The Blue-Eyed Sheiks,* in 1979. For 40 years

Peter has been an inspiration, a mentor and a friend. Most importantly, Peter has taught me most of what I know and believe about economics and has provided tremendous insights into human and political behavior.

The person who shaped everything else is Alessandra, my wife for over 35 years and my best friend for 38. While this is my first book, it is not my first crusade. All my life I have attacked whatever I was doing next with energy and enthusiasm, whether it was family, business, politics, skiing, family or even hobbies like woodworking. Alessandra has been by my side offering advice, support and encouragement. This book is no different. The only person happier than me to see it in print is her.

A special thanks to Don (Mac) Wishart and his wife Wendy. "Mac" is a friend dating back to grade school and a retired pipeline executive. With Alessandra, in retirement from day-to-day employment the four of us travel. During our excursions Mac and I spend hours talking about business and politics. Thanks to Wendy and Alessandra for enduring our many discussions and period outbursts of utter frustration.

The book took a lot of editing my many people. The editors at Friesen Press have been very helpful. But special thanks go to my sister, Susan Andre, a ferocious editor and career student and defender of the English language.

Thanks to the folks at Colborne Communications Ltd. of Toronto who helped guide me through the world of copyrights and copyright permissions.

Every attempt has made to get every fact right and every quote properly reproduced and attributed. This story simply needed to be told, not modified or embellished.

All monetary figures are in Canadian dollars unless otherwise noted.

Introduction

AS CLIMATE CHANGE CONCERNS GROW, ALBERTA FACES AN UNCERTAIN FUTURE

It is the hot, dry, flaming, and smoky summer of 2018 and the world is cranky. Forest fires rage in Canada, the US, and Europe. Hot weather is blamed for fatalities on several continents. September's Hurricane Florence on the US east coast and Super Typhoon Mangkhut in Southeast Asia got the same publicity, and both occurred for the same reason. Commentators were virtually unanimous: the weather is the worst it has ever been because of climate change.

Climate change warnings have been growing for years. Many of the media reports carried the same theme. If only politicians had "done something" years ago, it wouldn't be like this. The implication is not only that mankind can change the weather, but it already has. If decisive action isn't taken immediately, every problem we have with the weather today will be much, much worse in the future.

Carbon is the fourth most common element in the universe by mass. Human beings are 20% carbon. It is everywhere and in everything. Because they contain the carbon component of dead plants and animals, coal, petroleum, and natural gas are called "fossil fuels." Scientists claim their use is the main driver of climate change today. The carbon energy that powered the Industrial Revolution, and the incredible evolution to society as we know it today, has gone from miracle to menace.

Therefore, the developed world has increasingly declared war on carbon. To prevent further and more severe climate change, in October 2018 the Intergovernmental Panel on Climate Change (IPCC) warned the world yet again of the impending disaster of continuing down the present path. Carbon emissions must be cut by 45% by 2030, only 11 years from now. Otherwise, volatile weather events caused by climate change would most certainly cause even greater devastation.

What then for Alberta? Can an economy and society that took a century to build retool itself in a decade?

Finding, extracting, processing, and consuming carbon resources is a huge global business. Driven primarily by Alberta, Canada is the fifth-largest combined oil- and gas-producing country in the world. The carbon-based energy sector is 10% of Canada's Gross Domestic Product. Depending how you measure it, the carbon-resource business is bigger than agriculture, forestry, mining, and the automotive industries combined.

Alberta's massive, essential, and valuable oil, natural gas, and coal resource industries have faced challenges before. But they have never been threatened with legislated and moral extinction for mankind's greater good. We're continually told the negative economic consequences of keeping it in the ground are insignificant compared to the benefits. For thirty years, the growing chorus has been that if we don't stop putting carbon in the air, the only possible outcomes are bad, awful, and worse.

But it is glaringly obvious to anyone who thinks this through that we can't live without it. Carbon-based energy and by-products have become as essential to modern civilization as water, food, and air. Those who espouse a future without carbon energy or the myriad of derivative consumer and industrial products are preaching the impossible without the creation of alternatives that do not currently exist.

Advice about how our coal, oil, and gas industries should operate is not new. Over the years these resources have faced market access challenges, foreign owner-ship restrictions, "fairness" in pricing and revenue sharing, high taxes and royalties, jurisdictional conflicts, protectionism, export controls, mountains of regulations, environmental controls, and general disdain, derision, and distrust.

But now carbon is a game-changing threat. Hardly a day passes without this issue in the news. In a society that has always been preoccupied with the weather, today every bad climatic event is unquestioningly blamed on climate change. There are few suggestions of possible positive outcomes.

Low-cost carbon energy has fuelled economic and population growth for centuries and improved the standard of living of billions. But with the miracle now the menace, the public debate is dominated by carbon taxes, carbon protests, carbon offsets, carbon politics, carbon conferences, carbon alternatives, carbon guilt, anti-carbon lobbyists and, most recently, carbon litigation. All leading to an impending carbon catastrophe. Many say it is already too late.

It gets worse. The basic scientific issue of how the atmosphere's composition affects the climate has become intertwined with investment, employment, economic growth, politics, politicians, non-government organizations, environmental crusaders, jurisdictional sovereignty, Indigenous land claims, the internet, social media, fake news, anti-business agendas, and even religion.

It is inconceivable that there exists a future for carbon that will satisfy everyone.

The financial impact of the anti-carbon movement has already damaged the Alberta and Canadian economies. In the past 10 years fierce and successful opposition by environmentalists and sympathetic politicians to new oil export pipelines and liquified natural gas (LNG) terminals has curtailed investment. Carbon development capital that once flowed into Alberta is leaving. Canada has the lowest market-set oil and gas prices in the world. Investor confidence has been crushed by political decisions that have raised taxes, lowered commodity prices, expanded regulation, banned crude oil tankers, and cancelled pipelines.

Regulatory paralysis has made Canada a joke among international energy investors. Current federal government plans to fix the pipeline review and approval process, Bill C-69, has left the remaining stakeholders—the ones stuck here—deeply concerned. The message to global capital markets is that Canada wants no part of expanding the industry that helped make the country what it is today, and in many ways the envy of the world. One or two new high-profile projects will not reverse this perception.

A breaking point occurred in the fall of 2018 when domestic oil prices collapsed, creating a made-in-Canada financial and political crisis. The spot price for Alberta crude fell to a fraction of international levels costing producers and all stakeholders from workers to governments billions. It was caused by a lack of pipelines, insufficient market access options, increased supply from recently completed multi-year oil sands projects, and the temporary shutdown of US refineries.

The tall foreheads at the IPCC should be pleased. They tell us repeatedly this must happen all over the world, and soon. Oil's market value plummets and production is stopped because the cost to produce it is greater than the market will pay. Drilling rigs are racked, programs cancelled, budgets slashed, and layoffs follow. That's what happened in Alberta.

The reaction was unprecedented; mass public protests were held by oil workers demanding pipelines and fair treatment from the rest of the country. Thousands of people took to the streets in downtown Calgary. Thousands more driving truck-mounted surplus oilfield equipment obstructed roads and highways in operations centres like Drayton Valley, Grande Prairie, Nisku and Medicine Hat. A convoy drove all the way to Ottawa to plead for attention and fairness.

Oil workers were frustrated that their jobs, and families and companies were being targeted in the name of climate change by people who believed they were "doing something" by opposing Alberta oil. But these same critics were not prepared to make any personal sacrifices or change their own fossil fuel consumption patterns. Worse, if the oil they needed didn't come from Alberta, they would gladly buy it from somebody else without making any environmental demands whatsoever on other suppliers.

In response to the planet's lowest oil prices, the Alberta government took the controversial step of ordering the withdrawal of 325,000 barrels of oil production per day from the market commencing in January of 2019.

Into this emotionally charged environment the new premier of New Brunswick, Blaine Hicks, raised the issue of revisiting the cancelled Energy East pipeline to supply his province's huge refinery with Alberta crude to replace foreign imports. In response and clearly oblivious or indifferent to the economic catastrophe in Alberta, the new premier of Quebec, François Legault, told the country a new oil pipeline in his province was not socially acceptable and the country should buy more of Quebec's clean hydro, not Alberta's "dirty oil".

Saving the modern world from the fossil fuels that created it will be a nasty piece of business, and it will be accompanied by massive financial and worker disruption that will ultimately affect the lives of billions. The future is already unfolding in Alberta. And only Alberta.

ABOUT THIS BOOK

This book is the story of carbon and its impact on Alberta and Canada—past, present, and future. It's about the people, the economy, the governments, and Canada's relationship with the rest of the world. It's about how enormous deposits of carbon in a once sparsely populated part of the country changed everything, as well as the incredibly diverse strategies, policies, and public opinions on what should happen next.

Carbon resources have caused some of Canada's most divisive political battles at different times for different reasons. Last century it was security of carbon supply, ensuring consumers never ran out of affordable energy. This century these same resources have been declared a threat to the future of the world because supplies appear limitless and don't cost enough.

Every time carbon becomes a political issue, the public demands that governments "do something." But regardless of what the public wants to hear, Canadian governments are powerless to solve or influence global problems. Despite continuous expressions of deep concern by almost everyone in or campaigning for public office, there is no evidence that Canadian governments alone can ever solve this problem.

At the same time, the internet and modern media have dramatically altered communications and attitudes. Public policy used to reflect what people needed. Now it is driven by what they want and how they feel. Facts no longer matter. There is no other way to explain the seemingly altruistic but utterly disingenuous demands for the immediate end of fossil fuels in the absence of zero-carbon substitutes for most of their applications.

Now what?

This book is not about the theories, research, and computer models supporting how the composition of the atmosphere has changed and will continue to change the climate. After all, "the science is settled." The carbon-menace ship has sailed. The question is where will it moor next?

Nor is this book a commercial for the benefits of coal, petroleum, and natural gas. They are so essential and ingrained into our lives that people don't even understand how integral they have become to life in the twenty-first century. The real issue is the absence of practical alternatives.

But as beneficial as carbon resources have been to Alberta and Canada, decarbonization will have the exact opposite effect.

The book is divided into three sections: carbon development, carbon politics, and carbon future. It concludes with what makes sense and some realistic common ground between two highly polarized positions. There is a solution that will make a difference and Alberta can be a huge part of it. But it is not considered or even mentioned in today's climate policy debate.

FROM MIRACLE TO MENACE

Carbon energy has powered civilization to unprecedented heights in only 200 years. More people than ever are living better than at any time in history. But in this century there have been major changes in what people and governments believe about fossil fuels. The multi-pronged assault on Alberta's oil, gas, and coal industries is well underway, the damage is already done and getting worse and it shows no sign of slowing down. The anti-carbon movement's success will be Alberta's economic failure.

Famous environmentalists like David Suzuki tell us we must leave the oil sands in the ground. Decarbonization is essential. Climate change scientists have issued warnings about "what" is happening many times over. The issue now is "how" we fix it.

After 30 years of increasingly alarming predictions about the impending global climate catastrophe, ignoring or denying the issue will no longer work. Whether climate change predictions are right or wrong, malicious or benign, scientific or alarmist, the movement to reduce then eliminate the use of fossil fuels—the world's most plentiful, powerful, versatile, and economical energy source—continues unchecked. The carbon-menace genie will be impossible to stuff back into the bottle.

Alberta cannot solve this problem alone, nor can it be solved without including Alberta. Alberta must and will help.

Moving mankind from cheap carbon energy to something else will be the greatest financial dislocation in history. Because for every human being whose life might be better tomorrow, billions must pay more today for what we in the west consider the essentials of life. And hundreds of thousands of people in the carbon-resource industries must lose their jobs, their houses, or their companies.

Climate change is no longer about warnings, theories, protests, virtue signalling or endless political expressions of concern. In Alberta it is about real people, real money and real suffering. In Canada, the future is here today, and it is ugly. But for very different reasons than unpredictable or extreme weather.

There will be no progress without a better understanding of the real issues and a complete rethinking of how the world's transition away from fossil fuels can realistically take place in a way that ensures that the cure isn't worse than the disease.

Section 1

CARBON DEVELOPMENT

Introduction
THE EARLY HISTORY OF CARBON DISCOVERY

"As civilization progresses, the pivotal role played by carbon is forgotten, ignored, or not fully understood."

There is so much carbon resource in the world in the form of coal, oil, and natural gas, discovering it was simple.

The easiest carbon-rich resource to discover was coal. Most often jet-black, coal was easy to differentiate from the rock around it and was found either on the ground or exposed in sloped terrain like a hill, cliff, or river valley where erosion or geological thrusts had exposed multiple layers of otherwise buried rock. There are reports that coal was mined and used in China as early at 3490 BC. Carvings made from coal date back another 3,000 years. Coal's use as a heat source to work metal goes back to 300 BC. In Wales, coal was used for cremation 2,500 years ago. Coal was mined by the Romans in Britain in the second century AD, and it was used for multiple domestic and industrial purposes in Germany and Britain beginning in the twelfth century.

The Industrial Revolution—which involved boiling water to create steam to drive powerful engines, which replaced people and animals of husbandry to move and lift things—was a major breakthrough in the development of modern industry and civilization. While wood worked, coal was much more powerful, generating more heat from much smaller quantities. Think progress? Think coal!

Oil was discovered by surface seepages whereby liquid petroleum from subterranean reservoirs found its way to the surface by following fissures and cracks in the earth. It was also black or dark brown, a liquid or a tar, and clearly much different than the surrounding dirt or rock. Noticing it was very easy, and this occurred in many places around the world. Historical research indicates Neanderthal man was using bitumen—heavy oil or naturally occurring asphalt—as long as 70,000 years ago. By the fifth century BC, natural seepages of petroleum were ideal for waterproofing boats made of natural materials such as reeds. In construction it was employed as an adhesive and sealing compound for the earliest indoor plumbing.

Oil was used for illumination and mummification in ancient Egypt. There are many more examples of the ancient uses oil for a multitude of applications, including medical treatments.

Natural gas or methane seepages were also common but nowhere near as easy to identify. Methane in its purest form is odourless and colourless. But a lightning strike or other accidental spark or flame would ignite it. It was probably not obvious to early man that the terrifying discovery of flames emitting from the earth from an unknown source would ultimately be a huge asset, not a mystifying liability. By 900 BC the Chinese were burning natural gas from wells (man-made holes in the earth) to accelerate the evaporation of seawater to obtain salt. The Romans knew what natural gas was and what it could be used for. There are stories from early Russia where churches were built around flaming natural gas seepages, so worship could take place in close proximity to an "eternal flame."

By the time European explorers ventured across the Atlantic to explore the New World, the three major sources of carbon energy were known, and their usefulness understood. According to the *Canadian Encyclopedia*, Canada's first coal mine was opened in New Brunswick in 1639. Mining was underway at Cape Breton (Cow Bay) in 1720. This mine was soon exporting coal by water to Boston and supplied the militia in Halifax. Having both iron ore and high-quality coal in Nova Scotia allowed that province to begin producing steel in 1870. Nova Scotia coal and steel proved very useful when construction of the Canadian Pacific Railway across Canada began in 1881. In the First World War three-quarters of the coal supporting Canadian industry came from Nova Scotia.

The massive oil sands of northeastern Alberta were also known to Europeans in the early days of the formation and expansion of Canada. In his 1978 book *Black Gold With Grit: The Alberta Oil Sands*, J. Joseph Fitzgerald recounts the first oil sands exports took place in 1719 when an Alberta Cree fur trader brought a sample of the oil-laden sand to the York Factory Hudson's Bay Company trading post in what would later become northern Manitoba (Fitzgerald 1978).

Fort York manager, Henry Kelsey, had recorded in his official journal that a sample, "of that Gum or pitch that flows out of the Bank of the River" was in his possession. Fitzgerald concluded Kelsey's description referred to the oil seepages near Fort McMurray, on the banks of the Athabasca River. These deposits that were already known to Fitzgerald. Because the Hudson's Bay Company was headquartered and founded in Britain, the fact that oil leaked out of the ground in Canada was known throughout the British Empire, three hundred years ago (Fitzgerald 1978).

The first European to witness the seepages that would two centuries later lead to the development of the third-largest crude oil deposit in the world was Peter Pond in 1778.

In 1912 the Dingman #1 well began drilling alongside the Sheep River in what is now Turner Valley. The selection of that location was easy, as there were natural gas seepages—some of which were flaming—coming out of the river bank.

While Canada didn't yet exist as a country in 1842, that was the year the colonial administration created the Geological Survey of Canada (GSC) founded in what was known as The Province of Canada. Geologists were hired with a mandate to determine what minerals and potential wealth existed in the geographical and geological inventory of the territory. Enough exploration work had been done by 1881 that the GSC was able to advise governments and railroads on the geology of the terrain to be encountered on various routes.

The exploration and settlement of Alberta by non-Indigenous people began in the late 1700s. By the end of the century fur trading posts were operating in several locations on the banks of the South Saskatchewan River and its tributaries which, of course, flowed to Hudson Bay. Up the creek *with* a paddle is how they got there.

When immigration and settlement began to accelerate in the latter half of the 1800s and early 1900s, geology as a science was well advanced and the federal government was an enthusiastic supporter of turning rocks into products and money to advance the development and colonization of the country.

The explorers and settlers knew what to look for when it came to coal and oil and, as importantly, what to do when they found them. Thanks to earlier efforts, the setters and developers also already knew where it a lot of it was.

While the usefulness of carbon resources is generally understood—even if it must be periodically explained—the concept of carbon as a menace is relatively new and growing.

But before Canada takes any further steps to get out the fossil fuel business, everyone should understand what it is, how it got here, why it grew, where and how we need it, and how big a piece of the economy of Canada and Alberta it has become.

1

CARBON CREATION – HOW ALBERTA BECAME A GLOBAL CARBON WAREHOUSE

"Carbon dioxide, the source of carbon essential for plants and animals, has been identified as a major threat to the earth's climate if levels in the atmosphere continue to rise. But clearly too little carbon dioxide is a threat to the earth for different reasons. Can't live with it, can't live without it."

Alberta's carbon story starts about 300 million years ago when this part of the North American land mass was submerged by primordial oceans. The waters teemed with living things. One of the more common creatures was a tiny multi-celled animal called a "diatom," a creature related to today's plankton.

One day a baby diatom asked his father what the future held for him. While there was nobody taking notes or recording it on their cellphone, the conversation went something like this.

"Daddy, what am I going to be when I grow up?"

"Well son, it goes like this. When you die you will fall to the bottom of the ocean. Over time you and all your friends, family, and relatives will be buried, compressed, and heated. Millions of years from now you are going to turn into oil."

"Daddy, what's oil?"

"I'm not sure but I think it's going to be very, very valuable one day."

Encouraged to learn his life was not wasted, the baby diatom asked his father one more question.

"Daddy, where do we live?"

"It hasn't got a name right now but one day they will call it Alberta."

Carbon is an element found everywhere and in everything. It is number six on the periodic table of elements but is the fourth most common element in the universe, measured by mass, behind hydrogen, helium, and oxygen. Carbon is created when three helium nuclei merge through the "burning" process of nuclear fusion, which takes place on active stars. This process also creates neon, magnesium, and oxygen.

Britannica Online Encyclopedia[1] has a simplified summary of carbon, which is helpful because this is generally a very scientific subject. When comparing chemical elements found in the earth's crust, carbon ranks 19th in terms of weight compared to other elements. Carbon easily bonds with other materials to form common minerals such as magnesite, dolomite, marble, and limestone. Elemental carbon was discovered by the first human to touch the charcoal remaining from a wood fire. Carbon joins sulphur, iron, tin, lead, copper, mercury, silver, and gold as one of the elements first identified by ancient man.

Diamonds are also made of carbon, but it took some time for scientists to figure this out. French scientist Antoine Lavoisier first reached this conclusion in the late 1700s.[2] He and other chemists bought a diamond and put it in a closed glass jar. Using a powerful magnifying glass for heat, the rays of the sun were focused on the diamond and they watched it burn and disappear. But when the jar was weighed it was unchanged. The carbon in the diamond had bonded with oxygen to created carbon dioxide.

Burning charcoal and diamonds in a closed environment both achieved the same outcome—no change in weight. This is how it was determined charcoal and diamonds were made from the same element.

Mixed with sulphur and potassium nitrate, carbon in the form of charcoal was used to create gunpowder. This was first used by one man against another over 1,000 years ago. There are more compounds in the world that contain carbon than don't. Graphene, "the thinnest, strongest material ever known," is also carbon.[3] Car tires are black because they are about 30% carbon from carbon black, an additive that strengthens rubber and makes it resistant to ultraviolet radiation from the sun. Carbon is used in graphite for pencils, added to steel to make it stronger, and is the black pigment in ink.

Slightly radioactive carbon, carbon-14 (or ^{14}C), is today used for carbon dating. This was identified by scientists in the 1960s. Carbon-14 is stored by plants and

---

1   "Carbon," *Encyclopedia Britannica* (March 14, 2018). https://www.britannica.com/science/carbon-chemical-element.

2   "Carbon Element Facts," *Chemicool: Cooler than Absolute Zero!* (July 25, 2014). https://www.chemicool.com/elements/carbon.html.

3   Ibid.

passed on to animals when they eat the plants. It has become a fairly accurate method of determining the age of organic fossils. It has a short half-life of 5,730 years, the rate at which it decays. Its half-life is considerably more useful than that of uranium for dating fossils because uranium has a half-life of about 4.5 billion years.

On Earth, carbon and life are one and the same, which is why plants and animals are called carbon-based life forms. Carbon is key to the diversity and complexity of organic forms because of the ease with which carbon atoms bond with each other in different ring and chain structures in three-dimensional forms.[4] With more than a million different carbon compounds identified in chemistry literature, in the nineteenth century the study of carbon and life forms was given the special classification of organic chemistry. At that time, most carbon compounds were thought to have originated from living organisms.

Like most living things, the human body is comprised of about 20% carbon, extracted from the carbon dioxide in the atmosphere.[5] This is expanded by, "All organic compounds, such as proteins, carbohydrates, and fats, contain carbon, and all plant and animal cells consist of carbon compounds and their polymers. With hydrogen, oxygen, nitrogen, and a few other elements, carbon forms compounds that make up about 18% of all the matter in living things. The process by which organisms consume carbon and return it to their surroundings constitutes the carbon cycle."[6]

Thus, it is not surprising to learn, "carbon compounds form the basis of all known life on Earth, and the carbon-nitrogen cycle provides some of the energy produced by the sun and other stars."[7] The sun provides another key ingredient of carbon-based life: photosynthesis. This is defined as, "the process by which green plants and some other organisms use sunlight to synthesize nutrients from carbon dioxide and water. Photosynthesis in plants generally involves the green pigment chlorophyll and generates oxygen as a by-product."[8]

---

4   "Carbon," *Encyclopedia Britannica* (March 14, 2018). https://www.britannica.com/science/carbon-chemical-element.

5   "Carbon Element Facts," *Chemicool: Cooler than Absolute Zero!* (July 25, 2014). https://www.chemicool.com/elements/carbon.html.

6   "Carbon," *Encyclopedia Britannica* (March 14, 2018). https://www.britannica.com/science/carbon-chemical-element.

7   "Carbon," *Wikipedia* (December 10, 2018). https://en.wikipedia.org/wiki/Carbon.

8   "Photosynthesis," *Oxford Living Dictionaries* (2018). https://en.oxforddictionaries.com/definition/photosynthesis.

Carbon dioxide, the source of carbon essential for plants and animals, has been identified as a major threat to the earth's climate if levels in the atmosphere continue to rise. It is more frequently asserted the damage is already done.

But clearly too little carbon dioxide is a threat to the earth for different reasons. Can't live with it, can't live without it.

As the link between carbon energy emissions, global warming, and climate change has gained acceptance, one of the proposals to mitigate the problem is called carbon capture and storage (CCS). Carbon dioxide is an industrial emission created in large concentrations by burning fossil fuels for generating electricity, oil refining, or oil sands upgrading. From industrial sources, the carbon dioxide is captured (extracted) from exhaust gases and then stored underground. Pilot projects to determine both technical and commercial effectiveness have been conducted in multiple places around the world, including Alberta and Saskatchewan. CCS will be revisited in later chapters.

However, carbon capture and storage is also the fundamental process that created coal, natural gas, and crude oil in the first place. Living organisms, animals and plants, extracted carbon from the atmosphere or the ocean and, using energy from the sun, used it for growth. Carbon capture and storage is essential to the continuation of all life on Earth. Life on Earth is one massive CCS system.

Accumulations of massive quantities of dead plants and animals, combined with extreme climate change, are the conditions that created deposits of carbon-based energy, more commonly called fossil fuels.

Alberta's Department of Energy website provides a succinct explanation of how petroleum was formed.

> Crude oil is a naturally occurring mixture of different hydrocarbon compounds trapped in underground rock. These hydrocarbons were created millions of years ago when ancient marine life or vegetation died and settled into the bottom of streams, lakes, seas and oceans, forming a thick layer of organic material. Sediment later covered this layer, applying heat and pressure that "cooked" the organic material and changed it into the petroleum we extract from the ground today.[9]

---

9   "What is Oil?" *Alberta Energy* (2018). https://www.energy.alberta.ca/Oil/AO/Pages/WIO. aspx.

That oil-bearing reservoirs often contain fossils or that petroleum is sometimes trapped in ancient coral reefs has provided very strong evidence of oil's marine biological origin.

The creation of natural gas was similar but the outcome quite different. This explanation comes from the US Energy Information Administration website.

> Millions to 100's of millions of years ago and over long periods of time, the remains of plants and animals (such as diatoms) built up in thick layers on the earth's surface and ocean floors, sometimes mixed with sand, silt, and calcium carbonate. Over time, these layers were buried under sand, silt, and rock. Pressure and heat changed some of this carbon and hydrogen-rich material into coal, some into oil (petroleum), and some into natural gas.
>
> In some places, the natural gas moved into large cracks and spaces between layers of overlying rock. The natural gas found in these types of formations is sometimes called *conventional natural gas*. In other places, natural gas occurs in the tiny pores (spaces) within some formations of shale, sandstone, and other types of sedimentary rock. This natural gas referred to as *shale gas* or *tight gas*. Natural gas also occurs with deposits of crude oil, and this natural gas is called *associated natural gas*. A type of natural gas found in coal deposits is called *coalbed methane*.[10]

Coal is very different from crude oil and natural gas because it is solid as opposed to a liquid or a gas. However, its roots are the same.

> Coal is a solid, black, readily combustible fossil fuel that contains a large amount of carbon-based material – approximately 50% of its weight. The formation of coal takes a significant amount of time, with coal beginning to form 290-360 million years ago. As well, there are extensive coal deposits from … about 65 to 144 million years ago.

---

10 "Natural Gas, Explained," *US Energy Information Administration* (December 11, 2018). https://www.eia.gov/energyexplained/print.php?page=natural_gas_home.

The formation of coal begins in areas of swampy wetlands where groundwater is near or slightly above the topsoil. Because of this, the flora present produces organic matter quickly – faster in fact than it can be decomposed. In these areas, layers of organic matter are accumulated and then buried. It is these layers of organic material that then form coal. The energy in coal initially comes from the Sun, and is energy from sunlight trapped by dead plants.[11]

Today's deposits of coal, natural gas, and crude oil are dead, decayed, and long-buried life forms, both plants and animals. In their second exposure above the ground after being long buried, their combusted carbon (accumulated while alive) creates carbon dioxide in the atmosphere. But remember, this was the source of the carbon content of living things in the first place.

The Carboniferous period,[12] from 359 million to 299 million years ago, is when scientists believe the first extensive forests appeared on the planet. Part of the late Paleozoic era, it is named after today's huge coal deposits that were formed during this time.

When people look at Alberta, the last thing they see is a place fostering the abundant growth of plants and animals in sufficient quantities to create massive deposits of carbon in the form of coal, natural gas, or crude oil. Places that grow a lot of things—flora and fauna—are wet and warm and a lot closer to the equator than Alberta.

But the climate change we all hear so much about nowadays is not new. Climate change has been occurring as long as the earth has existed. The plot of land legally and geographically referred to as Alberta today was very different when the carbon deposits began to accumulate hundreds of millions of years ago. For the past 500 million years, Alberta has been submerged, drained, hot, cold, wet, dry, barren, and lush.

In relatively modern times, at least geologically, Alberta was buried by continental ice starting about 25,000 years ago. The only parts of Alberta not covered with ice were the peaks of the highest mountains, the Cypress Hills in the southeast

---

11 "Coal Formation," *Energy Education* (July 21, 2018). https://energyeducation.ca/encyclopedia/ Coal_formation.

12 "Carboniferous Period," *National Geographic* (2018). https://www.nationalgeographic.com/ science/prehistoric-world/carboniferous/.

corner of the province, and the Porcupine Hills southwest of Calgary. The Cypress Hills area has unique flora and fauna today for that reason.

The final retreat of the last major ice began 13,000 years ago. This is when the river systems and hills as we see them today were shaped. The Bow Valley through the Rocky Mountains is much different than the terrain around it because it was moulded by the force and erosion of receding and melting glaciers. The massive ice cap was finally melted by rising temperatures.

The last remaining glacial ice coverage in Alberta is in the mountains. These glaciers have continued to recede in the past 150 years. This has been well documented by explorers and more recently by photographers.

It is postulated today that the cause and end of this ice period from 25,000 to 13,000 years ago was related to a shift of the axis of the rotation of the earth. This would change the angle at which the sun's solar energy reaches various parts of the earth, resulting in regional climate change.

According to an article titled, *Ice Ages Blamed on Tilted Earth* from March 2005, the current angle of 23.5 degrees can vary by up to one degree in both directions in what scientists believe are 40,000-year cycles. They compared the timing of the fluctuations with the last seven ice ages. It states, "... the ends of those periods – called glacial terminations – corresponded to times of greatest tilt." Further, "The researchers speculate that the glacier period has become longer in the last million years because the earth has gotten slightly colder—the upshot being that every once in a while the planet misses a chance to thaw out."[13] The basis of the theories is studying the chemical composition of sediments dating back millions of years.

Scientists have also theorized the earth suffers from "polar wander" over millions of years, where the axis of rotation of the earth may have shifted by 30 degrees or even as much as 55 degrees. If true, this might be the reason for the clearly dramatic climate change that has occurred to create massive deposits of coal, natural gas, and crude oil in Alberta currently buried up to three kilometres or more below the surface of the earth.

It took quite some time for the future Alberta to again become a place where plants and animals could survive and flourish. But it might be another 65 million years before the area returns to its prior wet and warm life-generating and plant- and animal-growing climatic splendour.

---

13   Michael Schirber, "Ice Ages Blamed on Tilted Earth," *LiveScience* (March 30, 2005). https://www.livescience.com/6937-ice-ages-blamed-tilted-earth.html.

The study of post-ice-age Alberta is helped by the relative newness (in geological terms) of discovered mammal fossils and advances in carbon-14 dating techniques.

A leader in this field has been the Simon Fraser University Museum in BC. In 2003 the institution revealed it had discovered a massive mastodon tusk (2.8 metres [10 feet 6 inches] long, weighing 88 kilograms [183 pounds]) along the Yukon/Alaska border that carbon dating revealed to be about 25,000 years old. This dates back to the beginning of the last ice age. A theory has emerged there was an ice-free corridor of some kind that ran from Alaska to the southeast, towards central North America, extending through central Alberta. Further dating of buffalo fossils in and around Edmonton indicated this ice-free corridor opened up from the north and the south about 13,000 years ago, allowing buffalo to migrate from both directions. The buffalo fossils were found in Edmonton's Clover Bar area in the 1950s and 1960s.

It is also theorized there was also an ice-bridge across the Bering Sea, which permitted the migration of humans as the ice age ended and the ice-free corridor expanded. There is evidence that Indigenous communities populated both ends of the ice-free corridor as recently as 14,000 years ago.

The account of early settlement by humans has been well studied. In his book *A History of Alberta* by James G. MacGregor, first published in 1972, the opening sentence in the first chapter, titled "The First Albertans" reads, "Eleven thousand years ago on the banks of the Oldman River near modern Taber, a primitive hunter butchered a buffalo. He crushed the animal's skull with a crudely fashioned hammer which, when found recently, was still embedded in the bone" (MacGregor 1981).

MacGregor reviews the Bering Sea ice-bridge theory that the first humans crossed into what would later be called North American at late as 30,000 years ago, then migrated south through a temporary recession of the ice cap, which opened an ice-free corridor through Alberta and farther south. Other scholars believe this happened later. MacGregor writes that about the time of "Taber man," the ice was still very thick. Edmonton emerged from under the ice about 10,000 years ago. The land on which Fort Chipewyan sits today didn't see sunlight for another 2,000 years. At times the ice was 2,000 feet thick.

About 7,500 to 4,500 years ago, with the ice having retreated, the area was blessed with a period of warmer and drier climate called Altithermal. To ensure things would never be easy, this included a time when Canada's prairie provinces became too hot and dry to support significant populations of people. About 4,000 years ago the temperature began to moderate and attracted nomadic hunters living on buffalo on the plains and moose, deer, and caribou in the forests. There were four distinct groups of Indigenous people in Alberta when the first European— Anthony Henday on behalf of the Hudson's Bay Company—appeared in Alberta in 1754.

The Indigenous population of Alberta prior to the arrival of the first Europeans is unknown. The population of the entire North-West Territory in the 1891 census was 98,967. This area included all of what is now Alberta, Saskatchewan, the Northwest Territories, the rest of Manitoba not already declared a province, and a large portion of Nunavut.

By 1882 all modern-day Alberta needed was a railroad, some borders, and a name.

2

CANADA CREATES ALBERTA

"The attraction of Alberta was not the weather, water sports, more affordable housing, or better government services. As the famous line from Bill Clinton's 1992 US presidential campaign stated so clearly, 'It's the economy, stupid.'"

Unlike most political jurisdictions around the world, how the land mass called Alberta came into existence is boring in the extreme. It was created by elected politicians and surveyors, not tribal conflict, invasions or, with the exception of the Great Divide in the Rocky Mountains, geographic anomalies.

When Canada became a country in 1867, Alberta did not yet exist. It was still part of a large territory claimed by Britain. Britain gained undisputed sovereignty over what would become western Canada in two stages. The first was the Anglo-American Convention of 1818—also called the Treaty of 1818—which was signed on October 20 of that year. It decided all the lands from the western end of Lake Superior to the Continental Divide in the Rocky Mountains would be divided on the 49th parallel (49° north latitude). Lands to the north became Britain's, and those to the south, America's. The southern border of Alberta was set.

On June 15, 1846, Britain and the United States signed the Treaty of Oregon, which selected the 49th parallel as the border between American and British areas west of the Continental Divide. The lands under dispute were primarily on the Pacific Coast, which was more settled compared to the interior of the country. This agreement extended the 49th parallel to the Strait of Georgia.

The next major step took place in 1870, when the British government transferred control of the vast North-West Territory to Canada. The Hudson's Bay Company had sold its massive claim to Rupert's Land to the new country in 1868. Two hundred years earlier, in 1670, Britain granted the Hudson's Bay Company a commercial monopoly over all the lands in North America from which the rivers and streams flowed into Hudson Bay. This extended west to what would later be called the Continental Divide (the highest point in the Rocky Mountains from which water either flows west into the Pacific Ocean or east into Hudson Bay or the Arctic Ocean). Because of the Red River drainage system, this also included territory that would later become part of the United States.

Canada had great interest in settling the vast empty areas of the west to assert its sovereignty and thus dampen American interest in acquiring or settling it. A major step was the Dominion Land Survey, a federal government initiative, which began in 1871 to create an orderly geographical grid, which chopped western Canada into one square mile (2.6 square kilometres) blocks for agricultural settlement and development. What the survey would call the "first meridian" (different from the global meridian system) was actually at 97° 27' 28.4" west longitude. It was set just west of Winnipeg because at that time it was the western end of settlement. As the survey continued, five more meridians were added. Because of the enormity of the land mass and the relatively tiny parcels into which it was to be delineated, properly surveying the entire area took years.

This is the origin of the Legal Sub Division (LSD) system oil and gas mineral leases are leased by today. It is common vernacular in the oilpatch in Alberta to describe industry exploration and production activity as being "west of the fourth" or "west of the fifth," the fourth and fifth being meridians created by the Dominion Land Survey. Oil and gas deposits in Alberta are deeper from east to west. Therefore, references to the Dominion Survey meridians are a broad regional reference to production depth or cost.

Oil company CEOs who say, "Our core areas are located west of the fourth" simply mean they don't intend to drill very deep. If they explore "west of the fifth," they are searching for deeper, larger, and more expensive deposits.

In 1905 the population of the prairie provinces had risen to the point that the region deserved provincial status, in part to set up regional governments with sufficient autonomy that the huge area would no longer be governed exclusively from Ottawa. There was pressure from the increasing population to move more government power closer to the governed.

The British North America Act of 1867, or "Confederation" as it is more commonly known, involved former colonies merging into a single country. Former colonies, now called provinces, were given unique and specific powers under law. Manitoba became a province in 1870, but it was only a fraction of its current size. British Columbia abandoned colonial status and joined Confederation as a province in 1871 with its current boundaries. British Columbia's eastern boundary would run along the Continental Divide until the 120th meridian

(120° west longitude) then proceed straight north to the 60th parallel (60° north latitude), which would be the northern border. The eastern border heads due north from the Continental Divide at Intersection Mountain at 53°48' north latitude. Everything in between British Columbia and Manitoba was still called the North-West Territory, demarcated into the districts of Athabasca, Assiniboia, Alberta, and Saskatchewan.

Alberta's name comes from the wife of the Governor General of Canada who served from 1878 to 1883. Her name was Louise Caroline Alberta, the fourth daughter of Queen Victoria. Lake Louise, the village of Caroline, and Mount Alberta are also named after her. They certainly got good utility from her first, middle, and last names.

As the west was settled, pressure grew to give the region more autonomy to be able to make more decisions locally instead of in Ottawa. Frederick Haultain, at the time considered the "premier" of the North-West Territory, was of the view it should be one large province to offset the political power of the large provinces of central Canada, Ontario and Quebec. The proposed name of the new province was "Buffalo".

When the decision was made to create two more provinces in addition to Manitoba—Alberta and Saskatchewan—federal determination to make three provinces instead of one large province appeared to be political. Manitoba was of the view it could manage the entire region. Ottawa disagreed, preferring to retain the power of the larger provinces, Quebec and Ontario.

All of this would become extremely important 70 years later when carbon-based energy would become one of, if not the, most valuable resources in the world.

The shape of Alberta was simple. The 110th meridian (110° west longitude) became the Alberta/Saskatchewan border—a perfectly straight line from the 49th to 60th parallels—which were the southern and northern borders, respectively. The western border was shared with British Columbia. The area was 661,185 square kilometres or 255,200 square miles. Once Canada was completed when Newfoundland-Labrador joined Confederation in 1949, Alberta would occupy about 7% of the total area.

With over 300 million years of carbon-creating history behind it, the last major step to create the foundation of today's Alberta was the Natural Resource Transfer Acts of 1930. This transferred control of resources to the provinces from the federal government. That this would one day become a major political issue that threatened the future of the country was, of course, unknown.

The fact that Alberta, Saskatchewan, and Manitoba were the only provinces that didn't control their natural resources had grown increasingly unpopular since their creation. This was addressed by the federal legislation in 1930 and all resources became property of "the Crown": the legal vernacular for provincial government ownership. Prior thereto, however, when people or corporations purchased or were granted surface land from the federal government, the mineral/resource rights were included in ownership. These would become known as "freehold" lands and the resources were not included in the federal/provincial transfer.

The three prairie provinces were very similar by design. In terms of area, Saskatchewan was given 651,900 square kilometres (251,700 square miles) and Manitoba ended up at 647,797 square kilometres (250,900 square miles). Technically Alberta was a bit larger, but only by an average of 1.5%. This may be an early example of a Canadian government central planning masterpiece.

But Alberta would leave its sister provinces behind economically. They were very similar until the big oil discovery at Leduc in 1947. This was confirmed by the 1941 census. The populations of Alberta and Manitoba that year were 796,169 and 729,744 respectively. Because agriculture was the biggest business at that time, Saskatchewan led the way with 895,992 identified citizens.

In 2016 Statistics Canada reported Manitoba had grown by 75% to 1,278,365. Saskatchewan's population was 23% higher at 1,098,352. Alberta, however, exploded to 4,067,175, an increase of 411%.

There is only one reason. Carbon—the commercial exploitation of oil, gas, and coal. The attraction of Alberta was not the weather, water sports, more affordable housing, or better government services. As the famous line from Bill Clinton's 1992 US presidential campaign stated so clearly, "It's the economy, stupid."

A commitment made to British Columbia when it joined Canada in 1871 was the construction of a national rail line. Construction began in 1881 and the Canadian Pacific Railroad reached Alberta in 1883. A branch line north to Edmonton was completed in 1891.

Between 1896 and 1914 over 2.5 million people emigrated to Canada and one million of them settled on the prairies. The growth was fairly evenly distributed, as demonstrated in the 1911 census when Manitoba had a population of 455,614, Saskatchewan 492,432, and Alberta 374,663. Because settlement went east to west with the construction of the railroad, and Saskatchewan had much more fertile and wide-open prairie suitable for agriculture, the fact that Saskatchewan had the largest population should surprise no one.

In terms of an economic base, the early vision did not look far beyond agriculture. It was the fur trade that opened the west for the Hudson's Bay Company. Hunting vast herds of buffalo to the point of near extinction followed the railroad. The subsurface resource potential was unknown.

Today, over a century after Alberta came into existence, its carbon-resource base is well defined. And it is world scale. An open question is the degree to which it is understood. Although Alberta does not have the world's largest reserves of coal, natural gas, or crude oil compared to other political jurisdictions with constitutional control over resources, it has very significant deposits relative to its area and population.

For crude oil Alberta has the third-largest recognized proven and recoverable petroleum reserves in the world behind, in order, Venezuela and Saudi Arabia. Alberta has 167 billion barrels while Venezuela and Saudi Arabia hold 301 billion and 267 billion barrels, respectively.

But in terms of crude oil density by area, Alberta moves up to second place at 253,000 barrels per square kilometre, behind 341,000 barrels per square kilometre in Venezuela but more than double 124,000 barrels per square kilometre for Saudi Arabia.

On a per-capita basis, Alberta is off the charts. Alberta holds 39,000 barrels of oil per citizen. Venezuela and Saudi Arabia, with much higher populations, hold 9,300 and 7,900 barrels per capita, respectively.

For natural gas deposits by area, Alberta holds 0.09 billion cubic feet (bcf) per square kilometre. Iran holds the largest reserves of natural gas in the world, which is about 20 times that of Alberta. But on a per-square-kilometre basis, Iran has about eight times as much as Alberta. On a per-capita basis, the two countries are very close to the same. Alberta has less natural gas by area than Venezuela and Saudi Arabia but close to twice as much per capita.

Other countries with substantially more natural gas than Alberta include the United States, Russia, Qatar, Nigeria, China, and Australia. The United States, Qatar, and Australia have established themselves as LNG exporters. Landlocked Alberta has yet to enter this market.

Alberta also has significant proven coal reserves, 398 tonnes per square kilometre. Natural Resources Canada reports 90% of Canada's coal lies in Saskatchewan, Alberta, and British Columbia. This is based on 40% of the total. On a per-capita basis, this works out to 61 tonnes per person.

The world's largest proven coal reserves are in the United States, and they are 100 times that of Alberta. But this equates to only seven times greater per square kilometre and about 11 times per person. Alberta's coal reserves dwarf those of Venezuela and Saudi Arabia by every measurement.

Countries with more coal include Germany, Poland, Russia, Australia, China, India, and Indonesia. However, of all the jurisdictions listed above, Alberta is the only one currently planning to completely phase out using coal for electricity generation.

While they both contain carbon, comparing coal to petroleum or natural gas is difficult. Oil and gas are commonly compared using an energy equivalent basis (BTUs or British Thermal Units). Gas is regularly measured in Barrels of Oil Equivalent, or BOE. Many producing companies report their reserves on a BOE basis.

The foregoing figures and calculations are based up what is called "proven" reserves, meaning they are recognized by governments and financial regulators as economically recoverable using current technology and market conditions. These are only a fraction of total reserves, the geological estimates of how much of each resource is in the ground - assuming it could all be discovered and recovered.

Currently, recoverable oil sands are only about 10% of what the industry calls Original Oil in Place, or OOIP. The actual amount of oil defined as oil sands in Alberta is in the range of 1.2 to 2.5 trillion barrels. The total proven oil reserves in the world were only 1.6 trillion barrels in 2015.

Natural gas reserves in Alberta are much larger than official estimates because they include "shale gas," which is contained in very tight rocks, as well as coalbed methane, which is natural gas trapped in coal. Estimates for Original Gas in Place run as high as 464 trillion cubic feet with another 506 trillion cubic feet contained in Alberta's massive coal reserves. This is about 20 times greater than current estimates of proven reserves.

Proven and currently economically recoverable coal reserves are also a fraction of the total resource. While official estimates of Alberta's proven coal are 2,633 million tons, there are another 2 trillion tons too deep to be commercially exploited at the present time. This is the coal that contains massive amounts of coalbed methane.

It is clear Alberta is a carbon-resource jurisdiction with reserves that are immensely important and valuable to the people who live here. They are large enough to influence the price and markets for these resources well beyond its borders. Alberta is a global carbon warehouse by any measure.

As people talk about the need to decarbonize Alberta's economic future, hopefully they will consider this chapter and will put all parties on the equal footing for the discussion, implementation, and impact.

3
COAL FUELS THE NATIONAL DREAM

"Alberta's coal industry helped build the railroad ... Had Canada not had coal first on the east coast and later thousands of miles to the west ... it is hard to imagine how the country's history would have unfolded as it did."

Coal is the world's most plentiful and lowest cost source of energy. It was easy to find because there it was, sticking out of the ground—that unusual black stuff. After man discovered that it burned, coal was harder to ignite but emitted much more heat than wood, dung, or grasses. Coal didn't have to be chopped or dried. When burned, coal creates intense heat, which has proven applications in keeping houses and buildings warm, making metal and generating electricity. According to the International Energy Agency, coal supplies 29% of total world energy and 42% of its electricity.

However, coal is increasingly vilified as a major threat to health and the climate. Coal has a very high carbon content. Its biological geological origins ensure it is rarely pure. When combusted without treatment, coal releases carbon dioxide and other gases and particulates into the atmosphere.

Early this century, the Province of Ontario declared burning coal to be a sufficient public health hazard that coal-fired power generation would be phased out. This was completed in 2014. In 2015, Alberta followed with a plan to end power generation using coal by 2030. Air quality concerns and coal's contributions to climate change have been the justification for phasing it out, despite its utility as a cheap and reliable source of energy for electricity generation.

It is useful, however, to review the incredibly important, largely forgotten, and unrecognized role coal played in building the country.

The Alberta Energy Regulator (AER, the modern-day name for the better-known Energy Resources Conservation Board) indicates coal is prolific in Alberta but the larger deposits are confined to the southern half of the province. After the glaciers receded and the Indigenous peoples began to populate Alberta over 10,000 years ago, any of them travelling in many places in Alberta would have seen coal. Historians believe at this time it was not regarded as anything other just another rock, just softer and black. When they needed fire, Alberta's earliest people burned wood and grass.

Geology created and buried the coal. As they melted, the receding glaciers created enormous water flows, which cut and shaped the topography of modern Alberta, particularly its river valleys and drainage systems. This exposed a lot of coal. The coal deposits were impossible to miss in the Bow Valley, the Crowsnest Pass, or deep river valleys such as the Oldman River in southern Alberta, Red Deer River in central Alberta, or the North Saskatchewan River farther north. Exposed coal seams were also located and recorded on the Pembina and Athabasca Rivers.

The first published historical record of coal in Alberta was written in 1793 by Peter Fidler, a surveyor and mapmaker employed by the Hudson's Bay Company. He observed the outcroppings on the Red Deer River at Kneehills Creek, close to what is now the town of Drumheller. A year later David Thompson, doing the same work for the same reason as Fidler, reported coal in the banks of the North Saskatchewan River. In 1811 near Rocky Mountain House a coal seam nine metres (29.5 feet) wide was identified by a representative of the North West Company, the main competitor to the Hudson's Bay Company.

In 1841, near what would become Edmonton, the governor of the merged Hudson's Bay Company/North West Company, Sir George Simpson, observed significant quantities of coal in the banks of the North Saskatchewan River.

An Alberta government Energy History website writes of these early observations of coal: "It was these first discoveries reported back to Europe that set the stage for Alberta's rapid transformation from undeveloped landscape into industrialized energy powerhouse of the twentieth century."[14] With the coal-powered Industrial Revolution underway, moving to a big empty place with a lot of coal was significantly more attractive than it would have been otherwise.

After Canada the country was founded in 1867 (prior thereto Canada was a British colony, officially known as the Province of Canada), a priority of the new government was to attract immigrants to settle the west. The Homestead Act of 1872 made it a law that adult males could lay claim to 65 hectares (160 acres) of

---

14 "Coal: Alberta's First Discoveries," *Alberta's Energy Resources Heritage* (2018). http://www. history.alberta.ca/energyheritage/coal/early-coal-history-to-1900/albertas-first-discoveries/ default.aspx#page-2.

land for a $10 filing fee, construction of a dwelling, and a commitment to live on the land for at least three years.

About the same time large quantities of coal were identified in the banks of the Oldman River valley near what would later become Lethbridge. In 1874 American whisky trader and entrepreneur Nicholas Sheran started operating a ferry across the river. Sheran was soon mining coal and selling it to nearby RCMP posts as heating fuel. Drawn by the reports of coal, Sir Alexander Galt travelled west, viewed the location and resource, and staked out a nearby coal mine in 1879. The site of the mine was Coal Banks, now known as West Lethbridge (MacGregor 1981).

Back then the west was seen a massive amount of empty land the government could settle and populate. The downside was it was a long way from the capital of Ottawa and from the Quebec ports where settlers from Europe arrived in ships. As the old joke goes, "You can't get there from here." Mass [large quantities] and timely movement of anything was impossible using the rivers and trails of the day, the only means of getting around.

That's why the Canadian Pacific Railroad (CPR) linking Canada from the Atlantic to Pacific oceans became a national priority. Construction began in 1881. The Alberta government version reads, "Both immigration and construction of the railway were essential to the development of Alberta's coal industry in the nineteenth century. Population growth created a domestic market for coal, while immigration provided coal companies with their labor force."[15]

On the other hand, powering a railway locomotive required heat to make steam. While locomotives could and did burn wood, coal was a much better fuel source. And once the railroad arrived at the vast treeless prairie that began in eastern Manitoba, wood was in very short supply. Hauling coal from Nova Scotia and New Brunswick grew increasingly expensive the farther west the railroad progressed.

There was very little if any coal in Ontario or Quebec because of geology. These provinces are largely built on the Canadian Shield and the rocks are classified as Precambrian. This era ended some 4.5 billion years ago and is thought to predate the first simple complex, multi-celled life forms about four billion years ago. Coal

---

15 "Coal: The Early Development of the Coal Industry: 1874–1914," *Alberta's Energy Resources Heritage* (2018). http://www.history.alberta.ca/energyheritage/coal/the-early-development-of-the-coal-industry-1874-1914/default.aspx#page-3.

(as well as oil and natural gas) is organic, so there is found in Precambrian rocks because they predate life on Earth.

The need for a new Canadian source of coal increased when the CPR was slated to go all the way to the Pacific Ocean, the farthest possible distance from current coal supplies on the Atlantic coast. When British Columbia joined Confederation in 1871, that new province's link to the east via a transcontinental railway became essential.

The Alberta government site continues, "The link between immigration and the coal industry was anchored by the appointment in 1880 of Alexander Galt to the position of Canadian High Commissioner to the United Kingdom. Galt was one of the leading figures in the development of southern Alberta's coal industry and, as High Commissioner, he was responsible for promoting emigration to Canada and trying to secure the British government's support for British colonization schemes in the Prairie West."[16]

Galt and a partner founded the North Western Coal and Navigation Company in 1882, mining near the early operations at Coal Banks. The word "Navigation" was in the name because the company would need to use the rivers to move coal to its main customer, the railroad.

The construction of the railroad and the growth of the coal industry were symbiotic. The fact that vast quantities of high-quality coal were available in western Canada—not just from Nova Scotia—made the economics of building the railroad significantly more attractive for two reasons. The logistics of supplying coal to the CPR's locomotives were simpler and, as importantly, considerably less expensive.

Doing today what Alexander Galt did in the 1800s would result in such vociferous accusations of conflict of interest it would not be permitted. Back then getting things done for the good of the country was regarded as nation-building, not blatant self-enrichment. Sir Alexander Galt was an elected member of the Province of Canada in 1867 and is considered one of the Fathers of Confederation. When he was in Britain working to populate western Canada and get the rail link constructed, he also owned a coal mine on the far end of the prairies. Capitalists call this, "enlightened self-interest."

---

16 "Coal: The Early Development of the Coal Industry: 1874–1914," *Alberta's Energy Resources Heritage* (2018). http://www.history.alberta.ca/energyheritage/coal/the-early-development-of-the-coal-industry-1874-1914/default.aspx#page-3.

The availability of coal to fuel the railroad once it reached the western end of the prairies was known when the project was conceived. Had the planners of the national railroad been forced to ship coal all the way from the Atlantic to the Pacific, things would have unfolded very differently. While coal could likely have been sourced from the western United States, there was no north/south rail link to provide it reliably or economically to the CPR.

By the time the CPR reached Medicine Hat in 1883, Galt had the coal there waiting. Work had begun a year earlier to supply coal from Galt's mines using boats on southern Alberta's river network. Access to the transcontinental rail line would expand the coal market east across the prairies and into Ontario and even Quebec. The railroad reached Calgary later in 1883.

In 1883 Galt's paddle wheel boat began to navigate the rivers from Medicine Hat to Coal Banks, with limited success. This mode of transportation was seasonal and depended upon water depth and lack of ice. However, two hundred tons of coal were delivered to the CPR. A year later the coal miners had three boats with barges delivering coal by water. But the hauling season was usually only a month long, confined to when the rivers weren't frozen over and the water levels were high enough.

By 1885, when construction of the CPR had moved well into the Rocky Mountains, a year-round solution was completed; another railroad to support the railroad. A narrow-gauge rail line was built 107 miles from Coal Banks to Dunmore, just east of Medicine Hat. With a link to the national rail line, Alberta's first major resource export industry was underway supplying coal to the CPR and anywhere else on the transcontinental railroad where its transportation costs were price-competitive. Alberta was now able to ship coal east to Regina, Winnipeg, and all other markets serviced by the CPR (MacGregor 1981).

The CPR then proceeded farther west up the Bow River valley on its way to the surveyed route through the Kicking Horse Pass, Rogers Pass, and the west coast. Coal seams were visible in Canmore and identified by the GSC. Mining began shortly thereafter. The railroad opened its own mine at Bankhead just outside Banff in 1904, mining coal seams on the eastern flank of Cascade Mountain. Canmore was established and grew up as a significant coal town. It shuttered its last mine in 1979. Bankhead closed in 1922. It had been operating within Banff National Park, which was created in 1885. But back then industrial activity in a national park was overlooked as the price of nation-building and not considered reckless environmental degradation.

History reads that the national railroad built Alberta's coal industry. Reviewing the foregoing, it could as easily be said that Alberta's coal industry helped build the railroad. Certainly, had Canada not had coal first on the east coast and later thousands of miles to the west, it is hard to imagine how the country's history would have unfolded as it did.

The combination of Alberta coal and the national railroad created the foundations for massive growth of the coal industry and the province in the years ahead. "The development of the coal industry is thus part of much larger story of nation-building and economic development in southern Alberta. Immigration and railway development laid the foundation for the industry, which attracted investment from Eastern Canada, Europe and the United States. By the early 1890s, Alberta was set for one of the greatest natural resource booms in its history."[17]

And boom it did. The AER has a database that records every coal-mining operation in Alberta since the Sheran mine at Lethbridge was permitted in 1874. In the next 140 years, a whopping 5,474 coal mines were registered with the federal government until 1930. At that point the control of the province's natural resources was transferred to Alberta. Sorting through the permit issue and expiry data reveals that 51% of these mines likely never went into production because their permits expired either the year they were issued or the next calendar year.

However, that still leaves 2,690 mines that lasted two years or longer, meaning they likely produced some coal. Twenty mines operated for more than thirty years. Seventy-five mines were on the record as licensed for 20 or more years, 375 for more than ten years, and a whopping 1,963 were registered for five years or longer.

The big years for coal-mining expansion were to follow in the twentieth century. From 1880 to 1900 an average of less than 10 new mines were registered each year. During the last decade of the nineteenth century, coal production averaged 214,000 tons per year.

For the first decade to 1909, the average number of permits was for 39 new mines each year. While the growth in permits accelerated annually from 23 in 1900 to 74 in 1909, Alberta's coal mining was just getting going. In that decade Alberta produced nearly four times as much coal as in the previous 10 years, 966,000 tons

---

17 "Ibid.

annually. A rising population and growing industry increased demand for coal for heat and to fuel an increasing variety of steam-powered industrial operations. The expansion of the railroad network north and south in Alberta and the opening of a new northern national line (ultimately called the Canadian National Railroad [CNR]), ensured that the market for coal for locomotive fuel continued to expand.

In the next decade from 1910 to 1919, a total of 1,170 new mines were permitted, an average of 117 per year. During the war years from 1914 to 1918, the average rose to 132 new mines annually, over 1,300 in total. Coal production quadrupled yet again to an average of 3.9 million tons per year.

In the roaring '20s, 1,459 new coal mines were licensed. Permits peaked at 202 in 1922 but was down to half that number as the decade closed. Annual coal production nearly doubled during the 1920s to an average 6.5 million tons annually.

The coal business slowed down a bit in the 1930s, likely a factor of the Depression and the extended drought in agricultural western Canada. These were the legendary "dust bowl" years. In the 1930s expansion was steady with 1,030 new coal mines registered, still about 100 per year. But average annual coal output fell for the first time to 5.2 million tons.

Larger communities were emerging around the coal-mining operations. Some of the major mines of the day included the Atlas Mine at East Coulee near Drumheller and the coal mines at Coleman and Blairmore in the Crowsnest Pass. Supplying coal to the Grand Trunk Pacific Railroad (later Canadian National Railroad ([CNR]) spurred the development of the Coal Branch west of Edmonton and south of Edson and Hinton. Towns like Mercoal, Coal Valley, Robb, Cadomin, and Luscar sprouted up but today populate the list of Alberta ghost towns.

These larger mining centres were supplemented by hundreds of smaller mines serving residential communities. The best example was Edmonton. Under what is now the community of Beverley, it was estimated there were 50 coal mines operating from 1900 to 1950, but this figure is probably low because not all the mines were registered. Other parts of the city were literally undermined from the 1880s to 1940. Surface structures, new and old, tilted or collapsed as the integrity of the mineshafts below deteriorated. A CBC News article from 2016 said the records for one site showed seven mineshafts. Further investigation revealed 21.

In the 1940s new mine permitting fell by almost half to 531, an average of 53 per year. Regardless, annual production continued to grow, reflecting the impact the large, more industrialized mines had on total provincial output. Annual output rose again to average 7.5 million tons, up about 50% from the 1930s and back above the level of the 1920s. The record year in Alberta coal production took place in 1949: 8.6 million tons.

But by the 1950s growing supplies of crude oil, a growing local refining industry, and the discovery of large supplies of natural gas began to make a material dent in coal consumption for heating buildings and homes. Beginning as early as the 1920s, railway locomotives were being refitted with ever larger and substantially more efficient diesel engines, shrinking the market for coal-fired steam locomotives. Over the next 20 years North America would make almost a complete conversion from coal-powered railway locomotives to diesel-electric.

The British Navy began to show a preference for petroleum over coal to fuel its armada, particularly after large discoveries were made in Persia (now Iran) in 1908. Oil packed a lot more energy punch volumetrically than coal, and in many ways was easier to handle.

In the 1950s only 271 new coal mines were registered with the Alberta government, the lowest figure in the twentieth century. The peak was for the decade was 56 in 1950; the lowest figure was only five in 1958. While the decade opened with production of 8.1 million tons in 1950, by 1959 it was down to only 2.4 million tons, the lowest level since 1911.

The 1950s claimed some major coal-mining casualties like Brazeau Collieries at Nordegg, which had opened in 1911. Twenty-three of the mines that had operated for 20 years or longer closed during this decade. The decline continued in the 1960s. While applications to operate 178 new mines were granted, 88 of these permits were cancelled within a year. The real number was only 90. Nine of the mines in operation for 20 years or longer were deregistered during this time. The output in 1961 was only two million tons, the lowest level in half a century.

From 1970 to 2014, the next 44 years, a total of 305 coal mining licenses were issued but 142—or 47%—were cancelled in one year or less. An internet search of ghost towns of Alberta reveals numerous abandoned, single-purpose former coal-mining communities. To this day, unknown, unregistered, or unmapped underground coal-mining shafts and operations continue to plague residential communities built over top of them. Obviously, unmarked excavations or holes in the ground are a hazard to everything. The new housing on the west side of Canmore is but one example.

As the coal industry in Alberta lost its traditional markets of heating and transportation fuel, new markets emerged. One was large-scale strip mining of coal near the surface for electricity generation. Coal was at, or close to, the surface in several locations across Alberta. A commercial strip mine was operated at Wabamun Lake starting in 1914. Mines near Cadomin and Coal Valley in the Coal Branch area were surface mining operations. As surface earth and rock-moving machinery improved and the cost of labour for traditional underground mining rose, it made more sense to look at cheaper ways to excavate coal on or close to surface than traditional mining through vertical shafts and horizontal tunnels.

The first large scale strip-mining operation to support power generation began in 1962 near Wabamun Lake. In 1970 the Highvale Mine opened in the same area. Highvale would later become the Canada's largest surface-coal-mining operation. In the decade from 1960 to 1970, the percentage of coal surface mined rose from 49% to 70%.

Large-scale surface mining to support power generation gave coal a new lease on life. By the end of the decade in 1969, output had doubled and was back up to 4.4 million tons. The demand for coal for power generation caused coal production to rise to unprecedented levels; way more coal with a fraction of the cost and effort. By 1976 coal production had nearly tripled to 12 million tons. It doubled again by 1980 to 24 million tons. Then it nearly doubled again in the next 15 years, reaching 46 million tons in 1995, ten times the level of coal produced in the mid-1950s.

Metallurgical or coking coal, used to make steel, had a market that was impervious to the move to natural gas or refined petroleum products. In 1969 a coal mine to support Japanese blast furnaces opened in Grande Cache, in the foothills four hours west of Edmonton. To be economic, the mine needed the market expansion for Japanese steel making and the completion of the Alberta Resources Railway link north from the Canadian National main line, which ran through Hinton and Jasper. The Alberta government created Grande Cache through an order-in-council and construction of the town infrastructure began in 1969. At one point the population exceeded 4,000. Grande Cache officially became a town in 1983.

At the end of the twentieth century, annual coal production had tapered off somewhat to about 36 million tons, a level that was sustained for a decade. The major market for western-Canadian coal that has disappeared this century was for electricity generation in Ontario, which shuttered its last coal-fired generating plant in 2014. The decline in domestic consumption was offset by exports. Since

2011 coal production has been gradually declining to about 28 million tons in 2016 and 2017. While this was a decrease from the peak of 20 years earlier, Alberta's coal output was still nearly ten times higher than 100 years earlier.

Both of these primary markets for Alberta coal have been under pressure. Environmental concerns have convinced some governments to reduce or eliminate using coal for electricity generation. Under the Climate Leadership Plan introduced by the New Democratic Party government in November 2015, Alberta is intending to eliminate the use of coal generation for power generation by 2030.

Metallurgical coal, which represents a small amount of Alberta's total output and mined primarily for international markets, has been subjected to competition and pricing pressure in recent years. Grande Cache has endured a boom/bust cycle based on coal and steel prices. To stabilize the economy and population, the Government of Canada opened a correctional facility and a wood-chip processing plant has also been established in the community.

There were nine mines listed as producing in Alberta in 2017. The AER reported the Grande Cache mine produced no coal in that year because the owner shut down the mine due to financial problems in 2015. The primarily production was subbituminous and thermal bituminous coal for the power generation market. Coking or metallurgical coal accounted for only 4% of total production.

In March 2018 the AER released a Coal Supply/Demand through to 2027. It sees coal production declining steadily until 2023, then levelling off at about 14 million tons per year, half of current output. The regulator forecasts the metallurgical market to remain stable.

4

THREE CENTURIES OF OIL SANDS DEVELOPMENT

"... in the period 2005 to 2008, there was a one-time spike in oil sands mineral rights leasing from the government of Alberta that had never been seen before nor duplicated since. That there might one day be a pipeline takeaway problem was not considered"

Over time, crude oil (petroleum) would prove to be the most valuable of the three main sources of fossil fuel extracted from the earth. Once oil refining processes were perfected, gasoline (for cars; trucks; airplanes), diesel fuel, and jet fuel would prove to be ideal and compact sources of portable energy. While the advent of steam to run vehicles was a vast improvement over horse or oxen power, petroleum was the transportation fuel that changed the world.

Electric cars look more promising with every passing year. But replacing oil as the primary fuel powering airplanes, ships, trains, and transport trucks appears impossible in the foreseeable future.

That Alberta contained at least some petroleum was known before the arrival of explorers, settlers, and colonization. The Alberta oil story, which predates the province by almost 200 years, begins in the oil sands. Because of the early reports of seepages of oil out of the banks of the Athabasca River near what would be called Fort McMurray, the GSC took interest as soon as the vast region of the west became part of Canada in 1867.

However, this resource was the last to be developed in significant quantities and was by no means the economic engine that drove the modern oil and gas industry in the province.

Many deposits of low-cost, high-quality, high-value light crude had been identified worldwide. Many of these had the advantage of being located close to oceans and were therefore easily exportable to a global market.

Making the case to tap Alberta's enormous oil sands carbon resources would require overcoming challenges others did not face. These challenges remain today.

In 1985 Barry Glen Ferguson wrote a book titled *Athabasca Oil Sands: Northern Exploration 1875 – 1951*. The period studied ends 16 years the before first large-scale commercial exploitation of this massive resource when the Great Canadian Oil Sands (GCOS, later Suncor) mine and processing facility went on stream in 1967 (B. G. Ferguson 1985).

Ferguson wrote, "The tar sands were known as a curiosity which had some limited uses as pitch for canoe repairs, or so (explorer and namesake of the great northern river to the Arctic Ocean) Alexander Mackenzie reported in the 1790s. In his written reflections on the fur trade, Mackenzie described the 'bituminous fountains' or pools he found on the Athabasca" (B. G. Ferguson 1985).

Mackenzie wrote of the pools, "… into which a pole of twenty feet long may be inserted without the least resistance. The bitumen is in a fluid state, and when mixed with gum, or the resinous substance collected from the spruce fir, serves to gum the canoes. In its heated state it emits a smell like that of sea-coal. The banks of the river, which are there very elevated, discover veins of the same bituminous quality" (B. G. Ferguson 1985).

In 1848 John Richardson visited the region and reported finding a "mineral pitch" consisting of a sand and bitumen mixture. Richardson noted the "country is so full of bitumen that it flows readily into a pit dug a few feet below the surface (Ferguson 1985)."

This identifiable seepage was of course seasonal. Early explorers visited the region in the spring, summer, and fall when the rivers were free of ice. The oil sands would not behave this way in the winter. But what emerged from the earliest observations was the theory the oil sands sat on top of a large deposit of petroleum, at least as crude oil was understood at the time.

The mandate of the GSC was to perform an inventory of the geology and mineral resources of the country. A GSC expedition led by John Macoun visited the Athabasca region in 1875. He observed and recorded "oil shales" and "tar sands" from his canoe as he toured the region of the Lower Peace River, Lake Athabasca, and Athabasca River (B. G. Ferguson 1985).

In his notes Macoun wrote, "When we landed, the ooze from the bank had flowed down the slope into the water and formed a tarred surface extending along the beach over one hundred yards, and as hard as iron; but in the bright sunshine the surface is quite soft, and the men tracking along the shore often sink into it up to the ankles" (B. G. Ferguson 1985).

This was Alberta's first recorded oil spill. It was also early speculation that the key ingredient to turning oil sands into money was heat.

A report from an 1882 GSC investigation, led by geologist Robert Bell, included the theory that the oil sands contained "vast quantities of somewhat altered petroleum." Bell wrote, "The enormous quantity of asphalt, or thickened petroleum in such a depth and extend of sand indicates an abundant origin. It is hardly likely that the source from where it came is exhausted. The whole of the liquid petroleum may have escaped in some parts of the area below the sandstone, while in others it is probably still imprisoned in great quantities and may be found by boring (B. G. Ferguson 1985)."

So drilling it was. In 1894 parliament approved an initial investment of $7,000 for "experimental drilling." A driller from Ontario, where oil had already been discovered at Oil Springs (later Petrolia) in 1858, set up a rig at Athabasca Landing. By October of that year had reached 1,100 feet. Two natural gas zones were encountered, one of which caused a problem for the drillers. But no deeper pool of source oil was identified. The endeavour was abandoned the following year at a depth of 1,500 feet (B. G. Ferguson 1985).

Athabasca Landing—today the community of Athabasca on the river of the same name—is 400 kilometres upstream of Fort McMurray.

Two more wells were drilled in the general Athabasca region before 1897. The first was at Pelican Rapids, 215 kilometres upriver from Fort McMurray, and the other on the North Saskatchewan River at Victoria Settlement near Smoky Lake, northeast of Edmonton. The Pelican River well revealed the existence of considerable hydrocarbons in the region when it encountered natural gas with sufficient volume and pressure that the well blew out and flowed gas, uncontrolled for 20 years. Neither well encountered a deeper source of oil that was feeding petroleum to the surface (B. G. Ferguson 1985).

The result of the GSC exploration was described by Ferguson as discouraging and, after the expenditure of $38,000, the federal government investigations were discontinued until 1912. In the interim some private exploration was done in 1906 to 1908. But in 1913, as Ferguson (1985) wrote, "... the federal government placed the oil sands region under a reserve, halting the speculative activity." British companies had discovered major supplies of oil in Persia (now Iran) in 1908 and realized it could replace coal as transportation fuel. Ferguson (1985) wrote, "The British Admiralty was realizing the strategic and economic value of petroleum and began casting about for Imperial (British Commonwealth member countries) sources of crude oil."

It would be another 54 years before the oil sands would yield a product in commercial quantities that was anything other than intriguing and frustrating.

The oil discovery at Leduc in 1947 proved Alberta had large quantities of petroleum, not just oil sands. Regardless, nobody had lost sight of the Athabasca oil sands during the excitement of what would later be coined "conventional oil development." All the non-oil sands crude in the world would eventually be labelled "conventional" because turning the perpetually frustrating oil sands into money required processes that were clearly "unconventional" compared to virtually every other oil deposit in the world.

In 1921 Alberta formed the Scientific and Industrial Research Council (SIRC), which went on to be called the Alberta Research Council and lives on today as Alberta Innovates. Ferguson (1985) wrote, "It resulted from the efforts of academics and politicians interested in creating an organization which would conduct studies on resources and processes which might permit diversification of the provinces economy, thus freeing it from its dependence on agriculture."

Alberta's first industrial diversification mandate took place nearly 100 years ago. For the past 40 years the mantra has been to get out of the carbon-resource and energy business. Again, the more things change the more they stay the same. Isn't history amazing?

A joint federal/provincial initiative to do more with Alberta's coal and oil sands began in 1919. The SIRC took over the province's role. In 1920, "Karl Clark was appointed to study the extraction of bitumen from the oil sands and to determine the value of bitumen as a road surfacing material" (B. G. Ferguson 1985).

The World War had I been the first major international military conflict powered by petroleum. The realization dawned that the world might run out of oil, so the development of new sources became militarily strategic as well as economically important. Turner Valley's deeper oil had not yet been discovered. Despite all the indications of opportunity, Canada was an oil importer, at the mercy of the United States and the Middle East.

It was time to do something useful with the huge but recalcitrant oil sands. But determination and patience would prove essential.

Clark began laboratory work in 1922 using earlier research into hot-water separation, which had some proven applications in mining and oil shales in California.

In 1927 Clark published a study with the stated objective of, "the commercial development of the bituminous sands." But the oil would have to be separated from the sand. A small-scale separator at the University of Alberta, in operation since 1923, processed 85 tons of oil sands. A larger facility was built in 1924 in north Edmonton, which processed 100 tons. It was expanded and in 1925 and 500 tons of oil sands were run through it. By 1926 Clark and his team were working on economic models from steam shovels to final output. For $200,000 they figured they could build a plant that would process 1,000 tons per day which, operating 200 days a year, would yield 27,000 tons (185,000 barrels or 925 barrels per operating day) of bitumen (B. G. Ferguson 1985).

It was time to find partners and money, so Clark visited established oil-producing companies and operating refineries. Meanwhile, an experimental 600-foot stretch of road was paved with bitumen in Edmonton as a commercial test. It failed. Soldier on.

In 1929 an agreement was reached between the renamed Research Council of Alberta and the Mines Department of Canada. Including the National Research Council, it was called the "Bituminous Sands Advisory Committee." The pilot plant in north Edmonton was redesigned, taken apart, and reassembled on the Clearwater River in Fort McMurray in the area now known as the subdivision of Waterways. By the source at last! By August 1930 it had processed 800 tons of oil sands, yielding 75 tons (513 barrels) of bitumen with the lowest particulate content to date. The work continued with a budget of $87,000 for 1930/31 and $72,000 for 1931/32 (B. G. Ferguson 1985).

During this period numerous private operators continued to probe the oil sands, trying to figure out what to do with it. Wells were drilled, oil sands mined, experiments conducted. In 1927 a company named International Bitumen Co. appeared on the scene. In 1930 at a place called Bitumount, downstream from Fort McMurray, hot-water separation began with some success. After struggling for several years, more money was raised to expand Bitumount in 1937. When it operated, it produced from 15 to 34 barrels of bitumen hourly. It was expanded in 1938 but was still not profitable. While technically successful, it could only sell product when the river was open. International Bitumount was declared insolvent that year (B. G. Ferguson 1985).

In 1936 another plant was constructed by a company called Abasand Oils Ltd. at Horse River, very close to Fort McMurray. Learning from the work of Karl

Clark and the Alberta Research Council, fundraising had begun years earlier. After multiple financial and operational problems, by 1941 Abasand was operating on a seasonal basis from May to November. The plant had been retooled to do partial refining of the bitumen to be able to produce and sell gasoline and diesel fuel locally. The output was 41,265 gallons (1,179 barrels) of gasoline, 70,700 gallons (2,020 barrels) of diesel, and 137,500 gallons (3,929 barrels) of fuel oil (B. G. Ferguson 1985).

Securing wealth from the oil sands would never be easy. Abasand burned to the ground in late 1941 and destroyed the $700,000 investment. However, it was rebuilt and was back operating the next year.

World War II attracted federal government money. Due to concerns about oil supplies, in 1942 more capital and attention was focused on the so-far limited success at Abasand. But intense competition for labour and materials during the war ensured little was done on time or on budget. Abasand continued to process and produce marketable petroleum products for the local market, but never turned a profit. Another fire occurred in late 1945. But with the war over and the investment of over $1.9 million, Ottawa declined further funds. The plant was never repaired and what could be sold was in 1948 (B. G. Ferguson 1985).

The Alberta government was less than enthusiastic about the federal government's direct investment in oil production and commercializing the oil sands. But for the province, it remained unfinished business. In 1944 Edmonton made the decision to, as Ferguson (1985) wrote, "… build its own experimental separation plant in order to sustain the private enterprise which, both levels of government agreed, must be behind any effective commercial operation."

Sixty-three years ago, the governments of the day apparently knew their limitations when it came to the resource development industry.

The plan was to revive Bitumount since it was located near some of the best quality oil sands reserves. International Bitumen, the original developer of Bitumount, had been revived as Oil Sands Limited in 1942. Looking for government money, it was in the right place at the right time. A deal was struck and Bitumount was retooled and rebuilt by 1947 at a cost of $500,000, double the original estimate. Short of capital, Oil Sands defaulted on its share of the expenses and Bitumount became a 100% provincial project. A 1949 government report was summarized by Ferguson (1985) this way: "Thus, the pilot plant at Bitumount was a necessary means of demonstrating to private investors the possibilities of the Athabasca sands."

By 1949 Bitumount was up and running and producing up to 660 barrels of processed oil per day. It shut down briefly because the high volume of oil-soaked tailings (sand) were spilling over into the Athabasca River because "... even in 1949 such emission of pollutants was considered unacceptable." By the fall of 1949 Bitumount had cost $750,000.

By 1950 Alberta's oil future had changed immensely following Leduc, Redwater, and other major oil discoveries. It was hard to justify further investments in experimental crude supplies when moving proven and more profitable resources to market was a larger and more pressing problem. The government's Bitumount investment had proven its point; the technical elements of oil sands mining and processing had been resolved. Now it was a matter of economics and volume.

And it worked. But not right away. In 1953 a group from Toronto incorporated Great Canadian Oil Sands Limited (GCOS) to one day make money on the oil sands (B. G. Ferguson 1985). Another strategic event that reminded people of the fragility of the now important global supply of oil occurred in 1956 when Egypt blocked the Suez Canal, the primary route to ship oil from the Middle East to Europe. Egypt had nationalized the Suez Canal, which precipitated a planned military response by Israel, Britain, France, and later the Soviet Union. As was common during the "Cold War", tensions escalated leading to speculation of a possible nuclear response.

As the world digested the first of many reminders of the uncertainty of oil supplies from the Middle East, Sun Oil Company Limited of Toronto and US parent Sun Oil Company began to examine more domestic supplies. Sun had been aware of the oil sands since 1944 and acquired a promising lease in 1954. GCOS retained Karl Clark as an adviser after he retired from the Alberta Research Council after 34 years (B. G. Ferguson 1985).

GCOS obtained a lease and production purchase agreement from Sun in 1958. In 1960 GCOS applied to the Alberta Oil and Gas Conservation Board (AOGCB) for a permit to build a $110-million oil sands production complex to produce 31,500 barrels per day (b/d). It was approved in 1962 and with permit in hand, the regulator gave GCOS a year to find the money. In 1963 the cost was up to $234 million but planned output was increased to 45,000 b/d. The Alberta regulator approved the project so long as construction began in 1964 and the plant was operational three years later. This included a 428-kilometre pipeline to Edmonton to hook up with the existing export pipeline infrastructure. The job was on time and on budget with an official opening held September 25, 1967. The oil sands were real (B. G. Ferguson 1985).

However, pipeline takeaway capacity from Alberta remained a challenge. In 1960 the consortium of companies behind what would eventually become Syncrude also applied to the AOGCB for a permit to build an even larger plant with an estimated cost of $356 million to produce 100,000 b/d. But among conventional producers still struggling with prorationing, having more oil come on stream in such large quantities would clobber their business. Consideration of the application for a second oil sands plant was shelved until 1968.

The problem with using heat to separate bitumen from oil sands is that once the plant was running, you couldn't shut it off. This made it different than a conventional oil well, which could be stopped, started, or throttled back. Cognizant of stiff opposition to unconventional oil, a condition of the approval of GCOS in 1962 was a government announcement that oil sands output would be limited to 5% of conventional crude production.

Feelings ran high. While lunching at the Petroleum Club in Calgary, *Black Gold With Grit* author J. Joseph Fitzgerald recalls being verbally attacked when a club member discovered he was one of "those guys from the tar sands." The member said he was going to table a resolution forbidding the Petroleum Club to ensure "none of your kind" would again be allowed in the building (Fitzgerald 1978).

Despising the oil sands has become a popular pastime this century. But it actually started over 50 years ago in Calgary, for entirely different reasons.

In 1968 the regulator, as promised, reopened the file. The consortium had been reorganized as Syncrude Canada Limited in 1964. The Alberta government also reviewed its output ceilings. Fitzgerald (1978) wrote, "… the government ensured that oil sand production could go ahead without limitation, to serve markets which were not met by conventional suppliers." A revised application for an 80,000 b/d plant was submitted in 1968. A year later "the Board" approved the project with production to begin in 1976.

The corporate structure of the Syncrude consortium changed as companies were bought, sold, or merged. In 1971 Syncrude asked what was then called the Energy Resources Conservation Board (ERCB) to approve raising output to 125,000 b/d. While this was approved, the partners didn't rush to spend any money. GCOS had now had four years of operations to examine and it still wasn't profitable. In 1974 major partner Atlantic Richfield, at 30%, withdrew from Syncrude in part because of the capital commitments required for its huge oil discovery at Prudhoe Bay and the construction of the Trans Alaska Pipeline (Fitzgerald 1978).

The remaining partners mined their Rolodexes to replace Atlantic Richfield but found no takers. The fact that the estimated cost of the project was now up to $2.3 billion was a big factor. But another strategic event had occurred on the other side of the world in 1973, when members of the Organization of Petroleum Exporting Countries (OPEC) announced an oil embargo on many Western countries to protest Israel's war with Egypt, the Yom Kippur War. The price of oil skyrocketed. Canadian governments again concluded intervention was in the national interest. Alberta, Ottawa, and Ontario split the available 30% at 10%, 15%, and 5%, respectively, in the so-called Winnipeg Agreement in 1975.

Having the government as an equity partner wasn't all bad. The price of oil had quadrupled thanks to OPEC and the federal government had intervened with an oil price cap and export controls. This was the beginning of Canadian carbon fairness, discussed later in this book. But Syncrude would be guaranteed the world price for its oil once production began.

Syncrude started mining and processing oil sands in 1978. Thanks in part to OPEC, GCOS began turning a profit in 1974. The oil sands were now a major part of Alberta and Canada's oil industry. By 1979, the first full year of Syncrude's operation, the oil sands were producing an average of 91,000 b/d. This was under 10% of Alberta's crude oil output that year, but a remarkable accomplishment after 200 years of struggling to figure out how to make money from this massive carbon energy resource.

With the technology improving with every passing year, oil sands recovery techniques began to improve. Imperial Oil, a perennial pioneer, began experimenting with steam for subsurface or *in-situ* (Latin for "in place") recovery of oil sands buried too deep to mine. It began commercial operations at Cold Lake in 1975. Back then cracking the code to develop more oil was something Canadians were proud of. Imperial Oil, which since 1936 had sponsored *Hockey Night in Canada*, initially on radio and later on television, advertised its project and pumpjacks on to the nation.

The next big step was a new technique called Steam Assisted Gravity Drainage, SAGD for short. Using the ever-improving technology of horizontal drilling, a steam-injection well was drilled above an oil-producing drain hole. Steam was injected into the top wellbore to heat the reservoir. The thinner oil drained into the lower well, which was pumped to surface.

By 1980 in-situ production was up to 9,500 b/d, a fraction of oil sands mining at 128,000 b/d. But most of the oil sands resource was deeply buried thus economic for surface mining. By 1990 the figures for *in-situ* and surface mining production were 135,000 b/d and 208,000 b/d, respectively.[18]

In the mid-1990s the Alberta and Canadian governments developed the Generic Oil Sands Royalty Regime to attract more capital to this massive resource, a combination of reduced Crown royalties and accelerated capital investment deductions. Early this century, lobbying by the Alberta government and Canadian oil sands producers persuaded the US Securities and Exchange Commission to recognize oil sands as reserves for financial accounting purposes.

It worked. The huge number of 173 billion recoverable barrels appeared in the world's proven crude oil inventory, making Alberta the home of the third-largest oil reserves in the world, behind only Venezuela and Saudi Arabia. Proven technical success established over decades resulted in continuously lower operating costs. Advancing technology. Steady oil prices. Growing global oil demand. At the turn of the century, oil sands output was 610,000 b/d, 289,000 b/d from in-situ thermal recovery and 321,000 b/d from mining.[19]

There was nothing but good news. The world's private sector oil industry was looking to invest in a country with lots of oil, where private ownership was legal, which had no state-owned oil-industry competitors, and with no threat of nationalization by the government. With the knowledge that Alberta's oil sands were now the real thing—and that only 25% of the world's reserves were available for development to private sector oil companies—Alberta's second great oil boom was underway.

The result was that in the period 2005 to 2008, there was a one-time spike in oil sands mineral rights leasing from the government of Alberta that had never

---

18   Canadian Association of Petroleum Producers, *Statistical Handbook for Canada's Upstream Oil Producers* (Calgary, AB: CAPP, 2018). https://www.capp.ca/publications-and-statistics/publications/316778.

19   Ibid.

been seen before nor duplicated since. In that period, $3.1 billion or 73% of the total $4.2 billion in oil sands leased collected from 1991 to 2017 materialized.[20]

With leases in hand, the twenty-first-century oil sands boom was of mammoth proportions in terms of investment and its impact on Alberta, its citizens, and the country. Alberta experienced yet another period of accelerated growth and prosperity as billions poured in from all over the world to develop the oil sands. The production growth this century was staggering. In 2017 total oil sands production averaged 2.7 million b/d. In-situ output was substantially larger than oil sands mining at 1.44 million and 1.23 million b/d respectively.[21]

When leases were purchased and investment and expansion plans made earlier this century, pipeline takeaway capacity was not an issue. Conventional production was declining, which freed up transportation space. New pipelines were under construction. That Alberta's oil production might once again face pipeline takeaway capacity issues was not considered to be an insurmountable problem.

This was dead wrong.

---

20 Government of Alberta, Ministry of Energy. https://open.alberta.ca/opendata/ historical-royalty-revenue.

21 Canadian Association of Petroleum Producers, *Statistical Handbook for Canada's Upstream Oil Producers* (Calgary, AB: CAPP, 2018). https://www.capp.ca/publications-and-statistics/ publications/316778.

5

LEDUC OIL DISCOVERY TRANSFORMS A PROVINCE

"This was an oil boom by any definition, the scope of which dwarfed Turner Valley. Alberta had world class crude oil reservoirs, and the world took notice and showed up. The inflow of investment capital would indelibly change the people, the province, and the country"

In the far southwest corner of Alberta near Waterton, the search for oil had been underway since 1902. In a 1968 collection of essays titled *Dusters and Gushers*, author George Lonn recounts how surface seepages had been identified in 1888 in the mountains by two trappers assisted by Stony Indian guides. The oil had originally found in a stream called Oil Creek, which is today known as Cameron Creek. It was soaked up in gunny sacks, squeezed into containers, and sold locally as a lubricant. Lonn wrote that Alberta's early oil entrepreneurs "... built a sluice box which worked on the principle used by miners attempting to obtain gold ... When it floated to the top, sometimes as deep as six inches, it was skimmed off with a shovel. The production was five or six gallons a day and during that summer." The entrepreneurs "sold hundreds of gallons of oil" (Lonn 1968).

By 1901 this attracted commercial interest. A year later a rig was assembled and drilling began. The first well flowed oil. A 1958 book titled *Dynamic Decade*, written by University of Alberta associate economics professor and author Eric Hanson, focused on the changes in Alberta from 1946, the year before Leduc, and 10 years later in 1956. Hanson wrote that the Oil City well discovery flowed at 300 barrels per day (b/d) and produced 8,000 barrels in total. More wells were drilled but every aspect of the operation was hampered by the remote location, transportation challenges, and poor drilling equipment and techniques. A pump was installed in the one steady producer but mechanical problems capped production at only 40 b/d. The operation was abandoned in 1904 (Hanson 1958).

While the Turner Valley discovery of 1914 is often cited as the next big event in the history of Alberta oil, it was in fact a natural gas discovery. The next area where crude oil was discovered was in the Lloydminster/Wainwright area of east central Alberta. This was "heavy oil," thinner than the oil sands but not the light crude discovered in the Middle East in 1908 or at Spindletop in East Texas in 1901.

In *The Roughneck* magazine in 1982, your author wrote, "People suspected the Lloydminster area had oil or gas producing potential as early as 1919 when Imperial Oil drilled 18 holes and although the company laid claim to no discoveries, favorable geological reports were issued. In the 1920s traces of high-grade oil were discovered in a well when cattle refused to drink the water. The farmer … skimmed samples of the oil from the surface of the water and sent them off for analysis. Tests indicated it was light oil, and … did further indicate the region had hydrocarbon producing potential."

Oil was discovered at Wainwright, southwest of Lloydminster, in 1925. The area was expanded and still produces today. A small refinery built at Wainwright would later supply the CNR with diesel fuel. In the 10 years from 1927 to 1937 oil and gas would be discovered near Lethbridge, Del Bonita and Moose Mountain near Turner Valley and Taber.

More oil would be found in 1939 at Vermillion, Princess/Bantry, and Lloydminster. This was the year that the potential for heavier-gravity crude in the general area of Vermillion, Lloydminster, and Wainwright became apparent. Including Provost to the south and Cold Lake and Bonnyville to the north, this remains a major oil-producing and development region to this day. History has proven up major deposits of heavy oil in the region.

But back to Turner Valley. The natural gas in the discovery that sparked the first Turner Valley boom contained intriguing quantities of hydrocarbons more liquid than gas, but not crude oil. When samples collected from surface seepages were sent to laboratories for analysis, they were found to contain natural gas liquids (condensate), which often indicates a geological association with petroleum.

In 1924 by drilling deeper, gas and condensate were discovered in Royalite #4 in such quantities that it blew wild for three weeks before it was brought under control. The condensate (also called naphtha, a by-product of which is used for fuel in camping stoves and lanterns) was of sufficient quality it could easily be turned into gasoline. While explorers kept searching for oil, it wasn't quite crude in the classic sense. It did, however, spark continued investment and development.

Unfortunately, only the naphtha had commercial value, so the natural gas was flared, giving the area the title of "Hell's Half Acre" because of dozens of wells burning huge quantities of the worthless methane. It was so warm beneath the flares the homeless sought refuge during the winter during the Great Depression. This wastage of the hydrocarbon resources caused Alberta to pass legislation in 1938 to create the Alberta Petroleum Natural Gas Conservation Board (APNGCB), predecessor of the Alberta Oil and Gas Conservation Board (AOGCB), Energy Resources Conservation Board (ERCB), and today's Alberta Energy Regulator (AER).

As the world learned more about the nature and structure of oil and gas reservoirs, it was noted that natural gas and lighter liquids were often found on top of crude oil after being separated naturally because of specific gravity. In 1936 a local enterprise called Turner Valley Royalties (later Home Oil) struck crude oil at the unprecedented depth of 6,828 feet, the deepest well drilled in Alberta to that date. This was light-gravity, high-quality crude oil that had the greatest economic value because of its ability to be refined into a suite of commercial petroleum products, from motor oil to diesel fuel.

This precipitated yet another drilling boom in the next five years; welcome economic activity during the economic depression that was gripping the Western world. The Turner Valley gas/naphtha/oil field would become a huge asset for Canada and its allies during the World War II. The high-quality crude was refined into gasoline and aviation fuel in Calgary at record volumes and used by industry and the military to keep the economy functioning and the weapons of war produced and fuelled. But gasoline was tightly rationed for consumers.

In his landmark 1991 book *The Prize*, a history of the world's oil industry written by Daniel Yergin in 1991, the author asserted that a major factor in the Allied forces prevailing to defeat Germany and Japan in World War II was that the United States and Canada had significant supplies of domestic and strategically secure crude of oil. As the war progressed Germany resorted to synthetic fuel made from coal as oil supplies from Baku (Azerbaijan, then part of the USSR) and the Middle East were cut off. Japan, with no carbon resources, was reduced to distilling fuel out of pine roots after it was cut off from regional supplies from Burma and the Dutch East Indies (Indonesia). One reason Japan ordered "kamikaze" pilots to commit suicide by intentionally crashing their planes into US navy warships towards the end of the war was because the country didn't have enough fuel for the aircraft to make a return flight (Yergin 1991).

There were no notable discoveries during the war years or immediately thereafter. During World War II the primary development efforts were helping to build the

Alaska Highway and the CANOL (an acronym of the words "Canada" and "Oil") pipeline from Norman Wells oilfield in the NWT to Fairbanks, Alaska. The former was completed and 70 years later is used daily. The latter was never finished and never carried oil. A refinery was constructed in Whitehorse that was later moved to Edmonton. At the time they were considered the largest construction projects in the world, requiring the deployment of significant equipment and labour assets.

Turner Valley and other discoveries, primarily gas, had put Alberta in the business. But the economic impact was not materially significant. It was of more interest to the people that it employed or who bought its products than anybody else. In comparing the estimated percentage of personal income generated in Alberta by petroleum, agriculture, and other industries in 1947, the contribution of oil and gas was only 5% in 1947 (Hanson 1958).

The discovery of oil just outside Edmonton at Leduc on February 13, 1947, changed everything for the industry, Alberta, and the country. While Turner Valley had a significant impact on the fledgling industry, it was still a regional business. Major markets in Central Canada, the west coast, or the Midwest United States were out of reach.

Dynamic Decade, under the title "Alberta's Transportation Handicap," read "… without pipeline transportation the Alberta oil and gas potential would be largely untouched and applied mainly to serve a relatively small prairie market" (Hanson 1958).

The more things change, the more they stay the same. Sixty years later Alberta is still seeking more pipe. Market access would dictate what was developed and when. This has always capped the rate of growth and size of western Canada's oil and gas industry, a challenge that continues today.

Until 1946 Canadian oil and gas investment totalled about $200 million on everything except royalties, 80% of which was in Alberta. It was estimated that the industry had recovered its costs in the previous 30 years, but not much more. Many developers made no money while many more went broke (Hanson 1958).

According to the Alberta Petroleum and Natural Gas Conservation Board (APNGCB), pre-Leduc only 157 million barrels of recoverable reserves had been discovered in Alberta, over half of which—85 million barrels—had been produced. As the west continued to grow and oil production did not, finding more oil before Alberta ran out of it was a sound commercial decision (Hanson 1958).

The oil sands looked promising but remained a technical enigma. Natural gas was interesting but at this point entirely local. What now?

By the mid-1940s exploration theory and techniques had advanced to the point subsurface reservoirs that might contain oil could be identified from the surface using seismic sound waves generated at surface then recorded and analyzed after being bounced back from deep in the earth. Imperial Oil, formed when 16 small refiners merged in southern Ontario in the 1880s, had sold control to America's Standard Oil in 1898 to obtain expansion capital. Imperial discovered oil at Norman Wells in 1920 and helped with the CANOL project in 1942.

Imperial was determined to explore the potential of western Canada using seismic and test drilling. In late 1946 the company "spudded" (started drilling) Imperial Leduc No 1. It was "wildcat" (purely exploratory) well number 134. The previous 133 had all failed to locate commercial quantities of oil or gas.

Ten weeks later, at about 5,000 feet, traces of oil began to appear at surface. It was time to test the well. "The company took a chance and invited government and company officials, journalists and radio reporters to come out to the well site on February 13" (Hanson 1958).

An interesting choice of words but Imperial most certainly knew there would be oil coming out of the ground. The question was how much?

Although a number of people had already gone home—it was mid-winter—at 4:00 p.m. the combination of reservoir pressure and mechanical coaxing brought the well to life. Rig operator Verne Hunter (nicknamed "Dry Hole" Hunter after drilling 133 consecutive exploration failures) would later write, "Then with a roar the well came in (started producing), flowing into the sump (surface fluid storage pit) near the rig. We switched it over to the flare line, lit the fire, and the most beautiful smoke ring you ever saw went skyward" (Hanson 1958).

The rest is history. The second well, one and a half miles south, was drilled to 5,370 feet and struck oil. Imperial Oil drilled 33 more wells that year in the field. In 1948, 147 successful producing oil wells were drilled in the field. One, Atlantic #3, blew out in spectacular fashion and gushed millions of barrels of oil before it caught fire and was finally capped nine months later. While the original Leduc discovery attracted significant investor attention, Atlantic #3 captured the imagination of the world by proving Alberta had so much oil it gushed out of the ground in uncontrollable quantities.

Alberta's oil boom was underway. Leduc was the key to unlocking the geology of the Western Canadian Sedimentary Basin (WCSB), the vast region extending from Yukon and the NWT through northeastern BC, most of Alberta, southern Saskatchewan, and the southwest corner of Manitoba. Leduc showed everyone the type of underground geological structure that might contain oil.

Only a year later Imperial struck more oil at Redwater, northeast of Edmonton. Redwater was at about half the depth of Leduc but would prove to be over twice as big. In the next 10 years oil would be discovered in the same general area at new fields like Golden Spike, Wizard Lake, Acheson and Bonny Glen. Further south, discoveries were made at Fenn-Big Valley and Nevis. To the southwest of Edmonton, the massive Pembina discovery was made in 1955. By the mid-1960s oil was discovered near Valleyview at Sturgeon Lake, Kaybob near Slave Lake, Swan Hills, Virginia Hills, Judy Creek, Mitsue, Nipisi, and Rainbow Lake. The Rainbow Lake discovery was over 800 miles northwest of Leduc by road, except there were no roads, forcing the industry to move during the winter on frozen ground (Lonn 1968).

Proven reserve growth was eye-popping. In only ten short years, from 1947 to 1956, recoverable oil reserves discovered increased 2,300% from 157 million to 3.8 billion barrels. Crude oil production rose from 17,000 b/d in 1946 to 393,000 in 1956 (Hanson 1958).

It could have been nearly twice that volume except for pipeline takeaway capacity restraints. Government regulators invented a new term called "prorationing," a fair way of cutting back production from every well to ensure fair and equal access to limited pipeline space.

About 50 exploratory wells were drilled in 1946. This figure exceeded 400 in 1951 and stayed there for the next five years. Only 20 rigs were drilling in 1946. The number was 167 the year after Pembina was discovered in 1955. Investment figures were equally impressive. A total of $23 million was spent in 1946 by oil companies on administration, land, seismic, drilling, and development. By 1956 the figure for the year was $391 million, 17 times higher (Hanson 1958).

This was an oil boom by any definition, the scope of which dwarfed Turner Valley. Alberta had world class crude oil reservoirs, and the world took notice and showed up. The inflow of investment capital would indelibly change the people, the province, and the country.

Alberta's conventional oil production continued to rise through the 1960s and early 1970s, assisted by more discoveries and pipeline capacity. It peaked in 1973 at 1.43 million b/d but has declined almost continuously ever since. By 2017 the figure was down by 70% to only 446,000 b/d. Some of the great discoveries of the 1950s and 1960s are still producing oil, but only in marginal amounts. The recent development of light tight oil shales (LTO) using horizontal drilling and hydraulic fracturing has arrested the decline in recent years and conventional production should stay flat for the foreseeable future.[22]

There are two ways to make money in the oil business. One is volume. The other is price. Until OPEC began to change world oil markets in 1973, making money finding and selling oil had been a volume game. For Alberta, this was affected by how much oil could be found in a single well or pool and by pipeline takeaway capacity. Leduc was great because it had lots of oil that, once discovered and connected to buyers, was cheap and profitable to produce. While the oil sands contained enormous quantities of petroleum, it was not easy nor always economic to recover.

Price also helps. A lot. In inflation-corrected current dollars, the price of oil remained flat from 1946 to 1973, fetching about US$20 in 2017 dollars per barrel. In 1973 OPEC and other global events like the overthrow of the Shah of Iran in 1979 changed everything. By 1979 the price of oil had increased by about 600% in 2017 inflation-corrected dollars, reaching a price over US$120 per barrel. That was only duplicated when oil hit record levels of over US$140 in 2008.[23]

The OPEC-driven oil price spike of the 1970s was interesting because of the massive amount of wealth transferred from oil consumers to oil producers. Because carbon was always political, this caused some unusual and unprecedented responses from Canadian governments.

After prices collapsed in the mid-1980s, lower prices depressed exploration activity until early this century. Finally, the combination of low prices, a growing population driving demand, and a simultaneous decline in spending on the hunt for new supplies once again caused the world's oil supply/demand curves to cross. This resulted in a period of rising and higher prices beginning in 2003 that, with the exception of the global financial collapse of 2008/2009, lasted until 2014.

---

22 Canadian Association of Petroleum Producers, *2018 Crude Oil Forecast: Markets and Transportation* (June 2018). https://www.capp.ca/publications-and-statistics/crude-oil-forecast.

23 Capital Professional Services, *InflationData.com* (2018). https://www.inflationdata.com/.

During that period significant new supplies were developed such as the oil sands in Canada and light tight shale oil in the US.

It was the growth of North American production from non-traditional sources that once again caused the global oil supply/demand curves to reverse.

The decision by OPEC and oil-producing partners like Russia to restrain production in late 2016—assisted by a significant lack of investment in new supplies in 2015, 2016, and 2017 caused by collapsed oil prices—have again caused world oil demand to exceed supply. In 2018 the world price of oil had risen again above US$70 a barrel, significantly above the 60-year inflation-adjusted average price of about US$40.[24]

In the fall of 2018 due to a variety of factors, oil supply once again exceeded demand and weekly North American prices fell as much as 40% from a multi-year high in October to the end of the year. Although they are obviously different words, *petroleum* and *volatility* seem to have increasingly become synonymous this century.

While the price of oil has been volatile for the past 45 years and supply disruptions have been either anticipated or threatened, the world has never actually run out of the essential carbon-based transportation liquid fuels: gasoline, diesel, and aviation fuel.

Despite growing concerns this century about the environmental impact of fossil fuels like oil, oil demand continues to grow. Current forecasts indicate consumption will continue to rise for the next 20 years unless some amazing breakthrough in transportation technology emerges. World oil demand in in 2019 is forecasted to rise above 100 million b/d, which is 35% higher than the 74 million b/d mankind consumed when the Kyoto Accord was signed in 1997.

As for running out of oil, based on human ingenuity and technology, it will never occur.

---

24 Ibid.

6

NATURAL GAS – ALBERTA'S PLEASANT CARBON SURPRISE

"The recovery in Alberta's oilpatch in the 1990s and until 2008 was primarily powered by natural gas, as production from conventional oilfields continued to decline. During this period about 70% of the new wells drilled had natural gas as the target."

Alberta's most valuable carbon surprise was natural gas. Petroleum had been spotted in the oil sands 100 years earlier. Coal was sticking out of the ground everywhere. Natural gas was discovered by accident, but it was a pleasant surprise indeed.

Natural gas is a fossil fuel fraught with contradictions. While a plentiful, relatively clean, and powerful form of heat, gases are by nature difficult to transport and can be dangerous to handle - particularly if they are odourless and flammable. Solid coal and liquid oil are easy to move and far less hazardous.

Gas requires compression and plumbing to be rendered useful. The *Cambridge Dictionary* definition of resource is, "a useful or valuable possession or quality of a country, organization, or person."[25] Natural gas did not technically fit the definition of resource without significant investment in transportation and distribution infrastructure. However, once safely delivered to its destination, natural gas proved to be a remarkable fuel with a promising future.

Alberta has lots of natural gas. In the early days most of it was encountered when drilling for something else. When gas was struck at Langevin near Medicine Hat in 1883, the CPR was searching for water for its steam locomotives. When an exploratory well to better understand the oil sands was drilled at Pelican Rapids in 1897, it hit natural gas which blew out of control for nearly 20 years. In Alberta gas appeared in well water or bubbled through ponds. Leaks to the surface would burn, either by accident or design.

---

25 "Resource," *Cambridge Dictionary* (2018). https://dictionary.cambridge.org/dictionary/english/resource.

Finding gas often created problems. An uncontrolled flow, as often occurred during early encounters, was later called a "blowout." While pure methane is not poisonous, it is highly flammable and therefore easily ignited by any spark. Gas not physically connected in some way to a place where it can be used is of no commercial value. For years, when gas was discovered in remote locations when drilling for oil, it was technically interesting but economically worthless. Without field processing, compression to provide pressure for flow, pipelines, and a distribution network to the end user, gas was often trapped by economics and logistics.

Natural gas holds tremendous promise today as a lower-carbon alternative to coal for electricity generation. But without significant investment in transmission and distribution infrastructure, significant conversion from coal to gas on a global basis will take a lot of time and a lot of money.

The American Public Gas Association (APGA) website describes the early history of gas as commercial fuel. As early as 1626 there are records of Indigenous Americans igniting natural gas seepages near Lake Erie. Natural gas from coal was used for illumination around 1785. Baltimore, Maryland did the same thing in 1816. The first well with natural gas as the target was "dug" in New York in 1821. Philadelphia owned a municipal gas distribution utility in 1835.

By the time settlement in Alberta was underway following the completion of the railroad to Calgary in 1883, people knew what natural gas was and what to do with it when discovered. That Alberta had so much gas was not yet known. The commodity wasn't commercial until enough people were concentrated in one place to justify the significant investment in transportation and distribution infrastructure.

In 1969 George de Mille wrote a book titled *Oil in Canada West: The Early Years*. When revisiting the story of the CPR hitting gas in 1883, de Mille recounts how the gas caught fire and burned the rig down. This happened frequently. Early settlers and developers knew gas was handy stuff if you could safely harness it. The original Langevin well and another drilled a few feet away supplied gas to heat CPR buildings until 1954 (de Mille 1969).

The CPR needed coal for its steam locomotives, so in 1890 the company drilled to verify the existence and depth of the coal seams that had been observed in the banks of the South Saskatchewan River near what is now Medicine Hat. More gas. This well blew wild for a year because the drillers of the day had no idea how to contain or control natural gas in large pressures and quantities. But around

Medicine Hat, natural gas was everywhere. By 1899 gas was used for residential heating in the community. Two years later, in 1901, there were seven commercial wells in the city and rules about how gas would be transported and distributed were created to ensure public safety. By 1904 deeper drilling revealed more gas under more pressure.

Medicine Hat, which would later call itself "The Gas City," was the first example of what carbon energy resources would do for the economy of Alberta—before the province was even founded. In 1901 the town created a municipal gas system, which was operational a year later. This attracted industry. While early gas was used for home heating, de Mille wrote gas was used "… to burn lime for plaster [probably the first industrial use of gas in Alberta] and for processing in the meat packing plant." By 1904 Medicine Hat had enough gas for the whole community and a natural gas distribution system was created for street lights and home heating (de Mille 1969).

In terms of agriculture—the primary attraction for the west's settlers—the Medicine Hat region was a major flop. It was right in the centre of what would later be called the Palliser Triangle, a particularly dry section of southeast Alberta and southwest Saskatchewan, subject to frequent droughts. However, Medicine Hat had the combination of the national railroad and lots of water coming down from the mountains in the South Saskatchewan River. Add in seemingly endless supplies of cheap fuel and the community boomed.

Medicine Hat became a city in 1906. Gas made it a magnet for industry. Soon it supported woollen mills, flour mills, greenhouses, foundries, breweries, glass manufacturing, and brick builders. The news of the day declared, "There is no doubt now but that Medicine Hat will become the manufacturing centre of the west."

To secure its energy future, in 1915 the city secured the mineral rights from the federal government to 12 sections of land surrounding the community. This was transferred to the province in 1946. Another 54 sections of mineral rights were secured in 1964.

Medicine Hat is today included in the Alberta Culture and Tourism website under the subheading "Gas." The text reads, "So effective was the city at propagandizing that it brought in settlers by the hundreds: between 1911 and 1913, the population practically tripled. The Alberta Bible Society found it necessary to distribute Bibles in fifteen different languages! By the end of 1913, the city claimed to be fueling thirty major industries by twenty natural gas wells. The gas was so plentiful that at time it was problematic; on more than one occasion a pocket burst as someone was digging a new water well."[26]

In the earlier days cheap heat and lots of clay created the business of bricks, tiles, sewer pipe and household items such as jugs, crocks and churns. In later years it would be petrochemicals, rubber and even tires. "Medicine Hat is not only Western Canada's first gas field but also one of Canada's greatest fields. Its productive sand extends over some 967,700 acres and contained, at the onset of production, at least 2.5 trillion cubic feet of natural gas" (de Mille 1969).

That was the 1969 figure. Today the National Energy Board estimates Canada has proven gas reserves of 1,225 trillion cubic feet and is the fifth-largest natural gas producer in the world by production volume.

Medicine Hat's success attracted a lot of attention and created a gas development boom of sorts. Gas exploration began in and around Calgary in 1895. Drilling moved west to Lott Creek in 1906 and to southeast Calgary in 1908 on the Colonel Walker Estate (now Inglewood). Gas from that well supplied the Calgary Brewing and Malting Company for years. Drilling for the CPR, more gas was found near Bow Island in 1909. The well was called "Old Glory" and apparently had initial production of 8 million cubic feet per day (de Mille 1969).

The Canadian Western Natural Gas, Light, Heat and Power Company was incorporated in 1911. In 1912 a 16-inch, 274-kilometre gas pipeline to Calgary was completed in 86 days. In creating Alberta's first natural gas utility serving multiple communities, the pipeline ran near Lethbridge which voted to grant a distribution franchise to Canadian Western. Gas would later also be supplied to Nanton, Okotoks, Granum, Claresholm, and Fort McLeod (de Mille 1969).

> The gas arrived in Calgary with great fanfare on July 17, 1912. A flare was set up and some 10,000 citizens showed up to witness its ignition. A report from a Canadian Western employee recounted the event. "Eugene Coste (founder of Canadian Western) and his wife were there and Whitey Foster was in charge of the valve control. At a signal from Mr. Coste, Whitey

---

26 "Natural Gas: Creation of an Industry – Accidental Industry – Medicine Hat," *Alberta's Energy Resources Heritage* (2018). http://www.history.alberta.ca/energyheritage/gas/creation-of-an-industry/accidental-industry/medicine-hat.aspx#page-2.

turned on the valve and he turned it on plenty, because coming out of the standpipe there was first a tremendous amount of dust, then stones and great big boulders, two or three pairs of overalls, piece of skids – almost everything came out. There was a tremendous roar the people started to back up, and there was almost a panic. Well, finally Mrs. Coste was standing by with Roman Candles and she was shooting these candles at the standpipe trying to light the gas. And finally away she went – with a terrible bang… And that was the commencement of the initial installation of natural gas in the City of Calgary" (de Mille 1969).

That day still stands as a momentous event for the city. Canadian Western Natural Gas, or CWNG, became part of ATCO Ltd. in 1980.

The next major milestone was the first Turner Valley discovery at Dingman #1 in 1914. This took place on the banks of the Sheep River near surface seepages that had been known to settlers as early as 1888. Samples were collected in 1911 and sent to United States for analysis. This confirmed not only the existence of methane, or pure natural gas, but small amounts of "wet gas," natural gas liquids such as naphtha. After a few teasing traces of hydrocarbons, de Mille wrote, "On May 14, 1914, the Dingman well suddenly came in with a roar. The depth was 2,718 feet and soon the well was producing 4 million cubic feet of natural gas per day. That gas was saturated with straw colored light oil 'so pure it was used as car fuel at the well site.'" (de Mille 1969).

This was greeted with tremendous enthusiasm in Calgary. While the value of natural gas was understood, the traces of petroleum-like liquids and the location of the discovery right outside of town caused great interest. Hundreds of Calgarians made their way to the Dingman discovery to see for themselves (de Mille 1969).

The Turner Valley discovery is considered Alberta's first classic oil boom because it involved speculation as well as hydrocarbons. De Mille wrote, "Within a few days more than 500 companies had been formed and there was a wild scramble to obtain shares in the action. Offices were opened in every available room along the business streets which were festooned with great cotton banners proclaiming the names and prices of newly issued stock" (de Mille 1969).

Over the years Alberta's carbon energy entrepreneurs would be regarded with a combination of shock, awe and bewilderment by their fellow Canadians. This is where and how it all started.

As gas consumption grew and output from the original gas wells in Bow Island declined (the fate of every conventional oil and gas well and reservoir in the world), Turner Valley was connected to the Canadian Western main distribution line in 1922. With production volumes higher and stable, it is estimated the Turner Valley wells also produced about 11,000 barrels of "natural gasoline." At the same time the significantly depleted Bow Island field was converted to Alberta's first storage reservoir using gas from Turner Valley. Gas was injected in the summer then produced during periods of much higher demand in the cold winter. Storing gas during periods of low demand to meet peak demand is a standard feature of the North American gas industry today.

Edmonton was about a decade behind Calgary in switching to gas for reasons that were not necessarily geological. Edmonton sat on top of significant reserves of coal, thus dwellings, buildings and industry had enjoyed ample supplies of cheap energy for decades. However, in 1914 gas was discovered in quantity at a field near Viking about 130 kilometres northeast of Edmonton. Edmontonians had been unsuccessfully drilling for natural gas since 1897 but were in no hurry to switch fuels. Converting to gas from coal was originally rejected by the citizens after it was proposed by city council.

However, after nine years of debate and consultations with outside experts, natural gas arrived in Edmonton in 1923. It took several more years to the citizens to complete the switch from coal (de Mille 1969). The final operating entity, Northwestern Utilities, was eventually born and became a unit of ATCO Ltd. in 1980.

Many communities joined the search for local supplies of natural gas. Between 1912 and 1915 Wetaskiwin, Tofield, Vegreville, Morinville, Castor, Camrose, Athabasca and Taber drilled for gas but didn't always find commercial quantities. Several of these communities, plus Lacombe, Ponoka and Red Deer, were added to Edmonton's integrated gas network (reservoir to burner tip) in later years.

Oil and gas seepages identified at Tar Island, about 20 kilometres downstream from the community of Peace River, were identified in 1893. Drilling for gas began in the Peace River area in 1909 but no commercial quantities were encountered. More seepages were identified near Pouce Coupe on the Alberta/BC border. An exploratory well completed in 1922 hit enough gas to blow out of control but was harnessed for local heating. That ended poorly when gas accumulating in the rig cellar caught fire resulting in multiple injuries. In 1923 drilling resumed and the well was deepened. It came in at 10 million cubic feet per day but was capped because of the lack of local markets or distribution infrastructure (de Mille 1969).

The First World War from 1914 to 1918 slowed development activity as capital and manpower were focused on the military and Europe. That Alberta had the largest known gas reserves in the British Empire at that time caught the attention of the British Admiralty. For safety reasons they were looking for supplies of helium to replace hydrogen in airships. Helium had been detected in small concentrations (0.33% to 0.36%) at Bow Island and Medicine Hat. But separation was expensive and never moved past the experimentation stage in Alberta (de Mille 1969).

In its ongoing quest for oil across western Canada, Imperial Oil discovered natural gas in significant quantities at Kinsella, near the producing Viking field northeast of Edmonton. After the Second World War, but before the 1947 Leduc discovery, Imperial had invested in determining the size of the Kinsella gas discovery with a view towards turning natural gas into gasoline to supply the local market. The Leduc discovery made that unnecessary and Kinsella was sold to Edmonton's gas utility ensuring secure supplies for continued expansion (de Mille 1969).

In 1944 a significant discovery of gas at Jumping Pound west of Calgary changed the business. Using seismic to identify the structure, the ongoing hunt for the next Turner Valley oilfield found "sour gas"; natural gas containing large quantities of hydrogen sulphide. Jumping Pound was not put on production until 1951. The original field plus offsetting discoveries would prove to be massive. The specialized processing plant built for the field moved Alberta into the business of extracting and marketing elemental sulphur for the first time. The primary market for sulphur was the growing pulp and paper business on the west coast (Hanson 1958).

Up until the Leduc discovery, explorers were of the view Alberta, with the exception of the oil sands, was primarily a gas-prone basin. Once the Leduc discovery unlocked the geological key to the type of reservoirs to be found at greater depths, a number of major gas fields were discovered by the mid-1960s. These included Pincher Creek, Cessford, Rimbey, Pembina, Harmatten-Elkton, Crossfield, Waterton, and Edson (Hanson 1958, Lonn 1968).

From a public-policy perspective, the Alberta government was enthusiastic about natural gas as a provincial resource but reluctant to consider exporting it. In the book *Promise of the Pipeline* written by Peter C. Newman under commission to TransCanada Pipelines Ltd., the author set the state for Alberta becoming a national supplier of gas.

"In the decade after Leduc, exploration companies hunting for oil stumbled onto more than 24 trillion cubic feet of natural gas. Nobody knew what to do with it. As early as 1949 the province's Dinning Commission had determined that Alberta did not have enough gas reserves to permit any exports. This attitude was perpetuated by the Gas Resources Preservation Act that set up the Alberta Petroleum and Gas Conservation Board to regulate gas removal" (Newman 1993).

Producers, however, had another view. Carl O. Nickle, publisher of the industry bible *Daily Oil Bulletin,* wrote in 1945, "The odds are that gas – long regarded as a relatively worthless by-product in the production of crude oil, will in the not-too-distant future rank with coal and oil among the most valuable of the province's natural resources" (Newman 1993).

As mentioned in the previous chapter, after the 1924 discovery of large quantities of naphtha at Turner Valley enormous volumes of gas had been flared to extract the more valuable liquids resulting in the creation of Alberta's conservation board and regulations. Of 1.3 trillion cubic feet of gas produced from Turner Valley, over 1 trillion cubic feet had been deemed worthless and flared. This happened all over the world and continues today (Newman 1993).

In explaining the market rationale for selling more Alberta gas, "Nickle pointed out that consumption was rising in the US, so that by 1945, 4 trillion cubic feet of gas were being sold, supplied by 60,000 wells, and that the US new had gas reserves estimated at 150 trillion cubic feet, with the substance being sold in 34 of the 48 states to 10 million domestic customers, 785,000 commercial and 45,000 industrial users" (Newman 1993).

During World War II natural gas was in short supply in the US but Canada wouldn't sell it any. Huge gas pipelines had been built in the US from Texas to markets in Chicago, Detroit, New York and California. Why not do this in Canada?

In 1947 the federal government took charge and dispatched its chief geologist from the GSC to investigate. Legislation was passed in 1949 giving Ottawa control over pipelines that crossed provincial boundaries. However, Alberta's Social Credit government of the day was, as Newman (1993) wrote, "… super-careful about in whispering the possibility of legitimizing gas exports." And, like everything else, it was political. "The Western Canada Coal Operators' Association had mounted a strong lobby claiming that a pipeline would replace the equivalent of 75 carloads of coal every day and some 3,000 miners would be thrown out of work, while Canadian railways would lose $7 million a year in freight" (Newman 1993).

Exporting gas was a high-profile issue in the 1952 provincial election. Only 40% of the province was using gas. The Edmonton Chamber of Commerce, late to the gas party only 30 years earlier, recommended the issue be deferred for five years. The Liberal Party opposed exports. Newman summarized, "Farmers were told that if the export of gas to outside industries was allowed, these industries would have no reason to come to Alberta. 'Before we allow export,' he [the Liberal leader] thundered, 'every Alberta hamlet must be served first.'" (Newman 1993).

The main proponents of turning Alberta's growing inventory of natural gas into money were the federal government and producers. The pipelines that were finally built, and the stories behind them, are contained in the next chapter. But by 1954 the case for exports was overwhelming and the Alberta government authorized the sale of 4.2 trillion cubic feet over the next 25 years.

Author Earle Gray wrote in the 1982 book *Wildcatters—The Story of Pacific Petroleums and Westcoast Transmission:* "The grime and soot that had blackened Vancouver were disappearing, and the dense killer fogs that had shrouded its streets were already lifting."

Of interest today as Alberta and BC argue over market access to tidewater for oil, it was natural gas that helped clean up the air in the lower mainland. Pushing through the Westcoast Transmission natural gas pipeline in the 1950s was, as usual, fraught with delays and drama. But in 1960 the *Vancouver Sun* newspaper commented on continued growth in natural gas heating installations. "Those who have ever tried to heat a home with wet cedar sawdust may have mercifully forgotten the soul-searching experience. But chances are they can still smell the fumes from the clogged hopper, gleaming evilly in a damp basement" (Gray 1982).

What followed was a massive investment in pipelines, gathering systems, gas plants, compression, and infrastructure. Cash flow from gas resulted in—you guessed it—more gas. The more pipeline infrastructure that was built, the more communities connected because the product was nearby and the hookup cost was lower.

What was not known at the time was the greatest gas development boom in Alberta would not take place for another 40 years. While exporting gas from Alberta to BC, Ontario, and Quebec, selling it to the US was another matter entirely. For years Canada had a natural gas surplus test. Producers had to prove they had a 25-year supply of proven reserves before they were able to export gas from the country.

After oil prices collapsed in the mid-1980s, gas prices and production volumes were deregulated in 1985. The export surplus test was modified to become market-based instead of reserve-based in 1987. This precipitated a natural gas boom in Alberta that lasted a quarter century until new supplies from shale gas in the US collapsed prices earlier this decade.

The National Energy Board (NEB) issued a report on deregulation 10 years later in 1996. The market performed perfectly as it tends to do when left alone. The NEB reported the price for customers was down 40%, gas replacement costs in Alberta had fallen by 50%, but "Cash flow from production was maintained by an almost fourfold increase in exports … These increased exports drove export revenue up from $2.6 billion in 1986 to $5.5 billion in 1995." Domestic gas sales had also increased by 30%.

The recovery in Alberta's oilpatch in the 1990s and until 2008 was primarily powered by natural gas, as production from conventional oilfields continued to decline. During this period about 70% of the new wells drilled had natural gas as the target.

Natural gas has entered an entirely new era, globally, for several reasons.

The first is shale gas. Starting in the US early this century, two new technologies have created a shale-gas revolution by facilitating the recovery of gas from rock formations that would not previously produce. The first is extended reach horizontal drilling (long wells drilled sideways). The second is multi-stage hydraulic fracturing, which shatters the rock with high-pressure fluid to create channels or fractures from which gas can flow from otherwise impermeable reservoirs.

Shale gas was originally classified as unconventional gas, but as production increases and extraction processes are refined, the "unconventional" handle is increasingly being removed.

The American opportunity is massive and similar reservoirs exist in Canada. The evolution of shale-gas extraction technology has collapsed the price of gas in North America. The greatest reduction in high-carbon-emitting coal burned for electricity ever achieved has taken place in the US thanks to ample supplies of cheap natural gas. North America has the lowest carbon- energy input costs in the world because of natural gas. Compared to competitors on other continents, access to cheap energy as an industrial input has been a great benefit for American companies.

The second is security of supply. Natural gas has been used as a political weapon in Europe by Russia, a country with immense supplies and exports of gas. Cutting off or restricting gas supplies in the winter is dangerous, at least for the people who need it for cooking and heat. The commercial imperative of continuing gas sales to sustain revenue is the default behaviour of gas producers. Using fossil fuels as a geopolitical weapon has happened before.

The third change is LNG exports. By liquefying natural gas, methane can be shipped across the ocean to new markets. This has been a growing business in Qatar and Australia, and most recently the US Canada has examined LNG exports but due to regulatory delays and interventions, none are on production.

However, on October 2, 2018, Canada LNG, a consortium involving Shell Canada and several Asian partners announced a $40 billion project to build Canada's first large-scale LNG export terminal. It is expected to be in production in 2023. LNG from North America is attractive in Europe for strategic reasons and in Asia to replace coal and nuclear power for electricity generation.

Forecasts are for global natural gas consumption to increase significantly in the next 20 years. Gas is increasingly being labelled a "transition" fuel. Because it is clean burning compared to coal and heavier grades of crude, it can replace these carbon energy sources and reduce carbon dioxide emissions until a large-scale, zero-carbon energy source available in large quantities at a lower cost is developed.

As recently as 2010 a new field—dubbed the Montney—was discovered straddling the border of northwestern Alberta and northeastern BC. With the application of new technologies the massive find is considered to be one of the most prolific natural gas/gas liquids reservoirs in North America and possibly the world.

The NEB, hardly oil and gas stock promoters, concluded in 2013 that the Montney reservoir contained 449 trillion cubic feet of marketable natural gas, 14.5 billion barrels of natural gas liquids and 1.1 billion marketable barrels of crude oil. The NEB wrote, "In 2012, Canada's total demand for natural gas was 3.1 tcf. The Montney's estimated 449 tcf of marketable natural gas would therefore be equivalent to 145 years of Canada's 2012 consumption."[27]

---

27 "Frequently Asked Questions - An assessment of the unconventional petroleum resources in the Montney Formation, West-Central Alberta and East-Central British Columbia," *National Energy Board* (2018). https://www.neb-one.gc.ca/nrg/sttstc/ntrlgs/rprt/ltmtptntlmntnyfrmtn2013/ltmtptntlmntnyfrmtn2013fq-eng.html.

Combined with all the other gas reservoirs ranging from the Beaufort Sea, to the Western Canada Sedimentary Basin (WCSB), to the east coast offshore, to the Utica Shales of southern Quebec (which are commercially exploited across the US border), Canada has secure supplies of natural gas for domestic and export markets for decades if not centuries.

7
PIPELINES TO PROSPERITY

"In just 20 months, oil producers and the governments of Alberta, Saskatchewan, Manitoba, Minnesota, Wisconsin, Canada, and the United States agreed this oil pipeline was useful and important and cleared the way for financing, construction, and commissioning."

Canada's prairie provinces are hardly an ideal location from which to export large quantities of anything. And unlike solid resource-based products like coal, grain, or wood, liquid petroleum and gaseous methane have unique transportation challenges, even when the supply is close to market.

Let's define challenges. The carbon-rich WCSB is on the eastern flank of the Rocky Mountains, which run parallel to the Pacific Ocean. This leaves the western fringe of the WCSB about 1000 kilometres from the coast. Mountainous terrain between the resources and tidewater is the most rugged in the country.

The most underpopulated regions of North America are north, south, and east of the WCSB. Unfortunately, the biggest markets of Central Canada, the US Midwest, California, the Gulf Coast, and the east coast are farther away than the ocean.

Geology is the good news, meaning geological history put a lot of hydrocarbons in the WCSB, most of it in Alberta. The bad news is location, because the resources are in the wrong place, the customers are in the wrong place, or both.

Today, getting more pipe built to move product remains one of the province's greatest challenges. Pipelines have been in the news for most of this decade, and for all the wrong reasons. This problem is not new, just different. The anti-carbon movement has capitalized on the public-pipeline-review-and-approval process to turn previously routine and boring regulatory hearings ground zero for carbon energy's opponents. The strategy is clearly working.

As early as 2700 BC in Asia, enclosed tubes were invented to carry water and waste such as sewage. This process, which we call plumbing, was created by the early Romans who used lead or clay for the tube and asphalt as a sealant. Iron pipe for plumbing was increasingly common in the US in the 1930s. The word "pipeline" (as opposed to "pipe" or "piping") was coined specifically to reference moving hydrocarbons.

Exactly when and where the first hydrocarbon pipeline was constructed is unknown. Oil was discovered in many places in the world and moved around by whatever means was available. However, one record indicates that a short wooden pipeline was built from an oilfield in Pennsylvania to a railroad terminal in 1862. More oil in this part of the US saw an oil trunk line supplying Buffalo, Philadelphia, Cleveland, and New York by 1879.

Liquid petroleum could be moved in closed vessels like barrels and tanks, so the horse and wagon and railroads were the immediate transportation solutions. Natural gas was more complicated and could only be transported by sealed pipe. As written in Chapter 6, natural gas was first distributed by pipeline in the US in the early 1800s.

Pipelines quickly became the preferred method of safely and economically moving hydrocarbons from the source to the customer. Under "pipelines," the website of the US Central Intelligence Agency (most data from 2013) reports 124 countries have a total of about 3.5 million kilometres of pipelines, 75% of which are in the US, Russia, and Canada. These countries are three of the largest oil and gas producers in the world in terms of production volumes, and three of the largest nations in the world by geographical area.

The CIA number is probably low because it doesn't include field-gathering and consumer-distribution systems. For example, the CIA reports that Canada only had 110,000 kilometres of pipeline in 2017 while the US had 2.2 million kilometres of pipe in 2013. Meanwhile, the AER reports it oversees 426,000 kilometres of pipe in Alberta alone and this figure does not include final-distribution pipelines operated by utilities that are regulated by a different government department.

Every single detached home and building has several buried pipelines on the property on which it sits. Having any of them leak causes problems. Water and natural gas are inbound, sewage outbound. Electricity, cable TV, telephone, and fibre optics are increasingly encased in a hermetically sealed tube, a pipeline by any definition. The neighbourhood has bigger pipelines for all of these, plus another pipeline system for storm water sewer drainage.

A buried pipeline map of a major urban centre would resemble the knitting of a piece of cloth. Pipelines are absolutely essential for a modern society—particularly an increasingly urban civilization—to exist in what people would consider a normal fashion.

Yet somehow the main pipelines that carry new or increased supplies of oil or gas have become lightning rods for the decarbonization movement. If you live in modern society with heat, water, light, and sewage—all carried by buried pipeline—opposing one new pipeline while being utterly dependent on hundreds of others is like opposing air while breathing deeply to explain why.

As written earlier, the earliest pipelines in Alberta were to move natural gas and oil from the well to the consumer. Medicine Hat had natural gas plumbed in at the beginning of the twentieth century. Calgary had natural gas arriving by pipeline in 1912, followed by an oil pipeline after the deep oil discovery at Turner Valley in 1936. Edmonton received natural gas by pipeline in 1923. Smaller communities along these routes were connected over the years.

Before the Leduc discovery of 1947, Alberta's oil and gas industry catered to local populations because of the province's geographical isolation. "Transportation costs have a great effect on the degree to which the resource potential of Alberta is used ... without pipeline transportation the oil and gas potential would be largely untouched and applied mainly to serve a small prairie market (Hanson 1958)."

It is remarkable and, at least for Albertans, tragic how little has changed in 50 years when it comes to pipelines.

Before pipe, all markets were regional. "Before Turner Valley became a producer of crude oil in 1936, most of the crude used by Alberta and Saskatchewan refineries came from the US, much of it from the Cutbank field in Montana. By 1941 Turner Valley had replaced imported crude in these two provinces. Prior to 1950, the only oil pipelines in the country were in Alberta and near Sarnia. Oil was discovered in southern Ontario in 1875" (Hanson 1958).

The 1947 Leduc discovery changed everything. Redwater, twice as big, was discovered in 1948. Two things became obvious. Sufficient reserves had been identified to warrant pipelines to larger markets and construction had to begin immediately to justify continued investment. After multiple markets, including Sarnia and Chicago were studied, the proponents concluded Superior, Wisconsin, would be suitable. On the western end of Lake Superior, the eastern terminus would connect Alberta to the US Midwest and allow oil to be shipped by tanker (seasonally) to Sarnia through the Great Lakes.

The Interprovincial Pipe Line Company (IPL) was formed in early 1949 and Imperial Oil assumed a minority interest of one third. The Canadian Parliament passed the Pipe Lines Act in April 1949, giving the Board of Transport Commissioners jurisdiction over pipeline routes, operations and tariffs. Shortly afterward the Interprovincial Company was created by a special act of parliament. The Board of Transport Commissioners approved construction of the Edmonton-Regina section in June and the section to Gretna, Manitoba, on the international border, in September. A wholly owned subsidiary of Interprovincial, the Lakehead Pipe Line Company, was formed to build and operate the American section from Gretna to Superior, Wisconsin. (Hanson 1958)

Imperial Oil was the discoverer of Leduc and Redwater. The Board of Transport Commissioners was not replaced by the NEB until 1959.

IPL finished construction of the 1,817 kilometres pipeline in late 1950. It would have been done sooner except for a pipe shortage. Capacity was 95,000 b/d. The line also supplied refineries in Regina and Winnipeg. An extension to Sarnia was completed in 1953. This added another 1,020 kilometres of pipe, bringing the total from Edmonton to Sarnia to 2,837 kilometres. By 1956 the system had been expanded to a capacity of 265,000 b/d.

Let's review what you just read. In just 20 months, oil producers and the governments of Alberta, Saskatchewan, Manitoba, Minnesota, Wisconsin, Canada, and the United States agreed this oil pipeline was useful and important, and cleared the way for financing, construction, and commissioning. IPL is now Enbridge and the original IPL route has been expanded several times.

This is why we have a wealthy country and an oil industry to argue about in the twenty-first century.

But there was still more oil than pipe. Expansion to west-coast markets was also attractive. While the BC market was only 40,000 b/d, the Pacific Northwest states were consuming 250,000 b/d, but with little refining capacity.

If only refineries could be built in Washington, it seemed entirely reasonable that Alberta crude could supplant a large part of the tanker shipments from California. The outbreak of the Korean

War created a critical crude oil supply in the Pacific region. This gave impetus to proposals to pipe Alberta crude westward. On March 21, 1951 the Trans Mountain Oil Pipe Line Company was incorporated by a special act of the Parliament of Canada. (Hanson 1958)

The route was Edmonton to Burnaby, 1,155 kilometres via Jasper, the Yellowhead Pass, and appropriate river valleys. Construction began in February of 1952 and was completed 19 months later, in September 1953. Refineries in Washington were expanded and a new one built. Initial capacity of 150,000 b/d grew to 200,000 b/d a year later.

BC's support was unequivocal. A *National Post* article dated April 11, 2018 quoted the BC Board of Transport Commissioners in an article from the *Surrey Leader* in the 1950s which read, "It is needless to state that the province of BC is desirous that no time be lost in establishing the pipe line. It is of great concern to the people of British Columbia that the line should be an all-Canadian route ... to ensure that the people of British Columbia shall enjoy to the fullest extent the benefits to be derived from the development of Canada's natural resources."

Wow. But in the 1950s, Canada was still a country, not a collection of increasingly isolated provincial nation-states more concerned about local voter support in the short term than the greater good of the province and the country in the long term.

Despite the construction of pipelines east and west—in lightning speed compared to modern times—oil discoveries were still outstripping takeaway capacity. Although actual production was up to 393,000 b/d by 1956, potential production was estimated by the APNGCB at 684,000 b/d had pipeline capacity been available (Hanson 1958).

Compared to today, the problem of more oil than pipe was much worse 50 years ago, at least percentage-wise.

The solution was "prorationing," the only fair way to deal with this very political problem. Under prorationing, all producers would be throttled back in a proportional manner. Reluctantly, it became the law of the land. No pipe and no better ideas. Shutting in oil because there's nobody to buy it is not uncommon from time to time in the oilfields of the world. Shutting in oil because there's no way to transport it is unique to western Canada. For the last seven decades every other petroleum-producing region of the world has run out of oil before they've run out of the capacity to transport it.

In most jurisdictions where there is more oil than pipe, like in the booming Permian Basin of West Texas, they build more pipeline capacity in a hurry. As this book goes to press, there are five new oil pipelines either recently completed or under construction in this region of the US alone. The total capacity will be as much as 2 million barrels per day.

Meanwhile, since 2015 Canada has seen four pipelines cancelled or postponed due to political intervention, insurmountable regulatory hurdles, or legal challenges.

If the production takeaway problems of the conventional industry were not sufficiently onerous, along came the oil sands, described more fully in Chapter 4. As GCOS came on stream in 1967 and Syncrude in 1978, pipeline access and production prorationing regulations in the name of fairness had to be rewritten to accommodate this non-conventional production. Once a thermal recovery mine and plant are up and running, it can't be turned off or even throttle it back like a conventional well.

When GCOS was approved in 1962, Alberta decreed that oil sands production would never exceed 5% of total output. This was to acknowledge pipeline transportation challenges and keep the peace in the rest of the industry. But by the time Syncrude came on stream in 1978, new pipeline capacity had been added. The production ceiling for oil sands was increased so long as that it could be sold in new markets (Fitzgerald 1978).

The natural gas business also needed pipelines to expand beyond local markets. As was written in Chapter 6, gas exports beyond Alberta's border were restricted by law in 1949 until it was proven all domestic demand had been satisfied. Perhaps the remarkable early twentieth-century economic success of Medicine Hat—located in the middle of nowhere—was because large quantities of cheap gas got people thinking this resource would attract more people and industry to Alberta, and therefore it should be saved or protected.

But there was continued pressure to connect Alberta gas to new markets from producers who had found enormous quantities of it. Another proponent of Alberta natural gas export was the federal government, which believed gas was a superb heating fuel for the long Canadian winters. In his 1993 book about TransCanada Pipe Line, author Peter C. Newman wrote about the US: "... America's post-war economy was enriched by an unprecedented burst of gas pipeline construction. Three lines then ran from Texas to Chicago and in 1947, the record-breaking Biggest Inch, a 30-inch pipeline, was engineered from Texas to Southern California. Two years later, the 2,475-kilometer Super Inch, a 34-inch line, connected Texas and northern California and the same year another 2,474-kilometer pipeline was rushed to augment natural gas flow to New York City. The longest pipeline ran from the San Juan Basin in New Mexico to the Canadian border" (Newman 1993).

The US experience proved a concept many still don't believe today - selling non-renewable resources results in more resources, not less. Ottawa put more pressure on Alberta and the industry to sell gas to Eastern Canada and the Alberta government acknowledged gas exports would help, not hurt, Alberta's economy.

So began the story of TransCanada Pipe Lines Inc., which was incorporated in February 1951. Like most of the early investments and initiatives in Alberta's growing oil and gas industry, it was sponsored by American interests. Liberal federal "minister of everything" Clarence Decatur (C. D.) Howe was pushing for the quick development of an all-Canadian pipeline route for gas to Ontario and Quebec as far as Montreal. Howe was even prepared to offer financial incentives for construction, but the fact TransCanada was not Canadian-controlled made this complicated. That was fixed by a merger with a Canadian company; not that difficult since the major asset of both companies was only an idea.

By 1954 TransCanada had a plan to build 2,897 kilometres of 30-inch pipe to Toronto with a 22-inch extension to Montreal. The total distance was 3,589 kilometres. Gas supply contracts and Alberta government export approvals in hand, construction was ready to begin in 1956.

Selecting the route over the north side of the Great Lakes required federal funding since the line was more expensive than it would have been otherwise. When the government tried to ram the legislation through the House of Commons in May 1956, what occurred would later be called the Great Pipeline Debate. It wasn't so much about the money or the project, but the behaviour of the Liberal government, which wanted the pipeline built, buried, and operating in time to campaign in the 1957 election on the party's nation-building prowess. Further, there were time-sensitive construction deadlines by TransCanada associated with the terms of the government funding (Newman 1993).

The Liberal government attempted to force enabling legislation through the House of Commons in the last 14 days of that year's legislative session. The opposition, enraged by the Liberal government's arrogance, was determined to ensure this didn't happen. The often-bizarre tactics employed by both sides became national news. Alberta, the natural gas industry, TransCanada, gas consumers, and all those who would benefit from its completion became collateral damage in a political battle.

This became the first of many political battles over carbon resources that would follow in the next 60 years.

Finally, the bill was passed in the House of Commons and was ratified by the senate a day later. Construction began in 1957 and a major Canadian carbon energy megaproject started flowing east from Alberta on October 27, 1958. Its original volume was 300 million cubic feet per day (Newman 1993).

Concurrent with the TransCanada project was the construction of a gas-gathering system in Alberta, sponsored by the provincial government in 1954. This was called Alberta Gas Trunk Line Company (AGTL), slated to become the exclusive collector and distributor of natural gas in the province. Shares were sold to Albertans, Alberta utilities, gas producers, gas exporters, gas processors, and directors. On top of the equity, the government purchased about three times as much in bonds to finance construction. It was completed in 1957 to ensure TransCanada had access to gas when its line was completed a year later. AGTL was renamed NOVA, an Alberta Corporation in 1980. When it merged with TransCanada in 1988, it operated over 3,000 kilometres of gas gathering pipelines and associated compression facilities across the province (Hanson 1958).

As was the case with oil pipelines, a similar project was undertaken in the other direction at about the same time. The promoter in this case was Westcoast Transmission Company, a sister company of Pacific Petroleums Ltd. Pacific had discovered large quantities of natural gas in the Peace River region of northwestern Alberta and northeastern BC west coast wanted to ship gas to Vancouver, then south into the US, which created regulatory issues on both sides of border (Gray 1982).

After six years of applications and hearings, in late 1955 Westcoast was finally given permission to build its new pipe west. The 30-inch west coast line was completed in August 1957, a total distance of 1,046 kilometres from Alberta to Vancouver. Another line ran to the US border.

When the gas line was opened for service it was a big event. BC Premier W. A. C. Bennett said, "This is the greatest event for British Columbia since the completion of the Canadian Pacific Railway united the province with the rest of Canada. All British Columbians will always be indebted to the McMahons (founders of Pacific Petroleums and Westcoast Transmission)" (Gray 1982).

It is unlikely anyone can remember a BC premier saying something affectionate about a hydrocarbon pipeline, at least in the twenty-first century.

Politics has always factored into Canadian oil and gas pipelines. In the late 1950s, there was producer interest in extending the IPL crude oil pipeline network to Montreal. This precipitated another round of classic Canadian political energy introspection, which is discussed in detail in the Carbon Politics section. Canadian oil to Montreal was never allowed in the 1950s and 1960s, essentially because imported oil cost less once transportation costs were taken into consideration.

However, the OPEC oil price shock of 1973 caused the federal and Quebec governments to think about security of supply instead of cost. Therefore, the federal government precipitated the extension of the IPL line from Toronto to Montreal to put more of Canada on secure supplies of Canadian crude. This was completed in 1976.

Of interest is that after the world oil crisis ended in the 1980s, the line sat idle. In 1997 it was reversed to permit the shipment of offshore oil from Montreal to Sarnia. In 2012, after being approached by Alberta customers with production in the west and refineries in Montreal, the application was made to reverse Line 9 yet again to begin shipping Canadian crude oil Montreal. By this time the same oil pipeline that had been declared a strategic asset in the national interest in the 1970s became a political battleground for the anti-carbon movement. The second reversal was approved in 2014 and began operations in 2015. This supplied lower-cost western Canadian crude to refineries in Montreal that had previously operated on imported crude oil at much higher prices.

To handle Alberta's ever-growing crude production capacity, the IPL/Enbridge main line to Superior, Wisconsin was expanded several times in 1962, 1963, 1965 and, to accommodate the oil sands, 1967.[28] Conventional oil production from the four western provinces peaked at about 1.7 million b/d in 1973 and declined annually for the next 30 years. Conventional crude output and pipeline takeaway capacity were in harmony after decades of prorationing.

---

28 Interprovincial Pipeline Company, *Mileposts: The Story of the World's Longest Petroleum Pipeline* (Calgary: Interprovincial Pipeline Company, 1989).

The growing gas business was another matter. The OPEC-induced carbon energy shortage of the 1970s resulted in a series of ambitious natural gas pipeline projects from Alaska and the Mackenzie Delta to move large quantities of strategically secure North American natural gas to expanding and demanding markets in the US and Canada. These projects enjoyed significant political support at the time with the thesis that increased supplies of natural gas from North American sources would allow the continent to become more self-sufficient in carbon energy by using domestic gas to displace imported foreign oil. Every route passed through Alberta.

During this period massive pipeline megaprojects were in the news regularly as producers proposed the Mackenzie Valley Pipeline to bring gas to market from the Beaufort Sea and the Mackenzie River Delta in the Northwest Territories and the Alaska Highway Gas Pipeline to connect the enormous oil and gas reserves of Alaska's North Slope to the United States. Frontier pipeline development created a new era of issues, including Indigenous peoples' impact, land-claims participation, environmental protection, and consideration of the impact of large-scale industrial development over long distances where none had taken place before.

Neither of these projects were ever built. The Berger Inquiry, public hearings initiated by Ottawa in 1974, concluded in 1977 that nothing should be built for at least 10 years while all the social and political issues were reviewed and hopefully resolved. But by the late 1980s, the markets had changed entirely. This pipeline was only officially declared dead by its main commercial proponents in 2017, 40 years later.

While the Alaska Highway Gas Pipeline was never completed in its entirety, the "prebuild" section called Foothills Pipe Lines Ltd., was completed in 1981 and began carrying gas from Alberta to the US border. In 2000 a new project called the Alliance Pipeline was completed to carry natural gas from northeastern BC and northwest Alberta to Illinois in the US.

Adding pipeline capacity this century has been both successful and impossible. As conventional crude output continued to decline, oil sands production grew. Total crude production (conventional plus oil sands) from western Canada rose from 1.9 b/d in 2000 to 3.7 million b/d in 2017. This required a lot of pipe and still does.[29]

---

29 Canadian Association of Petroleum Producers, *Statistical Handbook for Canada's Upstream Oil Producers* (Calgary, AB: CAPP, 2018). https://www.capp.ca/publications-and-statistics/publications/316778.

Under Conservative Party of Canada (CPC) Prime Minister Stephen Harper, a lot of export pipe was approved and completed while he held office from 2006 to 2015. This included the original Keystone line in 2010 (600,000 b/d), Alberta Clipper in 2010, (500,000 b/d), and some modifications to the existing Enbridge system to carry more oil in 2016 (300,000 b/d).

But thanks to fossil fuel opponents and sympathetic politicians, more pipe capacity has been rejected than built. Northern Gateway was approved by the Harper administration in 2014 but officially killed by the current Liberal government in 2016. The original decision on Northern Gateway was overturned by the Federal Court of Appeal when it concluded the pipeline had been approved without sufficient consultation with First Nations impacted by the project. After the Justin Trudeau administration announced its intention to ban oil tanker traffic on BC's north coast and stated it would refuse to approve Northern Gateway under any circumstances, the sponsoring company Enbridge gave up.

Former US President Obama terminated Keystone XL in late 2015. TransCanada Corporation abandoned Energy East in 2017 after new federal approval processes made the costs, timelines, and outcome unworkable and unknown.

Had they been allowed to proceed, those three lines would have carried over 2 million b/d. Assured, safe, reliable, and low-cost pipeline-takeaway capacity would have significantly changed the decisions of oil sands developers in the past five years. The anti-carbon movement, supported by US-based environmental organizations, has been very successful in inflicting significant economic damage to oil producers, but only in Canada.

Because of pipeline construction challenges, Alberta is once again faced with the problem of more production than pipe. When the original investment decisions were made to purchase oil sands mineral rights mid-way through the last decade, it was not anticipated or even contemplated that pipelines would become political battlefields for the anti-carbon movement. Market-access challenges, combined with rising costs and lower oil prices, have resulted in multiple oil sands project delays or cancellations. Several international operators have sold their oil sands properties to invest their money in places without Canada's self-inflicted obstacles.

In late 2016 Prime Minister Justin Trudeau approved the Trans Mountain Pipeline and Enbridge Line 3 expansions at the same time as he announced the termination of Northern Gateway. After the 2016 US president election, newly elected US President Donald Trump reversed the decision on Keystone XL made by President Obama a year earlier. This caused TransCanada to drop the lawsuit for damages it had filed against the US government to recover all or part of the $2 billion it had invested since 2007 trying to get the pipeline completed. The Enbridge Line 3 replacement/expansion is scheduled to be shipping oil in the latter half of 2019.

The Trans Mountain project, with its western terminus in Burnaby, BC, became the new battleground for the anti-pipeline, anti-carbon movement. In the 2017 BC election, the new minority New Democratic Party government of John Horgan was elected on a platform that included opposition to Trans Mountain, although the project had been approved by the province's previous Liberal government. Horgan remains premier of BC because of the support of Green Party Leader Andrew Weaver, also a fierce Trans Mountain opponent. The BC government has vowed to use every legal means possible to prevent the Trans Mountain expansion from being completed.

The uncertainty caused by intentional delays by the BC government and the municipality of Burnaby, plus numerous court challenges, caused Trans Mountain sponsor Kinder Morgan to announce in April 2018 that unless certainty was established by May 31, 2018, the company would withdraw from the expansion project.

On May 20, 2018 the federal government announced it was purchasing the pipeline from Kinder Morgan for $4.5 billion to ensure the expansion was completed. It seems clear that the economic damage of cancelled oil sands projects, cancelled pipeline projects, and highly depressed domestic oil sands prices caused by transportation issues were the motivation. Problem solved.

Or so everyone thought. On August 30, 2018, the Federal Court of Appeal ruled the Trans Mountain approval was illegal because of inadequate Indigenous consultation and marine protection of killer whales (Orcas). This was the fourth oil pipeline either cancelled or postponed in less than three years. When or if Trans Mountain will proceed is unknown as this book goes to press. If it is approved, it will not carry a barrel of oil before 2020 at the earliest.

TransCanada is attempting to complete Keystone XL 12 years after it was first conceived. The company has gone back to its customers and received the commercial support and backing required to finance construction. TransCanada remains committed to complete it. But on November 8, 2018, an American federal judge in Montana ordered activity on Keystone XL to be halted yet again pending clarification of the legitimacy of the existing federal environmental approval permits. This the second time this judge has intervened in the Keystone XL construction process. On January 8, 2019 media reports indicated TransCanada was working through the legal obstacles to try and commence construction in June 2019 to have the line in service by early 2021. On February 1, 2019 the US federal government announced its intention to appeal the Montana judge's ruling. The resolution of this issue remains unknown as this book goes to press.

Geography has always been an obstacle for western Canada's oil industry. Oil's enemies figured out how to exploit the pipeline-approval process. Climate change alarmists continue to obstruct output growth long before the industry fully understood their enemies might be successful.

Meanwhile, the vast majority of people opposed to new pipelines to get Canadian carbon energy resources to market are prepared to live without some sort of fossil fuel energy, which is essential to keep modern society functioning. Nor is anyone in the developed world.

8

PETROCHEMICALS – ECONOMIC DIVERSIFICATION AT LAST

"Oil and gas are, and always have been, more than heat, fuel, and lubricants. The hydrocarbon-product business is massive and creates products that are so ubiquitous many people don't even know their source."

Because of our heavy reliance of primary resource industries, Canadians are often prone to describing ourselves as "hewers of wood and drawers of water." By exploiting the vast expanse of land and waters, we catch (fish), dig (minerals), cut (forests), and grow (agriculture) things and sell them to others. While the rewards are obvious, so are the risks, which range from the weather, to commodity prices, to scarcity when overproduced or improperly managed.

As a result, one of the great Canadian preoccupations of politicians and pundits is to encourage "industrial diversification" to liberate the citizens from our primary resource dependency:

> 'Hewers of Wood and Drawers of Water' – A biblical term (Joshua 9:21) that the Oxford Dictionary defines as 'menial drudges; labourers.' For Canada, it has been pejorative short-hand for the country's historic reliance on natural resources – versus 'value-added' manufacturing – for economic growth. The biblical phrase was first used by Minister of Finance Leonard Tilley[30] in 1879: 'The time has arrived when we are to decide whether we will be simply hewers of wood and drawers of water… or will rise to the position, which, I believe Providence has destined us to occupy.'[31]

---

30 Leonard Tilley was Conservative cabinet under Prime Minister John A. MacDonald, prior thereto the Lieutenant Governor of New Brunswick after the MacDonald government was defeated in 1873. Tilley was Canada's first Minister of Customs and later Finance, post-Confederation.

31 "Hewers of Wood and Drawers of Water," *The Dictionary of Canadian Politics* (2018). http://www.parli.ca/hewers-wood-drawers-water/.

For resource-dependent provinces like Alberta, doing more than producing fossil fuel resources has remained a preoccupation of politicians and political commentators since it became a province in 1905. It continues today.

What is not as well known about Alberta's immense carbon-resource industries is the degree to which turning raw resources into something more valuable has been an integral part of the business since the beginning. Oil and gas are, and always have been, more than heat, fuel, and lubricants. The hydrocarbon-product business is massive and creates products that are so ubiquitous many people don't even know their source.

This chapter is not about refining, which is turning crude oil into transportation fuels for land, sea, and air, or lubricants for things that rotate and reciprocate. This is reasonably well understood. Without processing, crude oil does not burn cleanly. When spilled, it makes a mess. The focus here is on how crude oil and natural gas can modified to create new substances for other purposes.

The process of breaking crude oil down into more useful compounds has been around since the mid-1800s. The most popular petroleum product at the time was kerosene, used primarily for illumination. The first people to literally "save the whales" were oil producers, who, by replacing whale oil with kerosene, spared the lives of countless cetaceans. To obtain whale oil for lamps, the animals had been overhunted to the brink of extinction. (This is ironic, considering the Trans Mountain pipeline was rejected in August 2018 by a court for not doing enough to protect whales). When the electric bulb was developed, electricity replaced kerosene for illumination. Over time the oil-fuelled internal combustion engine rendered the steam engine all but obsolete.

Many natural gas deposits also contain natural gas liquids (NGL) like ethane, propane, butane, and pentanes-plus (NGLs with five or more carbon molecules). Propane and butane have known heating properties and are common products in this application. Methane and the other NGLs have also become a building block for many other industries beyond fuel and lubricants, a business broadly classified as petrochemicals.

The decarbonization movement seeks the complete replacement fossil fuels for transportation and electricity generation. The popular and promoted alternatives are biofuels, wind, and solar power, commonly called renewables. Their promoters, sympathetic politicians, and governments speak enthusiastically about the economic and environmental benefits of conversion from fossil fuels to renewables.

Whether or not a practical alternative for airplanes, ships, trains, and trucks exists will be explored later.

What is not as frequently mentioned by the anti-carbon activists is the substantial business of petrochemicals and plastics which, at this point, have no known or practical alternatives. Those considering becoming an unquestioning supporter of the decarbonization movement should consider the following.

The petrochemical industry and its primary compounds are very technical. In this chapter every attempt will be made to write more about the final products than their chemical composition and ingredients.

Turning crude oil into more useful commercial products began in the 1800s, mostly by accident. The petrochemical business is well described this way. "This industry and the products it makes play an enormous role in our daily lives. Imagine life without gasoline, cosmetics, fertilizers, detergents, synthetic fabrics, asphalt, and plastics. All of these products—and many more—are made from petrochemicals—chemicals derived from petroleum or natural gas."[32]

The oil refining process is a major source of base compounds and ingredients for the petrochemical business. Refining crude oil releases "off gases" such as ethane/ethylene, propane/propylene, butane/butylene, and aromatics such as benzene and toluene. None of these have any use direct commercial use in the transportation fuel and lubricant business.

America's Energy Information Administration (EIA) describes the refining process. A US 42-gallon barrel of crude oil yields about 45 gallons of petroleum products in US refineries because of refinery processing gain. The volumetric gain comes in part from the addition of hydrogen molecules, which is why the process is called "hydrocracking."

According to the EIA, a refined barrel of oil yields 20 gallons of gasoline, 11 gallons of low-sulphur distillate (diesel fuel), four gallons of jet fuel, two gallons of heating oil and bunker fuel, and six gallons of "other products" including asphalt, lubricants, waxes, and petrochemical feedstock.

By the end of the 1950s, Alberta's early petrochemical industry was launched creating the building blocks for a wide range of products from fabrics to fertilizer; plastics to packaging. Nowadays the general category of synthetic fibres alone is

---

32 Lynda De Witt, "Petrochemical Industry," *Encyclopedia.com* (2003). https://www.encyclopedia.com/history/dictionaries-thesauruses-pictures-and-press-releases/petrochemical-industry.

better recognized by the materials known as nylon, rayon, polyester, Spandex, Orlon, and Kevlar.

The petrochemical industry in the United States began to grow in the 1930s and expanded in the Gulf Coast of Mexico due to the proximity to oil, natural gas, refineries, and tidewater. It started using coal but quickly embraced oil and natural gas as superior feedstocks. Synthetic rubber came in very handy during World War II.

Alberta's petrochemical industry began in Calgary in 1941 with the opening of a plant creating ammonia and ammonium nitrate. The feedstock came from existing pipelines and gas fields, including Turner Valley. This was the first plant in North America to use natural gas as feedstock. The ammonia was converted to ammonium nitrate on site and at Trail, BC At this point in history, the primary market was supplying wartime munitions plants in Winnipeg, eastern Canada, and the US However, advances in explosives rendered ammonium nitrate obsolete for the military. By 1943 the plant was turning out ammonium nitrate fertilizer in pellet form. This product was sold in the US, Canada, Europe, Africa, and Asia. The ammonia not converted to fertilizer in Calgary was sold as feedstock to plants in BC (Hanson 1958).

This business still operates in Calgary today under the name Nutrien. It was incorporated as Consolidated Mining and Smelting Company of Canada in 1909, which went into the sulphur-based fertilizer business at its Trail, BC, copper mine as a way of reducing pollution from sulphur dioxide and sulphuric acid, a by-product of copper smelting. In 1966 the business became known by its abbreviation, Cominco Fertilizer. The name was changed again to Agrium in 1995, when Cominco exited the fertilizer business. When Agrium merged with Potash Corp. in 2016, the new company's chosen name was Nutrien, now a global fertilizer supplier.

Thanks to carbon-based petrochemicals, Alberta's natural gas industry has been helping feed North America and the world for over 75 years.

The next natural gas-based petrochemical plant was opened near Fort Saskatchewan by Sherritt-Gordon Mines Limited in 1954. Its primary product was ammonia, used for leaching, the chemical separation of metal minerals from the native ore. Concentrated nickel/cobalt/iron sulphate ore was transported by rail from a company-owned mine in Northern Manitoba to the plant in Fort Saskatchewan. The finished products were shipped to the US and Central Canada while the residual ammonia was sold into the fertilizer manufacturing business. Alberta's third ammonia processing plant opened in Medicine Hat in 1956 with the final product being nitrogen and phosphate fertilizers (Hanson 1958).

Alberta's first large petrochemical plant not dedicated to ammonia was built in 1953 on the eastern edge of Edmonton. The sponsor was the Canadian Chemical Company, a unit of Celanese Corporation of America, a manufacturer of textiles made from cellulose and other petrochemical products. To supply the cellulose component, Celanese built a pulp mill at Prince Rupert, BC, in 1951. The ample supply of oil and the growth of oil refining in the Edmonton area made Alberta a logical location for the other element of its burgeoning synthetic-fabric business. The fuel for the plant was natural gas from a local producer near Morinville, northwest of Edmonton (Hanson 1958).

In the same year, 1953, Canadian Industries Limited, or C-I-L, opened a plant. Its roots began in the explosives business in the 1800s and became C-I-L when five explosives manufacturers merged in 1910. It then expanded into other petrochemicals, including fertilizers, plastics, and ammunition. Its main interest in Alberta was producing polyethylene. C-I-L was affiliated with Imperial Chemical Industries Limited of England, which first developed polyethylene in 1937. By the time the Edmonton plant, on what is today called "refinery row" opened, the product was used for insulation, plumbing, housewares, kitchen products, wrapping electrical and communication cables, packing material, and the non-light sensitive component of photographic film (Hanson 1958).

In determining where to open in Canada, Imperial chose Edmonton over Montreal because it made more sense to be close to the raw materials than to end-user markets. The main feedstock, ethylene, came from natural gas from what was called a "conservation" plant at Imperial Oil's (no corporate relationship) Devon plant, extracted from the Leduc field. The ethylene was manufactured from the ethane or NGL that comprised about 20% of the field's natural gas production. This process would become a larger part of Alberta's petrochemical industry 20 years later (Hanson 1958).

The discovery of large reservoir containing high concentrations of hydrogen sulphide gas (so-called "sour gas") caused the concurrent growth of the elemental sulphur extraction business in Alberta. Growing markets for gas made construction of dedicated plants to extract the sulphur for a separate marketplace economical.

Canada's first sulphur recovery facility was located at the huge Jumping Pound gas field west of Calgary. This is referenced in Chapter 6. To this point most of North America's sulphur came from Texas and Louisiana. The market for the field's gas was Calgary, as the city's population grew and demand for gas expanded in both residential and industrial markets (see Nutrien above). Gas sales began in 1951 and sulphur recovery began in 1952. Sulphur was in high demand for the manufacture of paper from wood pulp, so the entire output of the early Jumping Pound plant was bought by a plant in Powell River, BC. Jumping Pound was expanded in 1954 (Hanson 1958).

By 1950 sulphur was in such short supply that the US began to ration exports. To exploit the high concentrations of hydrogen sulphide gas at Turner Valley, a sulphur recovery plant opened there in 1951. This operation was viable because the gas was being produced and shipped to Calgary. A fourth sulphur recovery plant at Pincher Creek was linked to the completion of the TransCanada natural gas pipeline, which went on stream in 1958. With a market for the natural gas, the economics of investing in sulphur recovery became attractive (Hanson 1958).

A half-century ago it was clear that Alberta had many of the key ingredients for a growing petrochemical industry. This included political stability, a growing and educated workforce, low taxes, and ample supplies of feedstock and fuel. But despite an early and encouraging start in the 1940s and 1950s, Alberta's petrochemical sector would face challenges that continue today. The realities of geography resulted in long distances to major markets and transportation costs that were higher than those of competitors on tidewater. This would always restrain expansion of this industry in the province.

The late 1950s and 1960s were much quieter for petrochemical expansion in Alberta. But with gas exports on stream east and west, as production in the natural gas business grew, so did field processing. Every field containing larger quantities of hydrogen sulphide gas was fitted with a sulphur-knockout-and-processing facility as the field was connected for production. Fields included Waterton in 1957, Carstairs in 1958, and Edson in 1962. As exploration continued, Alberta government regulators concluded about one-third of the gas reservoirs contained hydrogen sulphide, which typically accounted for 20% of all gas production.

Some of the bigger fields to be discovered that would yield significant quantities of sulphur were Crossfield north of Calgary, Harmatten-Elkton north of Cochrane, Kaybob near Fox Creek, Hanlon south of Edson, and Caroline northwest of

Sundre. When the Kabob plant came on production in 1972, Chevron Canada, the operator, described it as, "the largest sour gas plant in the world."

The website for The Sulphur Institute describes sulphur's major industrial markets. "Sulphur is the primary source to produce sulphuric acid, the world's most used chemical and a versatile mineral acid used as an essential intermediate in many processes in the chemical and manufacturing industries. Sulphuric acid is used by the fertilizer industry to manufacture primarily phosphates, and also nitrogen, potassium, and sulphate fertilizers. Sulphur is also used in many other industries including non-ferrous metals, pigments, fibers, hydrofluoric acid, carbon disulphide, pharmaceuticals, agricultural pesticides, personal care products, cosmetics, synthetic rubber vulcanization, water treatment, and steel pickling."[33]

When the Alberta Progressive Conservative government led by Peter Lougheed was first elected in 1971, it had an activist agenda to diversify the economy. One of the key initiatives was to expand the petrochemical industry despite the geographically induced cost challenges. Thanks to OPEC in 1973 and the first oil price shock, the sharply increased revenue collection by the province gave it significant financial clout to intervene in the direction of the economy. As importantly, there was increasing acceptance among voters that governments could and should take a more activist role in directing the economy.

Peter Lougheed's administration was able to execute a vision of creating what is now the world's largest landlocked petrochemical complex, NOVA Chemicals at Joffre. In 1973 NOVA, an Alberta Corporation (formerly Alberta Gas Ethylene Ltd., a unit of government-mandated AGTL created in 1954) committed to build an ethylene and derivative manufacturing plant. The NOVA Chemicals website reads that four years later, "Arrangements for long-term financing of the ethylene plant were successfully concluded ... after several months of negotiation."[34] It was game on and Lougheed was labelled a visionary. The original plant was expanded in 1981 with approval required from the province delivered through an order-in-council.

Government financial support to make the project economically viable came in the form of a supply of ethane feedstock at prices that assured the plant would

---

33 "An Introduction to Sulphur," *The Sulphur Institute* (2018). https://www.sulphurinstitute.org/.

34 "NOVA Chemicals Joffre Site: Milestones 1973 – 2013," NOVA Chemicals (June 2013). http://www.novachem.com/ExWeb%20Documents/joffre/Joffre%20Milestones%20Fact%20Sheet_final%20proof%204.pdf.

make money. While ethane was a known petrochemical feedstock, in the early 1970s rising gas volumes meant there was more ethane produced with rising gas production volumes than there was demand in local markets. Further, the fact that the petrochemical business was undergoing significant expansion in Sarnia, Ontario, with feedstock from Alberta was the type of economic issue that had become politicized at the time. The ethane was removed at the other end of the pipeline in Ontario and used as petrochemical feedstock outside of Alberta.

In Sarnia, the petrochemical business had started in the 1940s and had grown over the next two decades thanks in part to reliable and competitively priced supply of oil from Alberta and pipeline infrastructure to Central Canada constructed in the 1950s. Natural gas arrived in the late 1950s. In 1973 four companies joined forces to form Petrosar, the largest petrochemical investment in Canadian history.

As the supplier of the key ingredients, Alberta was determined to get into the petrochemical industry in a larger way. To capture the ethane feedstock before the gas was shipped east, so-called "deep cut plants" were built in places like Empress to extract ethane and supply Joffre. The combination of ethane extraction, the construction of processing and transmission infrastructure, and building the main plant at Joffre resulted in the greatest petrochemical construction boom in Alberta's history.

Today NOVA Chemicals at Joffre is still operating. It is owned by the International Petroleum Investment Company, a unit of the government of Abu Dhabi.

When natural gas prices rose significantly early this century, the competitive advantage that Alberta's petrochemical business was built upon—cheap feedstock—contracted. While the province enjoyed windfall royalty income from natural gas prices ten times higher than past levels, the petrochemical business was squeezed. This was the first major period of contraction for the petrochemical sector, as several projects were cancelled or postponed.

The growth in oil sands production in the twenty-first century resulted in great plans for expansion in Alberta's refinery row and petrochemical region, often called Alberta's Industrial Heartland. But expansion was tempered by the world economic collapse of 2008 and 2009, the oil price collapses of 2009 and 2014, and continued market-access issues due to intense opposition to new pipelines.

In 2008 there were tens of billions of dollars in new projects for oil sands upgrading and associated refining and petrochemical capacity expansions proposed for the region. In the end only one major project—the Sturgeon Refinery/Northwest

Upgrader—proceeded and it did so only because it received significant financial support from the province through guaranteed processing margins on feedstock supplied by Alberta's Bitumen Royalty In Kind (BRIK), a policy introduced in 2007 under which Alberta takes its royalty share of oil sands production in oil, not cash.

In 2016 the government of Alberta offered up to $1 billion in financial incentives for companies prepared to invest in petrochemical expansion in the province. This has helped generate investment commitments in several new projects. The development of new NGL-rich reservoirs like the Montney in northwestern Alberta have significantly increased the supply of NGL feedstock to the point of depressing prices. Thanks to new supplies of natural gas from the shale gas revolution in North America, gas is also cheap again. Financial incentives and cheap feedstock have combined to once again overcome Alberta's geographical market access handicap.

According to the Chemistry Industry Association of Canada,[35] Alberta produces 48% of Canada's industrial chemicals. The association reports that the total revenue in 2017 was $17.4 billion, the province's second-largest manufacturing subsector. Industrial chemicals accounted for 80% of the revenue. Exports were $8.4 billion. Direct employment was 7,400 and the estimated indirect employment multiplier raises this figure to 45,000.

In 2016 and 2018 Alberta announced financial incentive programs to help spur petrochemical investment. Under construction is a $3.5 billion facility to turn natural derivatives into plastic pellets. This is being built by Inter Pipeline Ltd. and will receive about $200 million in royalty credits against its feedstock costs under the 2016 program. Several projects to take advantage of the 2018 incentives are being reviewed.[36]

Leading up to the provincial election in the spring of 2019, Alberta's NDP government has been widely advertising a strategy to aggressively move Alberta into more heavy oil upgrading and petrochemical manufacturing. On January

35 "Competitiveness Scorecard, Chemical Sector: Alberta 2018–2019," *Chemistry Industry Association of Canada* (2018), http://canadianchemistry.ca/wp-content/uploads/2018/07/Alberta-Scorecard-2018-2019.pdf.

36 Dan Healing, "Petrochemical growth spurt expected despite rising desire for fewer plastics," *CBC News* (January 17, 2019). https://www.cbc.ca/news/canada/calgary/petrochemical-growth-spurt-expected-1.4981860.

22, 2019 the province announced $440 million in loan guarantees for a "partial upgrader," a $2-billion project that has been pushed along for years by Calgary-based Value Creation Inc. It would convert 77,500 b/d of bitumen into lighter crude. This would free up pipeline space because the feedstock would no longer have to be mixed with synthetic crude and natural gas liquids to create "dilbit," a thinner product that can be shipped by pipeline.

Value Creation still must find the rest of the capital but having the Alberta government provide up to 20% in what will surely be subordinated debt (because it is government guaranteed) should make finding the other 80% easier. The concept of finding ways to make Alberta oil sands easier to ship and refine was identified as an opportunity by a royalty review the NDP promised and conducted during the 2015 Alberta provincial election campaign.

On February 4, 2019 Pembina Pipeline Corp. and partner Kuwait's Petrochemical Industries Co. announced it would be proceeding with a $4.5-billion project to convert propane largely produced from the natural-gas-liquids-rich reservoirs of northwestern Alberta into polypropylene, a core raw material for the plastics industry. This project will qualify for $300 million in royalty incentives, part of Alberta's stimulus package.

These projects comprise the most significant expansion of Alberta's petro-chemical industry in years, proving yet again that cheap feedstock and government incentives can overcome the province's geographic and market access handicaps.

Alberta still holds the title of the world's largest landlocked petrochemical industry and, notwithstanding the determination of many to decarbonize the economy, probably always will.

9

OILFIELD SERVICES AND EQUIPMENT
– HIGH-TECH SECONDARY INDUSTRIES

"Domestically founded, owned, and operated oilfield service companies that supply knowledge, equipment, services, and personnel to the Canadian and international oil and gas exploration and production companies is a very big business. Much of the equipment designed, engineered, and manufactured in Canada is sold or utilized... around the world."

The old saying goes, "Oil is found in the minds of men." While indeed that's where carbon development starts, it is certainly not where it ends. Turning an idea into money is a complex, capital-intensive business that employs hundreds of thousands of Canadians. The knowledge, processes, procedures, equipment, services, logistics, and infrastructure required to create economic value for carbon resources is much larger and extensive than is understood or appreciated by most people not directly involved in the business. Including governments.

Another way of looking at it is that a dollar spent on carbon-resource exploration, development, or production travels a long way in the Canadian economy.

The general catch-all term for companies in the oil and gas business that don't actually explore for and produce the resources is oilfield services, or OFS. People recognize and understand the most visible symbols of the business, such as the drilling rig or pumpjack, the iconic "nodding donkey" surface oil pump. But the supply chain is significantly broader and more diversified than these two widely recognized symbols. This is an industry that also requires miles of pipe, tons of cement, electronics, heavy construction, transportation, chemicals, and a vast array of specialized mobile service equipment.

It is surely the largest and least-understood resource-support industry in the country.

When Alberta's oil and gas industry moved into its modern era after the Leduc discovery of 1947, the expansion was so rapid that significant incremental resources from exploration, to geology, to services, to equipment were imported from the

US While Alberta had some domain expertise and capacity from Medicine Hat, Turner Valley, and local natural gas development, Leduc and Redwater were of such scale that the province's existing OFS capacity was hopelessly inadequate.

In the early 1950s, oilfield folklore goes that the largest collection of Texans outside Texas was in Calgary. But there is no statistical evidence to support this. Alberta needed it and Texas had it. Houston has long been widely recognized as the oilfield service and supply capital of the world.

But as the industry grew and demand vastly outstripped supply, Alberta learned its population was comprised of a significant number of entrepreneurs. This is hardly surprising. Entrepreneurship is not unique to Alberta, but it is a characteristic of Canadians that is not widely appreciated. This was particularly true of those in the west, who were either original settlers or direct decedents of settlers and adventure seekers from other parts of the world more than willing to use their own brains and brawn to make something out of nothing. They had proven that.

With the land increasingly tamed and developed, Alberta's budding oil and gas industry was "in your face" and presented an obvious opportunity. It was another outlet for the pioneering instincts of the residents who had relocated to the frontiers. That's why they came here in the first place.

With demand for everything associated with oil and gas development vastly outstripping supply, more people and companies went into the oilfield service, supply, and support business at every level. While the growth of local oil and gas exploration and production companies has been well documented, the concurrent growth of the domestic oil service business is less well known.

As has been documented in previous chapters, the post-Leduc growth in the oil and gas industry was meteoric. While the macroeconomic statistical data is significant, on the ground, an oil boom is about manpower, logistics, equipment, and support. It's like the military. Take D-Day, the 1944 Allied invasion of the north coast of France in World War II. While the photographs and accounts of the events on the beaches are the visible history, the supporting logistical work is not. Manufacturing, transporting, mobilizing, and supporting all the soldiers with armaments, boats, tanks, airplanes, clothing, and food engaged a major portion of the North American economy.

This is also how the carbon-resource development industry operates. When the business is busy and growing, the impact on the broader economy reaches many people, products, and companies. When it goes in the other direction, history has proven on multiple occasions the negative economic impact is much broader than is widely understood.

Drilling a single well—a hole in the ground to search for and hopefully produce oil or gas—is a complex process. The early stage is exploration. Chasing surface seepages only went so far. First geologists would examine surface rocks to see if they were sedimentary, and if the layers likely covered ancient sea beds and forests. The industry invented seismic instruments, which generated sound waves at surface from explosives. The waves bounced back at different rates depending upon the shape of subsurface formations. In this way "traps," or structures that might contain hydrocarbons, were identified for drilling. This was the birth of the science of geophysics.

Once a promising site was identified and drilling rights secured, the next phase was construction; building a road and surface location. The winter was a good time to drill in Canada because the ground was frozen, and it cost less to move heavy equipment on the frost. No road was required. Of course in the spring and summer, the exact opposite took place. Lots of muskeg and soft ground in Alberta. Building and maintaining year-round roads has always been expensive.

Location built, the drilling rig was moved in. Drilling was supported by suppliers of drill pipe for the rig, steel casing to line the wellbore, cement to hold it in place and secure a seal, heavy steel drill bits to cut and crush the rocks, power tongs (giant pipe wrenches) to screw the casing together, electric instruments carried on electric cable to determine the type of rocks, and drilling mud and chemicals to clean rock cuttings from the wellbore. If the well was successful a wellhead, production, and surface valves called a "Christmas tree" would be installed. Each of these goods and services was highly specialized and supplied by a different company.

If the well was successful and production planned, a new batch of equipment and services was mobilized. This included production equipment for inside the well (packers), a downhole pump, a surface pump (pumpjack), a surface production facility, and miles and miles of pipelines to a central gathering system. Production equipment—tanks; vessels; valves; protective structures; gauges; instruments—spawned an entirely new business separate from drilling.

Electricity had to be installed at the wellsite to power the pumping unit and run the facilities. Finding oil in the middle of nowhere meant significant investment in support infrastructure beyond the direct equipment in the form of electricity, roads, and permanent housing.

In 1999 a University of Calgary business professor named Henry C. Klassen wrote a book titled, *A Business History of Alberta*. One of the very successful companies that solved an early housing problem that operates today is ATCO Ltd., founded in 1947 as Alberta Trailer Hire, which rented small trailers to tow behind cars and trucks to move things around. By 1950 the company rebranded as Alberta Trailer Company (ATCO), which sold portable housing primarily for oilfield workers in remote locations. With a partner, ATCO was building its own housing before 1960 and selling it in Alaska and other countries. The company was providing remote housing all over the world by 1970 and bought control of Canadian Utilities Ltd. (Klassen 1999). Today's ATCO operates all over the world in power, utilities, and infrastructure. Revenues in 2017 were $4.5 billion.

Another example is Vancouver-based construction giant Ledcor Group. It started out in construction building the access road to Imperial Oil's original discovery at Leduc in 1947. It grew with the industry for the next 30 years. In the 1980s Ledcor expanded across Canada and into the United States, opening offices from Toronto to Seattle. It diversified into seven construction divisions, including building civil, industrial, mining, pipeline, telecommunications, and highways. Still a private company, today's Ledcor has 7,000 employees working from 20 offices across North America.

The industry needed everything. Algoma Steel in Sault St. Marie in Ontario started building oilpatch tubing and casing coined OCTG or oil country tubular goods. Prudential Steel, now Tenaris, opened a pipe mill in Calgary and Stelco built a pipe plant in Regina. Western Rock Bit, founded by the drilling contractors, stuck a licensing agreement with Hughes Tool Company (Howard Hughes Sr. and Jr.) to assemble drill bits in Alberta. Global drilling contractor Precision Drilling was founded in 1951. Precision still had revenues of $1.3 billion in 2017, down from $2.4 billion in 2014 before the oil prices collapsed late that year.

Barber Industries started repairing drilling rigs in the Turner Valley oilfields in 1940 but was building Canadian-made drilling equipment, wellheads, and valves by 1960. Bruce Nodwell built his first off-road tracked vehicle to plow through the mud and rugged terrain of Alberta in 1952. This became today's Foremost Industries.

Based in Calgary, manufacturing company Standen's (An IMT Company) today states on its website that they are "… a Canadian manufacturing company making alloy steel products that are formed, forged, upset, threaded or heat-treated for a variety of markets. As one of North America's largest full-line leaf spring, suspension component, trailer axle and agricultural tillage tool manufacturers, Standen's prides itself on quality products and services." It was founded in 1924 as a blacksmith shop and has been integral in keeping Alberta moving for 94 years (Klassen 1999).

Within a few years after Leduc, many of the Americans who'd come to Alberta happily returned home. While the economic climate was attractive, the physical climate was oppressive from October to March. This gave more Albertans more opportunities. Over time the majority of the specialized equipment and service contractors were replaced by local companies using new, Canadian-made gear.

The Canadian ownership of the OFS sector has continuously increased in every aspect of the business. Today OFS equipment of all types is engineered, designed, and manufactured in Alberta and employed or shipped all over the world. Canada has a diversified geological basin ranging from shallow oil and gas to deep reservoirs with some of the highest concentrations of hydrogen sulphide in the world. Temperatures range from -40°C in the winter to +40°C in the summer. Experience has proven if OFS equipment will work in Canada, it will work anywhere.

As politicians and pundits endlessly opine about industrial diversification and the need for value-added industries, there has been one right under their noses that was growing and innovating continuously for decades without government incentives, assistance programs, grants, or special tax treatment. Public-policy blunders in later years would prove that for the most part, governments hardly knew what this massive industry was or how it operated. Canadian oilfield service is big, growing, dynamic, innovative, respected, and in demand in every oil- and gas-producing region of the world.

The oil and gas business is the largest resource industry on Earth and entirely dependent on highly specialized knowledge, equipment, services, and technology. The vast majority of the equipment and services have no application in other industries, nor can the business succeed without them. World oil production will soon reach 100 million b/d. At an average price of US$60 per barrel, annualized and converted into Canadian dollars, this totals $2.9 trillion in gross revenue. Natural gas sales total another about $1.3 trillion per year. Plus natural gas liquids.

The total revenue is a big number, greater than the Gross Domestic Product of all but four countries in the world—China, United States, Japan, and Germany.

By comparison, the global mineral mining business is only a fraction of that size. Annual revenue for the top 40 global metals and minerals mining companies—which account for the majority of production—peaked at US$731 billion in 2012 when commodity prices were higher. This figure was only US$600 billion last year.[37] This includes coal. Conservatively assuming that this is only half of the total world mining revenue and converted to Canadian dollars, the 2017 figure only rises to about $1.5 trillion, about one-third of the size of the oil and gas business.

Which makes the giant oilfield service sector supporting global oil, gas, and NGL exploration and production globally proportionately huge. The challenge is how to measure it. Private research companies place the figure at US$139 billion by 2025, but this only includes a slice of the industry using specialized equipment such as drilling rigs (offshore and on land), directional drilling services, pressure pumping (cementing and hydraulic fracturing), and so on. The top 10 international OFS companies, including pipeline operators, will have estimated revenue of US$221 billion for 2018.

This would only be 5% of total production revenue so it clearly does not include the majority of the support services required to keep this industry going. If capital expenditures and production operating costs are included, it is not unrealistic to estimate that oil and gas developers spend more like 50% of revenue on third-party suppliers of knowledge, goods, and services. Based on the foregoing estimate of total oil and gas production value, this would make the worldwide OFS business $2 trillion annually.

In 2009/2010 your author was Chairman of the Petroleum Services Association of Canada (PSAC), the umbrella group for Canada's OFS industry. Each year the chairman undertakes a specific project of his or her choice to enhance or broaden the knowledge of the OFS industry. The question that really needed answering was how big Canada's OFS business really is.

Following the disastrous New Royalty Framework project of the Ed Stelmach PC administration of 2006 to 2011, it became glaringly obvious that even the Government of Alberta didn't understand the total size, reach, or economic

---

37 "Total revenue of the top mining companies worldwide from 2002 to 2017 (in billion USU.S. dollars)," *Statistica: The Statistics Portal* (2019). www.statista.com/statistics/208715/total-revenue-of-the-top-mining-companies/.

footprint of the province's carbon-resource-support industries. Announced in the fall of 2007, the New Royalty Framework increased Alberta production royalties significantly to ensure the people of Alberta, through their government, received their "Fair Share."

Alarmed by rising taxes and reduced profits at a time when other jurisdictions like BC, Saskatchewan, and the US had more competitive economic terms and similarly promising geological opportunities, investment capital fled Alberta. Drilling and investment declined as companies spent their money elsewhere. Before long, the ripple effects began to trickle through the economy. It was described at the time as "unintended consequences." These included empty hotels and restaurants in rural Alberta, declining new-vehicle sales, and falling house prices in urban centres.

What was eye-opening about the slowdown in oil and gas investment in Alberta and the greater economy is that they took place at the same time as oil reached its all-time peak of US$147 a barrel in July 2008 and natural gas prices were still three to five times higher than in 2018.

Ignorance of the impact of public policy on the oil and gas industry's OFS supply chain was not new. When the National Energy Program was introduced by Pierre Trudeau's Liberal government in 1980, the impact on drilling and investment was debated on the floor of the House of Commons. As drilling rigs followed their customers as they moved their money to more investor-friendly opportunities in US, the opposition accused the government of introducing policies that were chasing capital, equipment, and jobs out of the country. How was Canada supposed to become self-sufficient in oil supply in the face of a global supply shortages when the money and expertise was fleeing the National Energy Program?

The government responded by wanting to know if the equipment leaving was oil rigs or gas rigs. Oil rigs leaving was bad because there was a shortage of oil. The exodus of gas rigs was not a problem because there was a surplus of that commodity. That there was only one type of drilling rig to drill any kind of well was not known by the government that had just completely rewritten the rules for the entire Canadian petroleum industry.

Understanding the ripple effects of oil and gas investment requires good data. Unfortunately, there isn't any. Perhaps the reason is because most of this growth took place in the hinterland of western Canada, not close to or under the watchful eye of Ottawa-based Statistics Canada. Or another reason might be that in the early stages, growth was financed by foreign-owned oil companies, not local capital.

Either way, the subcategories of industry to describe the oil and gas support sectors are very narrow. The primary measurements for the oil and gas resources industries are the volume and value of oil, gas, and NGLs produced. Where, how, and on what the exploration and production companies spent their money was not considered important, and therefore not tracked or studied.

The main source of tracking economic activity in industrial subsectors is the North American Industry Classification System (NAICS) Canada, last updated in 2017. Companies code their income tax returns using this system, which then makes the information available for accumulation and analysis by Statistics Canada (StatsCan).

One large NAICS category covers *Mining, quarrying and oil and gas extraction.* It becomes slightly more granular under *Oil and gas extraction.* While this includes some spending on processing such as "natural gas cleaning and preparation plants," the sub-category does not include "performing oil field services for operators, on a contract or fee basis." The sub-sector *Services to oil and gas extraction* is an expanded description of over 40 categories of specialized services to put new wells on stream and keep them producing. *Oil and gas contract drilling* includes drilling rigs and some related services like directional drilling, but not the equipment components of the well.

But none of these classifications include many other key elements of the upstream oil and gas industry including construction, trucking, logistics, equipment, pipe, and much of the infrastructure required to move the resources to market and turn them into money.

The Canadian Energy Research Institute (CERI) in Calgary has long been regarded as a credible and independent source of statistical data about the role of the carbon-resource industries and their place in the Canadian economy. To figure this out, PSAC contracted CERI to use expanded Statistics Canada data to include other key industrial inputs required for this industry to function but which were not included in the existing NAICS/StatsCan process.

The definition of OFS was expanded to include all services related to oil and gas exploration, development and field support except exploration, mineral rights acquisition, development, production, and in-house finance and administration. This was subjected to a cross reference based on revenue in Canada and international markets for 36 of the larger, Canadian OFS companies in multiple sectors.

OFS was divided into four distinct classifications.

The first was *Direct*, the business most commonly considered to support field spending including drilling, production, engineering, construction, and professional and scientific services. This is the best-known OFS sector because of highly visible assets like drilling rigs and hydraulic fracturing equipment.

The second was *Indirect Manufacturing*, one layer back from what was traditionally considered oil and gas, but essential services nonetheless. This included non-residential building construction (shops, warehouses, plants, facilities), power boilers and heat exchangers (a growing business in oil sands thermal recovery), metal tanks (heavy gauge for fluid handling and storage), iron and steel pipes from purchased steel (casing, flow line, pipelines), mining and oil and gas field equipment manufacturing (more specialized equipment being built in Canada with every passing year), and pump and compressor manufacturing (the industry moves a lot of gas and fluid). The allocation to OFS was fractional because Statistics Canada data reported these categories for the entire country. Only a portion of the economic activity was allocated to the upstream oil and gas industry, often as little as 5%.

The third was *Other Manufacturing Industries*, which was treated the same as *Indirect Manufacturing* with allocations on a fractional basis. This included transportation and engineering construction, ready-mix concrete manufacturing, industrial, commercial, and service industry equipment manufacturing, engine turbine and power equipment manufacturing (prime movers for everything), and communication and energy wire and cable manufacturing.

The last category was *Other Industries*, which included sand, gravel, clay and ceramic and refractory minerals, mining and quarrying (including fracturing [frac] sand), communication engineering construction, machine shops, support activities for transportation, warehousing and storage, and truck transportation.

The first report was released in October 2010 and covered the 2006 calendar year, a good year for activity but hardly the strongest of the decade. That year Canada's GDP was $1.353 trillion, of which OFS accounted for $65 billion or 4.8%. Total taxes paid by the OFS (corporate and payroll) were $9 billion. Calculated this way, total OFS employment was 800,000 or 4.8% of the entire Canadian labour force.

By this measure, OFS was the second-largest industry in Canada on a GDP basis behind only that of its customers, the oil and gas explorers and producers. At $65 billion, OFS was only slightly smaller than the combined figures for automotive ($25 billion), mining ($18 billion), and agriculture ($26 billion).

As a cross reference, PSAC studied the revenue of 36 publicly traded, Canadian OFS companies involved in drilling and well services, oil and gas services and equipment, and pipeline, infrastructure, and processing (more commonly called midstream). These companies had a total revenue in 2009 of $12.8 billion of which 35% or $4.5 billion was from markets outside Canada.

Like the slogan in the old (and now illegal) cigarette commercial used to say, "You've come a long way baby."

PSAC made this information public and began trying to explain it to governments. Even in Edmonton (which once had a sign at the south entrance to the city billing itself "The Oil Capital of Canada," only to have city council take the sign down several years ago), the reaction was skepticism and in some cases disbelief. Politicians, bureaucrats, and regulators had never seen information like this before. But when the spreadsheets showing the methodology and mathematical linkage back to the Statistics Canada GDP input/output model were produced and analyzed, resistance evaporated. The provincial governments of the western producing provinces and the federal government all agreed the concept and methodology was sound and acceptable.

This was not the only undertaking to explain to Canadians how a dollar invested in oil and gas development had ripple effects throughout the Canadian economy in businesses not typically associated with the carbon-resource industries. Working with the Canadian Association of Petroleum Producers (CAPP), CERI created similar models for oil sands development and production. The reports revealed that while Alberta was the greatest beneficiary of capital investment in this business, Ontario, Quebec, and in the Maritimes also benefited.

A CERI/CAPP study indicated that in 2014 and 2015, which were still strong years for oil sands investment, the "Oil Sands Supply Chain" did business with 3,400 companies across the country spending $738 million in BC, $462 million in Saskatchewan, $1.57 billion in Ontario, $385 million in Quebec, and $145 million in the four Maritime provinces. CAPP states, "In 2017, the oil sands supported and created more than 205,000 direct and indirect jobs across Canada."[38]

The most common measurement of oil and gas investment is capital expenditures, which usually involves drilling for conventional oil and gas and plant-and-facilities construction for oil sands. Wells drilled and capital invested are big numbers that always make the news. The health of the industry is most commonly evaluated by this information.

But the biggest number nobody hears much about is existing-production operating costs. This has been increasing continuously this century as production volumes rise, as do the associated units' costs of production because of the increasing penetration of non-conventional sources of carbon such as oil sands.

Operating costs for Canada's upstream petroleum industry rise with production levels. In 2001 $11.4 billion was spent to operate conventional and oil sands production. By 2017 this figure had risen to $41 billion.[39] By way of comparison, the total expenditure of the government of Alberta in that year was $53.1 billion. British Columbia spent $46.7 billion, Saskatchewan $14.5 billion, and Manitoba $13.5 billion.[40]

Pulling the direct operating costs of the oil and gas industry out of the economy would be similar to shutting down the governments of Saskatchewan and Manitoba entirely, plus about 25% of the spending of either Alberta or British Columbia.

Another element of the oil and gas business, and OFS in particular, that is seldom included in analyzing its economic impact is capital investment by the OFS sector. These are not included in client capital expenditures or operating costs because they are financed from the balance sheets and capital providers of the OFS companies. They build new equipment and new assets and over time hope to recover their costs and a profit on business they would not get otherwise.

In 2014, 21 Canadian publicly traded OFS operators invested $21.5 billion in new revenue-generating assets. This included five drilling contractors, three companies in the remote workforce housing sector, five providers of downhole

---

38 Canadian Association of Petroleum Producers, "Oil and Gas: Creating 6.6 Million Person-Years of Employment," *Context* (2017). https://context.capp.ca/energy-matters/2017/btn_ceri-study-6-million-years-employment.

39 Canadian Association of Petroleum Producers, *Statistical Handbook for Canada's Upstream Oil Producers* (Calgary, AB: CAPP, 2018). https://www.capp.ca/publications-and-statistics/publications/316778.

40 Provincial government budget documents.

services such as hydraulic fracturing, waste disposal, and chemicals, and eight "midstream" companies building pipelines, processing, and storage facilities. Not all this money was spent in Canada, but it consumed a lot of steel, trucks, pumps, engines, valves, welding supplies, and labour.

In 2015, after the price of oil collapsed, these same companies still invested $16.4 billion. These activities tend to fly below the radar of job creation and economic benefits in many cases. They certainly employ a lot of people, consume a lot of materials, and pay their healthy share of payroll, corporate, and sales taxes.

The switch from conventional vertical drilling to extended reach horizontal drilling required a substantial investment by OFS to stay relevant. In the eight years from 2010 to 2017, five Canadian drilling contractors alone invested $8 billion in new equipment, primarily new-generation drilling rigs. As the new rigs rolled out, an equivalent number of old drilling rigs were mothballed as obsolete.

PSAC macro OFS figures were updated in 2015. The major change in methodology was to include for the first time a component of oil and gas production operating costs and expand the inputs to anticipate significant self-funded capital investment by OFS providers for 2009. Canada's upstream oil and gas industry had changed significantly in only three years. One major change was that the growth of extended reach horizontal drilling and hydraulic fracturing to exploit oil and gas shale deposits. This had resulted in significant investments in new equipment by OFS operators. Increased oil production had required more investment in production infrastructure for processing and transportation. And the production had switched from lower-cost natural gas to higher-cost oil sands.

In 2009 Canada's GDP was $1.48 trillion but OFS had expanded to $75 billion, or 5.1% of total GDP. OFS accounted for 685,000 direct and indirect workers. OFS was larger on a GDP basis that year than agriculture, mining, automotive, and forestry combined. The bulk of the activity—$39 billion—took place in Alberta. The next two largest, $21 billion in total, were Saskatchewan and BC. But $7.6 billion of the business took place in Ontario and $5.4 billion in Quebec.

The cross reference to Canadian-controlled and -headquartered OFS companies was expanded to 42. These companies had total revenues of $42.5 billion in the 2014 fiscal year, 54% of which took place outside Canada. Most of this was in the United States. This was further proof that the almost non-existent OFS business the Americans found in Alberta in 1947 had not only "Canadianized" but was also sufficiently large and capitalized to expand significantly into the same country that had helped create it nearly 70 years earlier.

Domestically founded, owned, and operated oilfield service companies supplying knowledge, equipment, services, and personnel to the Canadian and international oil and gas exploration and production companies is a very big business. Much of the equipment designed, engineered, and manufactured in Canada is sold or utilized by Canadian OFS operators around the world. Through this large, dynamic, and vibrant industry, Alberta's oilpatch abandoned the title of "hewers of wood and drawers of water" decades ago.

10

TODAY'S CARBON-RESOURCE INDUSTRIES

*"Canada is the fifth-largest petroleum- and natural-gas-producing jurisdiction in
the world behind the United States, Russia, Saudi Arabia, and Iran. Those who
propose Alberta and Canada do something else without fully understanding and
publicly admitting the financial impact are economically delinquent, politically
reckless, or both."*

Enormous and world-scale carbon resources have provided significant economic
advantages for Canada in general and Alberta in particular. Canada's oil, gas, and
coal endowments have long been the envy of the developed world, particularly
when prices are high. Canada is the fifth-largest petroleum- and natural-gas-
producing jurisdiction in the world behind the United States, Russia, Saudi Arabia,
and Iran. Those who propose Alberta and Canada do something else without
fully understanding and publicly admitting the financial impact are economically
delinquent, politically reckless, or both.

Of the top 10 countries in terms of oil reserves, Canada is number three. The
top 10 in order of proven oil reserves are Venezuela, Saudi Arabia, Canada, Iraq,
Kuwait, United Arab Emirates, Russia, Libya, and Nigeria. At the time of writing,
Venezuela is on the brink of economic and political collapse. Canada and Saudi
Arabia recently experienced a diplomatic tiff over Saudi Arabia's human rights and
its recent alleged murder of an American-based Saudi dissident in Turkey. Iraq
remains politically challenged 28 years after the First Gulf War in 1990. Russia is
regularly in the news for all the wrong reasons. Libya and Nigeria are in varying
degrees of political turmoil.

Meanwhile, Canada is the only one of these countries currently politically
engaged in penalizing and punishing carbon-resource industries in order to save
the other nine and the rest of the planet. Canada is the only major oil producer with
a leader who has publicly mused that the country must "phase out" of this business.

In terms of crude oil and liquid lease production (natural gas liquids or condensate), Canada ranks fifth in the world behind the United States, Russia, Saudi

Arabia, and Iran. For the global top 10, Canada ranks ahead of China, Kuwait, Mexico, Nigeria, and Venezuela. This list is dynamic because of constantly changing political issues in Nigeria and Venezuela.

Canada is also blessed with significant quantities of natural gas. The BP Statistical Review of World Energy for 2018 reports Canada is the fourth-largest producer of this resource behind, in order, the United States, Russia, and Iran. The next six for the top 10, in order, are Qatar, China, Norway, Australia, Saudi Arabia, and Algeria.

For proven reserves of natural gas, Canada drops down the ladder. However, most of these countries with a lot of gas don't have to deal with winter weather so the use of gas as a heating fuel is not as developed. The 15 countries with more proven reserves, according to BP are, in order, Russia, Iran, Qatar, Turkmenistan, the United States, Saudi Arabia, Venezuela, United Arab Emirates, China, Algeria, Nigeria, Australia, Iraq, Indonesia, and Malaysia. That Russia is the major gas supplier to Europe has become a political issue in recent years.

NGLs, the lesser known but nonetheless very valuable sister of crude oil and natural gas, have become a big business particularly with the unique reservoir content of recent large discoveries like the Montney formation in northwestern Alberta and northeastern BC NGLs include ethane, propane, butane, and condensates. While propane and butane are used as heating fuels, all these products are the chemical building blocks of the petrochemical and plastics industries.

Natural Resources Canada (NRC) reports that in 2016 (latest figures available) Canada produced 642,700 b/d of all of these products, of which 133,400 b/d were exported. This was offset by the import of 76,000 b/d of petrochemical feedstock (see Chapter 8). NRC data indicates 87% of this production came from Alberta. Half went to industry as feedstock. About 10% was used residentially for structural heating and barbecues. Nearly 25% was used commercially, primarily for heating, about 7% for transportation and 6% was used for agricultural applications. Provincially, Alberta used nearly 40% and Ontario almost 30%.

On a barrel of oil equivalent (boe) basis, ARC Financial Research Institute of Calgary estimates total crude oil, natural gas, and NGL production in 2018 will be 7.49 million boe/day, up about 50% from a decade ago. This will consist of 1.01 million b/day of conventional oil, 2.98 million b/day of bitumen (unprocessed oil sands) and synthetic crude (upgraded bitumen), and 2.60 million boe/day of natural gas.[41]

The location of the carbon resources within Canada has influenced settlement, in-country migration, external immigration, and the economic success of the regions in which they have been discovered and exploited.

Events have proven that notwithstanding concerns about the environmental impact of the growing consumption of fossil fuels, for the better part of human history there have been no significant economic penalties incurred or experienced by countries with handy access to domestic supplies of coal, oil, and natural gas. There have been lots of issues associated with producing carbon resources over the years, but selling it to somebody at a profit has, for the most part, not been one of them. Nor has being self-sufficient in domestic energy resources.

As will be explored more fully in the next section of the book, the majority of the problems caused by fossil fuels past, present, and future have been related to those *without* carbon telling those *with* carbon what they should and should not do next.

As explained briefly in Chapter 2 of this section, the best way to measure what carbon resources have done for the economy of the provinces or region fortunate enough to be blessed with them is by population. With (mostly) open borders for movement within Canada, citizens and immigrants get to vote with their wallets and decide where they want to live. The statistics indicate they will follow the money. Left to their own devices, people make economic decisions in their self-interest. Finding steady and lucrative employment is certainly one of them.

The 1941 census indicated Canada's population was 11.5 million. Alberta, with only 796,169 residents, comprised only 6.9% of the country's population. Ontario and Quebec were the political powerhouses federally because they comprised 61.9% of the population, 3,787,655 and 3,331,882 respectively. The four Maritime provinces of Prince Edward Island, Nova Scotia, New Brunswick, and Newfoundland (still a British colony in 1941) contained 12.6% of the population, a total 1,448,610. The four western provinces of Manitoba, Saskatchewan, Alberta, and British Columbia held 28.2% of the population, 3,239,786 in total.

In 1947, the discovery of large quantities of crude in Alberta in 1947 changed everything. In the next 75 years until the 2016 census, the migration, immigration, birth, and population growth patterns for Canada would change significantly for

---

41   ARC Energy Research Institute, *ARC Energy Charts*, December 17, 2018.

the region with the greatest concentration of carbon resources and the country as a whole.

By 2016 the population of Canada was 35,151,728. But the growth was not evenly distributed. The historic political and economic powerhouse of Ontario and Quebec saw their combined population fall slightly as a percentage of the total to 61.5%, 13,448,494 and 8,164,361 respectively. In 75 years Ontario grew by 255% while Quebec expanded by only 145%.

Canada's Maritime provinces, the smallest in the country, clearly peaked early. They were settled first because they were the closest to Europe by sea, but did not have the land area, resource base, or proximity to the huge American market-place enjoyed by other provinces. By 2016 the percentage of the total Maritime population had fallen by almost half to 6.6%. None of the provinces kept pace because of an average growth rate of only 61.1% for the 75 years since 1941, less than one-third of the country as a whole. Their total population in this census was only 2,333,322.

The region that outpaced the rest of the country was the west. By 2016 these four provinces comprised 31.5% of the total population, a total of 11,091,947, up by 242.4% with 7,852,161 more people than 75 years earlier. The growth was not evenly distributed. Manitoba gained 548,621, or 7.0% of the west's total growth. Saskatchewan saw the inflow of only 202,360 extra citizens in 75 years, accounting for only 2.6% of the west's meteoric population expansion in seven and one-half decades.

Alberta and British Columbia grew by 3,271,006 and 3,830,174 respectively. Alberta grew by 411% and BC by 468%. These two provinces were responsible for 90.4% of the growth in the west and 30.0% of the population growth of Canada, a total of 7,101,180 people.

Except for the mountains in the southwest corner and as discussed earlier, Alberta, Manitoba, and Saskatchewan are similar geographically and in terms of non-carbon resources. As mentioned, their land masses are almost identical. Manitoba has seasonal access to tidewater on its northeastern border through Hudson Bay. This could become year-round if global warming and Arctic-ice-contraction predictions come true.

Manitoba is somewhat closer to larger US markets in the Midwest. But the three states on the other side of the 49th parallel with which the prairie provinces share a border are not large measured by population and industry. Of the 50 US states, North Dakota ranks 48th by population and 45th by GDP; Montana 44th by population and 48th by GDP; and Minnesota, which shares about 100 kilometres of its northern border with Manitoba, is the 22nd largest state by population and 17th by GDP.

Of the three prairie provinces, Manitoba is the clear winner in terms of physical market access to the rest of the world with tidewater (seasonally) to the north and Minnesota to the south. However, the only year-round major transportation link to Hudson Bay—a railway—washed out in 2017 and has not been rebuilt. It has recently changed hands and been refinanced and should be operational soon.

British Columbia is an entirely different province in many respects. With the exception of the northwest border it shares with Alaska, BC has significant access to the Pacific Ocean and all the resources and opportunities that come with it. In the past fishing has been a major industry and trans-Pacific shipping still is as Canada's western port.

BC's forestry industry has always been, and continues to be, a major economic contributor, as does mining of a vast array of minerals, including coal. The north-eastern section has significant natural gas resources and some petroleum and natural gas liquids. Ports in BC export a lot of coal that is sourced from its own province, Alberta, and the United States. BC is also planning to expand into the exportation of LNG.

BC has two other assets other three western provinces do not: tourism and retirement. Both are significant economic contributors. When people sell their worldly belongings and move to another place for their autumn years, they don't often select Winnipeg, Regina, Saskatoon, Calgary, Edmonton, or any of the smaller centres in Manitoba, Saskatchewan, or Alberta. They seek the comparatively milder climes of BC Alberta has a tiny portion of BC's tourism and retirement industry in its eastern flank of the Rocky Mountains, primarily Banff and Canmore.

By comparison, BC is a tourism and retirement mecca in its interior, lower mainland, west coast, Vancouver Island, and the Gulf Islands. This imports billions of dollars from other parts of Canada and the world, resulting in significant economic opportunities that hardly register in the other three western provinces.

Alberta's carbon resources have been the major economic and population differentiator between it and its two sister prairie provinces. All this becomes extremely important as Alberta and Canada contemplate a transition to a low-carbon economy as is increasingly suggested must be done.

On a GDP basis, Canada's carbon-resource industries are the third largest. But as discussed in the previous chapter, the industry, despite its considerable size and impact on the economy, has never rated sufficiently detailed analysis to give economists, policy-makers, and the business itself the type of information required to make the best policy decisions. What is needed is data on the ripple effects in the goods-and-services support sector when the profitability of primary resource production is changed by law or regulations. The consequences of this will be discussed in detail in the next section, Carbon Politics.

According to Statistics Canada the "GDP at basis prices by industry" of the entire country, seasonally adjusted on a monthly basis for October 2018 was $1,946 trillion in "chained 2012" dollars. "Chained dollars" means the recent figures corrected for inflation for a more practical and useful comparison of the current versus prior periods.[42]

The broad categories are summarized as follows. Non-business sector industries, or the public sector, is the largest single segregated unit at $370 billion, or 17.9%. The next largest is real estate and rental and leasing—the country's commercial building and housing market—calculated at $248 billion. Third was manufacturing at $203 billion, followed by energy as a whole at $182 billion. Oil and gas extraction, mining, and quarrying, was the fifth-largest economic sector at $150 billion.[43]

Other notable industries are construction at $141 billion, finance and insurance at $130 billion, transportation and warehousing at $89 billion, information and communication technology and services at $88 billion and utilities at $44 billion. Other resource-related sectors considered quintessentially Canadian are agriculture, forestry, fishing, and hunting. They total only $35 billion combined.

By this calculation the mineral-extraction business (carbon resources plus mining) is 4.3 times larger than the other major resource industries combined. Removing stuff from the ground, processing, and selling it remains a core component of the Canadian economy regardless of what people think we do, like to think we do, or think we should do.

---

42 "Gross Domestic Product (GDP) at Basic Prices, by Industry, Monthly (x 1,000,000)," *Statistics Canada* (2019). https://www150.statcan.gc.ca/t1/tbl1/en/tv.action?pid=3610043401.

43 Ibid.

Perhaps this is explained by the growing urban/rural split. By 2011 Statistics Canada reported only 18.9% of the population was classified as rural where people harvest resources.[4]

On an annual basis, Statistics Canada provides significant granularity by business subsector. While "GDP at basic prices by industry" in 2012 chained dollars for conventional and non-conventional oil and gas extraction in 2017 was $101 billion, mineral mining and quarrying (coal, iron, gold, silver, copper, lead, zinc, and other metals) totalled only $22.2 billion. Other industries often in the news and seen as critical to the Canadian economy, are motor vehicle parts manufacturing at $8.8 billion, motor vehicle manufacturing at $6,5 billion, aerospace product and parts manufacturing at $7.6 billion, and pulp, paper, and paperboard mills totalling $4.7 billion.[44]

Other sectors not included in oil and gas extraction, but which should be added back because they would not otherwise exist, are crude oil and related products pipeline transportation at $5.0 billion, and natural gas pipelines at $4.8 billion. (This assumes the largest pipeline investments and expenses are the production gathering systems in Western Canada and which move product east, south and west. Canada would still get oil and gas from somebody, and this would require pipelines). Then there is the entire upstream oil and gas industry supply chain discussed in the previous chapter that is not listed in the Statistics Canada monthly data.[45]

From a revenue perspective, oil and gas production is significant. In its December 17, 2018, weekly snapshot of the Canadian oil gas and NGL industry, ARC estimated the value of production for the year at $110.4 billion ($14.3 billion lower than only five months earlier because of the collapse of Canadian oil and gas prices in the fall of 2018), after-tax cash flow available for reinvestment of $46.8 billion, and reinvestment (commonly called capital expenditures) in oil sands and conventional production of $40.4 billion. While these are big numbers, the same figures for 2014 before oil prices collapsed were $149.8 billion, $72.2 billion, and $82.7 billion respectively.[46]

---

44 "Gross Domestic Product (GDP) at Basic Prices, by Industry, Annual Average, Industry Detail (x 1,000,000)," *Statistics Canada* (2019). https://www1150.statcan.gc.ca/t1/tbl1/en/tv.action?pid=3610043406.

45 "GDP, Annual Average," *Statistics Canada* (2019).

46 ARC Energy Research Institute, *ARC Energy Charts* (December 17, 2018).

While reinvestment this year is up from the recent low of $38.4 billion in 2016, it is only a fraction of the 2014 values. If you noticed some wailing from the oil industry about cruel and unusual punishment by commodity markets and various governments, these numbers will help explain why.[47]

Carbon is down but certainly not out. For all these numbers to mean something, comparisons with other industries are essential. The big six banks had a total revenue of $173.9 billion according to *The Globe and Mail*'s "Report on Business Top 1000" ranking of Canada's largest companies by revenue. But the two main railroads, Canadian National and Canadian Pacific, were less than 10% of oil and gas production at $18.4 billion.

The big outfits that move oil and gas around or perform field processing— Enbridge, TransCanada, Pembina, Keyera, AltaGas, and Inter Pipeline—had combined revenues in 2017 of $59.4 billion, although not all of it was in Canada. Adding this back to the value of the production would increase the size of oil and gas yet again. No figures on how much the petrochemical business alone adds to GDP or revenue are publicly available. While Canada might still have a petrochemical business if it did not produce oil and gas, it certainly wouldn't be based in Alberta without the feedstock close nearby.

Another important way to analyze today's oil and gas industry is by employment. Direct employment figures through Statistics Canada are subjected to the same large-sector analysis that make granularity difficult. Using NAICS codes, Statistics Canada reports that in 2018 there were 18.7 million people employed, of whom only 272,300 worked in the broad bucket of "mining, quarrying and oil and gas extraction."[48]

At only 1.4% of the workforce, what could all the fuss be about? This is a fraction of the people working in construction, manufacturing, transportation and warehousing, financial and insurance, professional scientific and technical services, or even information culture and recreation.

But this data doesn't match that from other sources. As referenced in the previous chapter, the broader model for oil sands development that goes further down the industrial supply chain developed for CAPP by CERI identified a much larger industry than that identified by NAISC and Statistics Canada. "In 2017, the oil

---

47 Ibid.

48 "Labour Force Characteristics by Industry, Annual (x 1,000)," *Statistics Canada* (2019). https://www150.statcan.gc.ca/t1/tbl1/en/tv.action?pid=1410002301.

sands supported and created more than 223,000 direct and indirect jobs across Canada."[49] Figures for the country using the same methodology and all oil and gas production is 533,000 jobs in 2017, more than double the official Statistics Canada number.[50]

That was in 2017, after capital investment had plummeted compared to previous levels. Using ARC Energy Research Institute data, oil sands capital investment in 2014 was $33.9 billion. By 2017 it went down by 49% to only $13.2 billion. CAPP reports operating costs for oil sands in 2017 were $19 billion, down from $24.3 billion in 2014. The reductions are from operating efficiency, lower natural gas prices, and reduced prices for products and services. Operating costs for conventional production in 2017 were $22 billion.[51]

But the oil sands figures don't include the rest of upstream oil and gas industry. What about those jobs? Can or should major policy decisions about the future of this core Canadian industry be made when the regulators and the regulated can't even work from a common data set about the economic impact of this industry?

CERI issued another report in August 2017. The executive summary opens with, "The Canadian oil and gas industry is a significant contributor to the provincial and national economies in Canada. For the foreseeable future, natural gas and crude oil will be important elements in many economic sectors in Canadian and North American economies." Because Canada exports so much of its oil and gas to the US, CERI wanted to highlight the degree to which the economies of the two countries are intertwined. The US is a customer but also a major supplier of goods and services.[52]

CERI breaks employment and investment into direct and direct, stating, "For every direct job created in the Canadian oil and gas sector, two indirect and three induced jobs in other sectors are created in Canada on average. For every Canadian

---

49   Canadian Association of Petroleum Producers, *Canada's Oil Sands*, (July 2018).

50   Canadian Association of Petroleum Producers, "How Many Jobs Does Canadian Oil and Natural Gas Support?" *Context*, (April 18, 2018). https://context.capp.ca/infographics/2018/infographic_533000-jobs.

51   Canadian Association of Petroleum Producers, *Statistical Handbook for Canada's Upstream Oil Producers* (Calgary, AB: CAPP, 2018). https://www.capp.ca/publications-and-statistics/publications/316778.

52   Canadian Energy Research Institute, *Economic Impacts of Canadian Oil and Gas Supply in Canada and the US (2017-2027)*, (August 2017).

million dollars invested and generated in the Canadian oil and gas sector, the Canadian GDP impact is $1.2 million."[53]

In the middle of the study period, 2022, CERI sees about 105,000 jobs in natural gas, 135,000 jobs in conventional crude oil, and nearly 410,000 jobs in oil sands. This totals 650,000, a much larger figure than reported by Statistics Canada using conventional reporting by industrial sector. Total taxes paid to federal and provincial governments, directly and indirectly, including personal taxes, will be $40 billion.[54]

While the cost of renewables is indeed declining, they are not yet at a cost at which they are competitive with fossil fuels plus pay substantial net taxes to the provincial and federal treasuries.

For Alberta, the dependency on carbon resources for economic activity and employment is significant. One way to look at this is to "decarbonize" the economy using Statistics Canada data. Oil and gas extraction industries are only part of the picture. As a major producing region, Alberta also has significant support industries that would not otherwise exist in the province, as is the case in Manitoba and, to a lesser degree, Saskatchewan (which both produce oil but not on the same scale as Alberta).

Support industries include Statistics Canada classification such as "support activities for oil and gas extraction," "oil and gas engineering and construction," "petroleum and coal manufacturing," "pipeline transportation of natural gas," "crude oil and other pipeline transportation," "petroleum refining," "transportation construction engineering," "basic chemical manufacturing," "plastic product manufacturing," "machine shops, turned product," and "other chemical product manufacturing." In 2017 the GDP contribution of these sectors was $108 billion, 35% of the provincial total of $305 billion.

The argument can be made that that the last five—transportation construction engineering, chemical and plastic manufacturing, or machine shops—could be in other provinces, but they only account for 3.6% of the total.[55]

---

53 Ibid.
54 Ibid.
55 Hossein Hosseini, PhD, Economics Researcher, Canadian Energy Research Institute.

Using this methodology and adjusting for the fact these businesses might exist but in a much smaller form, Alberta without oil and gas would see its 2017 GDP decline by 32% from $305 billion to $206 billion.[56]

Big numbers. What's a GDP, anyway? Statistics Canada also includes jobs using what is called the "employment IO (input output) multiplier." Using the same categories for employment in Alberta in 2014, the result would be the loss of 355,352 jobs. That's a lot of solar panel technicians and windmill installers.[57]

On a cautionary note, StatsCan IO models assume an unlimited supply of labour and capital is available and that wage and interest rates are fixed in the economy. If the oil and gas support industries were removed, it is not unreasonable to assume that at least some of the jobs might have existed in other industries.

However, the reality of the importance of this industry in this province is undeniable, and prognostications of what Alberta would look like without carbon resources are regrettably shallow and incomplete. Political platitudes of a great future doing an as-of-yet unknown "something else" make pretty thin soup for the unemployed.

While pipeline opponents have been successful in stopping pipelines and slowing production growth, according to CAPP, just keeping the industry going cost $83 billion in 2017. This includes direct operating costs for oil sands, conventional oil and gas, and production maintenance capital expenditures for both. The total is greater than the combined annual spending of the governments of Alberta, Saskatchewan and Manitoba which provide essential services for nearly 7 million people.[58]

Many Canadians assume the country can get by without its oil and gas industry. Nobody feels the same way about the entire governments of the three prairie provinces, even their harshest critics, depending upon which party is power.

The last aspect to be considered is when developing future carbon and energy policy is the scope of the existing asset base. Alberta produces oil, gas, and natural gas liquids from every part of the province except the extreme northeast (about 15% of the total area, which does not contain sedimentary geology conducive

56 Ibid.

57 Ibid.

58 Canadian Association of Petroleum Producers, *Statistical Handbook for Canada's Upstream Oil Producers* (Calgary, AB: CAPP, 2018), https://www.capp.ca/publications-and-statistics/publications/316778.

to carbon resource accumulation) and the along the southeast border with BC, which is within Waterton, Banff, and Jasper National Parks.

Producing, processing, upgrading, refining, and transporting all this resource requires a significant investment in surface infrastructure. According to the AER this includes over 30,000 separate field production facilities for oil and almost 21,000 for natural gas.

Each of these are of significant size and mass, are primarily made of steel, sit on industrial locations that may be contaminated after years of services, and will cost literally billions to mothball, decommission, and reclaim.

Those are the small ones. There are over 600 natural gas processing plants in Alberta, including those that recover sulphur and natural liquids (NGL). Of these, 11 are classified as fractionation plants, which break NGLs into specific components such as ethane, propane, butane, and condensate. Eight plants are called straddle plants, which remove NLGs like ethane from natural gas transmission pipelines for use as petrochemical feedstock.

There are six bitumen upgraders in the province that convert over 1 million b/d of oil sands production to synthetic oil. Four refineries process 462,000 b/d of oil providing products primarily for the local market.

The point of this chapter is to point out exactly how big this business is and how many new jobs and investment will be required to switch from carbon resources to renewables, incurring massive economic disruption and dislocation in a giant industrial sector not well understood by carbon's critics or government policy-makers.

This is just the jobs, revenue, and economic contribution. To decommission all these assets, properly dispose of them, and perform comprehensive site remediation up to modern environmental standards will cost tens of billions, if not more.

Three and one-half years after the oil price collapse, economic observers still report the obvious. Who is listening remains in question. On May 17, 2018, *The Financial Post* carried an article titled, "Oil Pays the Bills – We Owe Our Economic Prosperity and Relatively High Per-Capita Income to Trade – And Crude Oil Dominates That Trade". Written by former Vice-President, Economics and Commodity Market Specialist with Scotiabank in Toronto, Patricia Mohr, she stated:

> "... many Canadians appear unaware of how critically important
> the oil industry is to the national economy ... In 2014, before

the oil price downturn, crude oil alone generated a $70-billion trade surplus for Canada – excluding smaller surpluses in refined petroleum and natural gas – far outstripping any other export category ... Even at the bottom of the oil-price correction in 2016, crude oil remained the largest contributor to Canada's merchandise trade, generating a $33-billion surplus."

Mohr added, "In 2016, the latest year for which data is available, Alberta payments to Ottawa were around $22 billion more than it received, amounting to $5,100 per capita, helping to fund equalization payments to Quebec, the Maritime provinces and Manitoba. Alberta has consistently shared a portion of its resource wealth with the rest of Canada. Per capita, Alberta's net federal payments are five times bigger than any other province." This is because, says Mohr, "Oil industry and tax and royalty revenues have strengthened the overall fiscal position of governments across Canada, helping to fund social services."

Based on a research paper from the University Calgary's School of Public Policy by Tim Hearn and Robert Mansell, on January 29, 2019 economist Jack Mintz wrote in *The Financial Post* that between 1961 to 2017 Alberta contributed $611 billion (in 2017 dollars) more to Canada's finances that it got back, second only to Ontario. Carbon comes through with the big buck yet again: "... since 2010, Albertans have paid $180 billion in net fiscal transfers, more than any province. No other province has contributed so much to the rest of Canada via federal taxes and spending. Ontario has a population three-and-half times larger than Alberta, but it only transferred $45 billion to the rest of the country since 2010."

During the same period under review and using the same methodology, Quebec received $476 billion, Nova Scotia $306 billion, New Brunswick $203 billion, Manitoba, $175 billion, and Newfoundland and Labrador $172 billion. British Columbia is the only other net contributor at about 10% of Alberta's total.

Saving the world by blocking Alberta's access to world markets and decarbonizing Canada is going to be expensive for everyone, not just morally delinquent oil sands producers.

Before the opponents of fossil fuels assure everyone displaced by decarbonization a promising new career in "cleantech" or renewable energy sources, it would be helpful if analysts and policy-makers could agree on how big the carbon-resource business really is, how many people actually work in it directly and indirectly, and what sectors and regions of the Canadian economy will be affected if major policy initiatives in that direction are undertaken in a broad and aggressive manner.

As opponents of carbon resources preach their "leave it in the ground for your own good" message for the oil sands, it would be nice if, for once, they asked the hundreds of thousands of Canadians whose jobs are considered a necessary casualty of global environmental salvation what they think about it. This is on top of the hundreds of billions of dollars of capital investment in assets, facilities, infrastructure, and the equity value of the companies themselves. Much of the ownership of these assets is held by private and public-sector pension funds owned by millions of Canadians.

At some point carbon's army of selfless critics must start attaching numbers to their claims and demands.

The ongoing and unsupported claims that there are sufficient economic opportunities in non-carbon energy in Alberta to accommodate all those displaced from their current employment should Alberta completely decarbonize its economy is absolutely incorrect and utterly disingenuous.

Section 2

CARBON POLITICS

Introduction
CARBON RESOURCES HAVE ALWAYS BEEN POLITICAL

"When anything becomes important, strategic, essential, and allegedly dangerous, its politicization is inevitable."

Carbon resources—coal, oil, and natural gas—are surely the most politicized naturally occurring compounds contained in the earth's crust. Once early man learned they were of great value in making life easier, they quickly became sought after.

Towards the end of the steam-powered Industrial Revolution, carbon energy—today more commonly called fossil fuels—was in progressively more in demand.

When liquid carbon fuels combined with technology to power the machines of war (first ships; later trucks, tanks, and airplanes) the value of carbon was upgraded from a convenience to a strategic necessity.

Then people figured out how to do more things with coal, oil, and natural gas; they began making plastics, medicines, and chemicals. At this point, carbon and carbon resources became essential to modern life.

But now carbon is increasingly identified as a menace. Ironically the perception of the resource has come full cycle from where it started.

When a commodity becomes important, strategic, essential—and allegedly dangerous—its politicization is inevitable.

The world's first major political intervention into the extraction of carbon was in the Middle East after the British made a huge oil discovery in 1908 in what was then called Persia (now Iran).[59] Crude oil packed a lot more energy punch volumetrically than coal; hence, the then-dominant British Navy considered oil strategic because it fuelled its warships.

In a 100th anniversary article of the 1908 discovery, author Randy Alfred wrote, "At the instigation of First Lord of the Admiralty Winston Churchill, the British

---

59   The British company that made the discovery was Burmah Oil Co. After multiple reorganizations it become the Anglo-Persian Oil Co., and later British Petroleum. In 2000 the company dropped the word "petroleum" from its name to appear greener and is now known simply as BP.

government became a majority (and at-first secret) shareholder of Anglo-Persian during World War I."[60]

More was discovered in Egypt in 1910. For Europe, the Middle East and North Africa became increasingly interesting, particularly with oil fuelling modern war machines from 1914 to 1918. After World War I, the partitioning of the Ottoman Empire during and by Britain, France, and Italy demarcated the borders of Turkey, Iraq, Syria, Jordan, Saudi Arabia, and Lebanon. The creation of these countries' borders employed survey transits resulting in straight lines, just as in western Canada.

This was the start of a century of carbon politics when politicians and governments began to interfere and intervene with the ownership, price, availability, and future of carbon resources. The problems with this form of nation-building continue to plague this region today. It has been long documented that Western military and political intervention in the affairs of countries in the Middle East and North Africa has cost hundreds of billions of dollars and tens of thousands of lives, all to ensure the uninterrupted supply of carbon fuel for the economies of the Western world.

These were battles *for* carbon. Today we have a battle *against* carbon. Fortunately, nobody has mobilized the military just yet, but this fractious engagement is in its early stages. All manner of ideas have been proposed to control carbon in the atmosphere. Many will be explored in this book.

This will come as a huge shock, but the main thrust of carbon politics in Canada has been—I hope you're sitting down—political. The direction of the day from federal, provincial, and municipal governments has been driven primarily by what elected politicians believe will be most popular with their supporters. It is of course rarely stated as such, certainly by the sponsor. Any new policy will be purportedly be based on the greater good, and couched in lots of words riddled with selflessness, altruism, forward-thinking, fairness, and the betterment of the community/province/country/planet.

It is useful to repeat the famous quote from the George Orwell essay from 1946, *Politics and the English Language*. Clearly jaded from the propaganda of World War II, Orwell wrote that political language, "is designed to make lies sound truthful and murder respectable, and to give an appearance of solidity to pure wind."

---

60 Randy Alfred, "May 26, 1908: Mideast Oil Discovered? There Will Be Blood," *Wired* (September 26, 2011). https://www.wired.com/2011/05/0526mideast-oil-strike/.

Federally, Canadian carbon policy (most often called energy policy, but it has been mostly about carbon resources) has reflected the tenor of the times. Up until the OPEC oil shock of 1973, Ottawa had been largely supportive of the development and exploitation of all carbon resources. The Geological Survey of Canada (GSC) was founded in the Province of Canada (then a British colony; now Quebec and Ontario) in 1842. Its original purpose was to conduct an inventory of the mineral resources of the region. Its mandate expanded with the country. Its website reads, "The GSC is Canada's oldest scientific agency and one of its first government organizations. It was founded in 1842 to help develop a viable Canadian mineral industry by establishing the general geological base on which the industry could plan detailed investigations. Throughout its long and colourful history, the GSC has played a leading role in exploring the nation."[61]

After a lot of oil and gas was discovered in Alberta and later Saskatchewan, British Columbia, and Manitoba, the federal government was quite helpful. To get all product—coal, oil, and gas—to market, Ottawa first ensured a transcontinental railroad was built in the 19th century. Then in 1949 the federal government passed the Pipe Lines Act to give it control over the interprovincial transport of crude oil and natural gas. This was to expedite proceedings in the national interest and overrule provincial obstacles, opposition, and politics. This greatly assisted construction of the first petroleum and natural gas pipelines east and west from Alberta in the 1950s. The National Oil Policy of 1961, which assured the Ontario market for western oil, was very helpful at the time.

In reviewing national energy policies pre-OPEC, the primary objective of the federal government was to help provinces and industry turn resources into jobs, wealth, and prosperity. While there were problems along the way and industry didn't always get everything it wanted, up until 45 years ago coal, oil, and gas producers for the most part had a strong ally in the nation's capital regardless of which party was in power.

All that changed when OPEC quadrupled the price of oil in 1973. The relationship with the federal government has been different ever since. Two chapters in this section are devoted to federal energy politics, starting in the late 1950s with the fallout from the Great Pipeline Debate of 1956 and ending with the Liberal federal government under Justin Trudeau committed to combatting climate

---

61 "Geological Survey of Canada," *Natural Resources Canada* (2018). https://www.nrcan.gc.ca/ earth-sciences/science/geology/gsc/17100.

change and meeting the so-called "Paris Agreement" commitments of significant GHG-emission reductions by 2030.

While control of subsurface resources was originally denied to the Province of Alberta when it was established in 1905, in 1930 Ottawa finally transferred that control. Discoveries at Medicine Hat at the turn of the century and at Turner Valley in 1914 and 1924 were interesting, but they were hardly considered national in scope, and therefore not worthy of federal government attention. The railroads had supplied themselves and Central Canada with as much Alberta coal as required and all was well with the relationship between Ottawa and Alberta regarding energy policy.

Until Leduc, Alberta was extraordinarily ordinary. The 1929 stock market crash did not spare Alberta. The province had difficulty paying interest on its borrowings. It had to bail out the Alberta Wheat Pool. During the Depression, cars were parked because gasoline cost too much. Horses were put back into service, pulling motorless cars (with the engines removed to cut weight) (MacGregor 1981).

In 1932 a new political party was formed when William Aberhart combined religion with economics to form the Social Credit, or "Socred" Party. In 1934 sitting Premier John Edward Brownlee of the United Farmers of Alberta was so desperate for ideas that he invited Aberhart and the man who invented the concept of Social Credit, Britain's Major Douglas, to Edmonton to share their ideas. Nobody had a clue what they were talking about.

Famed Canadian humorist Stephen Leacock wrote *My Discovery of the West* in 1937 In a chapter titled "Debit and Credit in Alberta," Leacock states, "Social Credit in the sense of economic theory is mere wind, words and nothing else. It creates a vague ideal of 'purchasing power' and wishes that everybody might have lots of it. So do I."[62]

The new party contested the 1935 election. Thanks to Albertans who had endured too much misery for too long, the Socreds won a large majority government.

The Depression continued. In 1936 Alberta defaulted on $3.2 million in government bonds and passed a law cutting the interest rate in half. Desperate for cash, the Socreds started paying bills, including wages to civil servants, with "prosperity certificates." In theory, these could be exchanged for goods, which the merchant

---

62 Stephen Leacock, *My Discovery of the West: A Discussion of East and West in Canada* (1937). https://gutenberg.ca/ebooks/leacock-mydiscoveryofthewest/leacock-mydiscovery-ofthewest-00-h.html#chap07.

could later redeem for cash from Edmonton. Labelled "funny money," these were quickly judged by Canada's Supreme Court not to be legal tender and cancelled.

In 1946, pre-Leduc was described this way. "Alberta's prospects were grey. The number of farmers was bound to decline. No one seemed to want two of Alberta's well-known resources, its 25 million tons of mineable coal and its 171,000 square miles of forests…. It appeared that Alberta's prospects had reached their zenith and were destined to decline" (MacGregor 1981).

Then large quantities of oil were discovered in 1947. Alberta and its politics would never be the same.

The other major element of modern carbon politics is climate change. As former Alberta Premier Ralph Klein said many times, the secret to success in politics is to find a parade and get in front of it. In this century there has been growing interest and concerns among the media, environmentalists, and a growing number of voters about the global-warming phenomenon.

There has always been a strong core of public distrust of corporations and capitalism in general, and oil companies in particular. That big oil companies and human decadence were combining to destroy the environment and the world was a perfect justification. The politically ambitious, always with their ears to the ground, realized the potential of turning climate change into a political issue.

Of course their motives, like those of all politicians, were and continue to be altruistic and exclusively for the greater good.

Scientific study of the atmosphere and climate have been conducted for centuries. Linkages between carbon dioxide and global warming were published in scientific papers as early as 1896. In 1972, 30 years after its creation in the middle of the World War II, the United Nations sponsored the first of many global environmental-protection conferences. As the UN's focus shifted to global warming and then climate change, these conferences would eventually shape the carbon politics of Europe, the United States, and Canada. The stated objective is to change the future of the world's energy complex to save it from the ravages of climate change.

The following chapters will discuss the relationship of carbon and politics, carbon policies in Alberta and federally, where the anti-carbon movement came from, what it has evolved into, and what politicians and the concerned have done and say what must be done to try to control the weather. Which is surely the most ambitious worldwide political undertaking in history.

1

THE ROLE OF POLITICS IN SHAPING CARBON POLICY

"If elected as part of a majority government, can that seemingly ordinary person who rang your doorbell last election and asked for your vote accomplish what used to be the sole purview of science fiction? Make it rain more or less? Control the temperature? Stop the glaciers from melting? Reduce the intensity of tropical storms?"

The old adage goes, "The road to hell is paved with good intentions." It is not enough to mean well; one must also do well. Put another way, trying is not the same thing as succeeding.

When it comes the politics of carbon, on multiple occasions Canadian political leaders have tried a variety of initiatives and policies in response to public pressure and voter expectations. The result has been regulations, legislation, and programs intended to make things better, improve the situation, or right egregious wrongs.

However, an objective review of the outcomes will reveal politicians are not always successful in accomplishing their stated objectives. Often these initiatives have resulted in "unintended consequences," collateral damage, or materially different results than what was promised and planned. But no matter how much the outcome differs from the promise or intent—even when disastrous—those running for public office keep promising and the voters keep asking.

Most of the major carbon-resource-related commitments and policies of the last half-century have been changed, amended, or withdrawn. A large number of the current slate of climate change and GHG mitigation strategies, programs, and laws in Canada might be described as "dynamic," that is, unlikely to survive beyond the next election. Sometimes the short shelf life of certain government initiatives is because they were overtaken by international events, technological advancement, or changing markets. Other times they were highly partisan or simply ill-conceived from inception and proven to be so once implemented.

Despite continuing media and public pressure to "do something"—and assurances from those seeking to govern that something will indeed be done—in North

America public trust in government institutions has been in general decline for the past 60 years. The questionable success rate of what governments accomplish—beyond providing essential public services and keeping the peace—might explain why.

Edelman is an international public affairs and research company founded in the US in 1952. Its mission reads, "We are a global communications marketing firm that partners with the world's leading businesses and organizations to evolve, promote and protect their brands and reputations."[63]

What the public is thinking or believes has been a core element of Edelman's service offering since the company's inception. Edelmen has been conducting large-scale surveys into the issue of public trust on behalf of its clients since 2000.

2018's Edelman Trust Barometer revealed that only 46% of Canadians had confidence in their governments, lower than their trust for non-government organizations, business, and the media. In the US the figure is only 33%. This means over half the people don't believe their publicly elected institutions are acting in their best interests. This distrust has not been improving. In 2014 Edelman reported that only 15% of respondents believed their political leaders "make ethical and moral decisions" and only 16% were confident politicians could "solve social or societal issues."

This century, new carbon policies and programs across the country have been focused on climate change, and on attempts to mitigate or reduce the impact of fossil fuel emissions. Think of the dozens of programs, plans, subsidies, and regulations that have been introduced to address this challenge. They involve everything from government financial aid to replace your furnace or hot water heater, the legislated inclusion of plant-based ethanol in gasoline, cash subsidies for electric vehicles, or incentives to equip your house with solar panels.

These are increasingly proving to have a much shorter shelf life than anyone imagined when they were created. The new Progressive Conservative government of Ontario, elected in 2018, reversed carbon policies introduced by the previous Liberal administration that began in 2005. The policies were extremely unpopular because of higher-than-anticipated costs and questionable outcomes. A year earlier, prior to the 2018 election, the incumbent Ontario Liberals had actually

---

63 "About Us," *Edelman* (2018). https://www.edelman.com/about-us.

begun to undo some of their own climate change initiatives to attempt to mollify public dissatisfaction with high electricity prices.

Carbon taxes to reduce GHG emissions have become a major issue in federal and provincial politics. Elections since 2015 in Alberta, Ottawa, and British Columbia have been followed by the introduction of even more aggressive anti-carbon policies. They include emission caps, carbon taxes and, in BC, government opposition to the expansion of pipelines carrying Alberta oil for export.

Alberta and federal elections in 2019 will both include carbon taxes as a major issue. The carbon politics of the recent past, present, and future are as controversial as those from the 1970s and 1980s. They have been expensive for consumers and taxpayers, costly to administer, and increasingly complex.

An obvious problem is that there is no evidence that collective Canadian climate change policies have made any measurable progress in achieving their stated goal of reducing global CO2 emissions. This should surprise no one given that most of the countries outside of the EU producing significant CO2 emissions have no coordinated nor cooperative approach, if they are undertaking emission reduction initiatives at all.

Global warming needs a global solution. There are 195 countries in the world. Thirty-four emit more than 2% of global GHG emissions each, about two-thirds of the total. This includes China, the US, 28 EU members, India, Russia, Japan, and Brazil.

Is there a common policy these countries could possibly agree to? *What* they should do is easy, as we're told frequently. *How* to do it remains elusive, likely impossible.

Another 12% of emissions, those over 1% but below 2%, are released by Indonesia, Canada, Mexico, Iran, South Korea, Australia, Saudi Arabia, and South Africa. Four—Canada, Mexico, Iran, and Saudi Arabia—are major oil producers while Indonesia and Australia are smaller oil and gas producers. Is there some carbon-emissions strategy these eight countries would jointly ratify?

The hot, dry, and smoky summer of 2018 exposed the major challenge of the politics of carbon and climate change. Because when media, commentators, and pundits questioned the resolve and policies of the politicians of the day, the most important question was never asked; can politicians actually do anything meaningful about major, trans-border issues like those associated with fossil fuels?

The headlines were breathtaking, but so was the air.

On August 2, 2018 *The Economist,* the respected British weekly magazine, published an article titled, "The World is Losing the War Against Climate Change." It begins by saying:

> Earth is smouldering. From Seattle to Siberia this summer, flames have consumed swathes of the northern hemisphere. ... Fires that raged through Athens last week killed 91. Elsewhere people are suffocating in the heat. Roughly 125 have died in Japan as a result of a heatwave that pushed temperatures in Tokyo above 40 degrees C for the first time.[64]

The Economist suggested that since the activities of mankind were the reason the weather was bad, doing the opposite it would reverse the outcome. Had the climate change mitigation the world has been warned about for years been acted upon sooner, things could be different today. It didn't have to be this way.

Call this 20/20 climate hindsight. Because we're living the results of past inaction, climate change is not just about the future.

Closer to home, British Columbia had another bad forest fire season thanks to hot, dry weather. The smoke headed east across the Great Divide into Alberta, resulting in some of the worst smoke conditions and air quality anyone can remember. The media called it the worst of everything ever. Yes, forest fires have happened before in BC After all, the province is almost entirely covered by trees. Everyone knows that. But because of man-made climate change, never had the fires been this awful.

The Globe and Mail columnist Gary Mason put fingers to keyboard on August 21, 2018 and expressed the view of many. But his message was more focused. The headline read, "While the planet burns our politicians fiddle." After three paragraphs of explaining what everybody knew—that the fires were terrible and air was awful—Mason concluded, "Amid this potential calamity, you would think the world's leaders would have been shaken into action ... Despite the emerging crisis created by an ever-warming planet – heat that is causing an increasing number

---

64 "The World is Losing the War Against Climate Change," *The Economist* (August 2, 2018). https://www. economist.com/leaders/2018/08/02/the-world-is-losing-the-war-against-climate-change.

of fatalities each year, temperatures that are destroying crops, fires obliterating homes by the thousands – leadership is missing."[65]

This is but a small sample of what was published on this subject in the summer of 2018. The headlines appeared daily. "New research predicts heat waves in Canada could become more frequent – and five times more deadly," a *National Post* headline declared. "Global temperatures will be abnormally higher from 2018 to 2022: study," said the *Huffington Post*. NBC wrote, "Global warming can make extreme weather events worse. Now scientists can say by how much."

There are three remarkable things about the new stories' collective conclusion that the terrible weather in the summer of 2018 was self-inflicted.

First, nowadays any weather or environmental event outside the norm, which is usually the multi-year average, is exacerbated by climate change.

The second is that the severity of these events could be reduced if governments enacted the right policies and enforced them.

Third, and perhaps the most important, is that few if any elected politicians went on the record and admitted that this situation was possibly beyond their control. Or that this was terrible bad luck caused by the normal cyclicality and unpredictability of the weather.

Which is the essence of the future of carbon politics. Armed with the right policy tools, can world governments really control the weather? If elected as part of a majority government, can that seemingly ordinary person who rang your doorbell last election and asked for your vote accomplish what used to be the sole purview of science fiction? Make it rain more or less? Control the temperature? Stop the glaciers from melting? Reduce the intensity of tropical storms?

This is the surely the most important question that should be asked, but nobody is asking it. Can a collection of formerly ordinary individuals gathered together for a relatively short period of time to form a government solve what is identified as one of, if not *the*, most important issue ever facing the future of the world?

In democratic countries, the skill set of a president/prime minister/premier is not climate science, technology development, energy, or international treaty creation, negotiation, and ratification. Generally, the only test Canada's future leader

---

65 Gary Mason, "While the Planet Burns our Politicians Fiddle," *The Globe and Mail* (August 21, 2018). https://www.theglobeandmail.com/opinion/article-while-the-planet-burns-our-politicians-fiddle/.

must pass is to secure enough votes within a political party to become leader, and then hope his or her party wins enough seats in an election to form government.

Modern democratic politics is about ordinary people, regular folks elected by other regular folks. What climate change needs is not politicians, but a large number of extraordinary people with extraordinarily common views and the conviction to carry out what must be done. These individuals must be prepared to commit, force, and convince their respective countries to do it.

There is no cause for optimism in Canada. The country has devolved to the point that we can't even get neighbouring provinces to agree. This is accepted politically. After all, the policy position of any province reflects what Premier(s) (pick one or more) Notley/Horgan/Ford/Pallister/ Legault must adopt to be re-elected. Why would anyone expect these people to do something in the national interest if it might cost them local votes or the next election?

When it comes to politics, people simultaneously live in two diametrically opposing mental states, yet they still expect something good to happen. Everyone understands how saving the world will require unprecedented cooperation among politicians willing to put the planet ahead of their own future. Many demand it. But at the same time the public is aware the politicians can't be depended upon to do it. We, the cynical voters, realize that a politician's personal future is, and always will be, more important to that elected representative than the big picture.

Regarding the matter of carbon and global warming, it is not whether human-induced climate change is occurring. Again, the science is settled. The issue is governments claiming to be able to fix just about anything, whether they actually can or not. Politicians promise to create jobs, end poverty, prevent discrimination, house the homeless, feed the hungry, protect the oppressed, treat the sick, lower taxes, and balance the budget. Simultaneously. Without any of them having to demonstrate any skills in any of these areas other than winning an election.

Which explains, unfortunately, why every one of these major problems remains. Perhaps if the politicians weren't beavering away with such determination, things would be worse. That's not the point. In many instances where measurable progress is evident, technology developed by the private sector is often a significant contributor. But many environmental activists are also anti-business, and too many politicians have little or no understanding of how business works. As governments continue to fall short of expectations, some members of the public stubbornly conclude that the only way to fix the problem is more government.

If the future of the world's climate is going to unfold in the catastrophic fashion increasingly described by so many—and the politicians we have elected thus far are unable to change the world's current course—is more of the same really the solution? People should undertake an objective review of how politics works in twenty-first century before they make more demands on a system that has a terrible track record of accomplishing what it claims to be able to do. There are many lessons in Canadian history, if people care to read them.

Canada is a relatively young country so the evolution of the expectations of government and what they say they can do are relatively fresh.

In the early days when Canada was inhabited by proudly self-reliant First Nations and growing numbers of industrious immigrant settlers here to build something out of nothing, the role of government was very basic. It existed to ensure law and order, provide a system of currency and banking with which to conduct commerce, and assist with transportation infrastructure.

In his 1969 book *Canada*, by Edgar McInnis, he wrote that government was very important for resource development, which has been the key economic driver in getting Canada this far.

> In many cases the full task of opening a new region is beyond the resources of private capital. Government helps in various forms, from prospecting to railway building, and has played a significant part in development the mining areas of Canada. A working harmony between large-scale enterprise and the federal and provincial governments has been almost a necessity in the development of the Canadian economy (McInnis 1969).

No argument there. Governments employing public funds and marshalling shared resources have played a huge role in the discovery, development, and exploitation of Canada's carbon-resource wealth: coal, oil, and natural gas. The railroad. Oil sands. Hibernia. Pipelines to markets for oil and natural gas.

> Of all the aspects of government economic policy none [...] has been of greater importance than transportation. The development of transport and communications has profoundly affected the growth of all modern nations, but to none has it been more basic and essential than to Canada. Preoccupation with this topic is a persistent theme throughout the whole of her history.

In the absence of railways, indeed, the present Dominion might never have come into existence (McInnis 1969).

Just replace the last reference to "railways" with "pipelines" and revisit the previous chapters on coal and pipelines in the first section of this book.

Early in its creation Canadians expected little from government and probably got less. In the 1983 book *A Short History of Canada* by historian Desmond Morton, the author wrote in the introduction, "Most people, of course, lived their lives within the limits of family and community. They looked to government not for social or economic programs but for bridges, wharves, the post office, and, quite rarely, for a place on the public payroll" (D. Morton 2001).

With rapid settlement and expansion underway, the country grew, interrupted only by terrible events such World War I from 1914 to 1918. The 1920s were a period of a significant economic growth. Capital markets were not heavily regulated. This contributed to the stock market crash of 1929 and the subsequent Depression. This was the first time Canadians looked to their governments for hope and assistance, out of desperation. Aspiring politicians responded with promises to make things better.

Conservative Party leader and Calgary lawyer and wealthy businessman R. B. (Richard Bedford) Bennett made a lot of comforting commitments in the 1930 federal election on behalf of the Conservative Party. Bennett figured he could solve the problems of the Depression by adjusting tariffs and trade policy.

He failed of course. The Depression was a global, not local, issue.

The Depression was devastating. The economy shrank. Unemployment was over 20% as jobs evaporated. Morton concluded his chapter on the Depression with the suggestion that the solution to the failure of government to materially assist the economy in a period of contraction would be more and bigger government. "The Depression forced thoughtful Canadians to realize how a feeble central government and penniless provinces crippled any collective response to economic crisis. In time, Canada might have recovered. Instead, a worldwide tragedy became Canada's salvation" (D. Morton 2001).

The tragedy was the Second World War.

The postwar period for Canada was one of growth. By now governments had introduced basic social programs primarily as income-levelling tools for those unable to work. This included unemployment insurance and old age pensions. Public investment in education grew, as did infrastructure such as more roads, more electrification through expansion of generating capacity and the grid and, for Alberta, oil and gas pipelines east, west, and south. However, government as an agent of massive social change was still not high on anyone's political agenda, be they a voter or campaigning politician.

As World War II drew to a close, the concept of "progressive" politics captured the imagination of North America's increasingly urban, university-educated populace. Sensing a shift in voter expectations, in 1942 the Conservative Party of Canada changed its name to Progressive Conservative of Canada to attract more voters. In a 2007 essay by William A. Schambra and Thomas West titled, *The Progressive Movement and the Transformation of American Politics*, the authors explained why Canada's Conservative Party was keen to rebrand itself.

> Progressivism was the reform movement that ran from the late 19th century through the first decades of the 20th century, during which leading intellectuals and social reformers in the United States sought to address the economic, political, and cultural questions that had arisen in the context of the rapid changes brought with the Industrial Revolution and the growth of modern capitalism in America. The Progressives believed that these changes marked the end of the old order and required the creation of a new order appropriate for the new industrial age.[66]

While industrialization was indeed lifting more people out of poverty and creating a growing middle class, the benefits were not evenly distributed enough to satisfy the increasing number of people who, thanks largely to the Industrial Revolution, had enough time and money to ponder such things. By the late 1960s and early 1970s, the frugality and selflessness essential to get through the Depression of the 1930s, while not forgotten, certainly didn't appear to be required again any time soon. The Second World War had been over for nearly a quarter century.

While there was no outbreak of world peace, the areas of active conflict were half a world away in the Middle East, Africa, and Southeast Asia. There was the lingering threat of global nuclear annihilation courtesy of the Soviet/American

---

66 William A. Schambra and Thomas West, "The Progressive Movement and the Transformation of American Politics," *The Heritage Foundation* (2007). https://www.heritage.org/political-process/report/the-progressive-movement-and-the-transformation-american-politics.

arms race, but if it happened nobody would be forced into military service through conscription. Most would be dead. The truly worried built and stocked fallout shelters.

The change in Canadian politics in this period is evidenced by who got elected. Pierre Trudeau, thinker and philosopher, became the Liberal Prime Minister in 1968. Confessed progressive Peter Lougheed ended a 36-year Social Credit dynasty in Alberta in 1971 with the election of the Progressive Conservative Party (PC).

Trudeau talked a lot about creating a "just society," pushed for a new Canadian constitution, and opposed the early Quebec separatist movement. These initiatives had little to do with schools and roads.

Lougheed, as will be explained in upcoming chapters, felt revitalizing an Alberta government was essential. A PC MP would later describe Alberta's outgoing Social Credit government as "moribund"; thus, change was required. Thanks to OPEC, Lougheed would have the funds to radically alter and expand Alberta's government without borrowing any money.

Following the failure of public policy to improve things during the Depression, the next clear example of a government promising to "do something" about global economic forces that failed completely was the National Energy Program (NEP) of the Pierre Trudeau Liberal government in 1980. The first major carbon crisis Canada faced was when OPEC quadrupled the price of oil in 1973. This was followed in 1979 by the Iranian revolution, which forced prices even higher. By 1980 the world oil price was ten times higher than in 1973. This was a period of high inflation, high interest rates, and recessions. The last thing the economy needed was skyrocketing oil prices.

Western governments were under great pressure to respond. The NEP was intended to right a great number of wrongs through the heavy hand of new policy and big government. Ottawa could and would manage this situation.

In 1982 author Peter Foster wrote a book titled *The Sorcerer's Apprentices— Canada's Super-Bureaucrats and the Energy Mess*. In the introduction Foster wrote, "The Liberals' National Energy Program – introduced in the October 1980 budget – was one of the most revolutionary, controversial and, it can be claimed, disastrous policies in Canadian history. Quite apart from its adverse direct impact on the energy sector, the NEP helped to decimate stock markets, severely impact the Canadian dollar and destroy the confidence of foreign investors. It also caused the country's industrial future … to be held ransom, and injured Canadian-US relations more than any policy in living memory" (Foster 1982).

But when introduced, the NEP was very popular outside Alberta. As oil prices rose in the 1970s oil companies were vilified as amoral, colluding, shameless profiteers. There was no way those backwards countries in the Middle East figured out how to hold the oil consumers of the world ransom on their own. Through the NEP the Canadian government would knock them down to size and to save helpless consumers.

Early in his book Foster explored "why" the NEP so was popular before describing "what" the policy did and how it failed. This was a time when it was becoming increasingly common for political parties to conduct opinion polling to figure out what the public wanted, then give it to them to win the election and stay in office.

Consultant Charles Perrault gave a speech where he said, "Most disturbing to me is what I call Government by Gallup and Goldfarb [the leading public opinion pollsters of the day], the Machiavellian fashioning of policies which, as stated in their simplistic objectives, play on the emotions of Canadians and have their very broad support, while contributing directly to economic stagnation and unemployment" (Foster, The Sorcerer's Apprentices—Canada's Super-Bureaucrats and the Energy Mess 1982).

Perrault continued, "Ours is a nation of economic illiterates. Only one-third of Canadians can even choose the right definition of profit in a three-part question, remember. The average Canadian is ripe for the plucking – and Ottawa knows it … Most Canadians like the NEP and applaud Canadianization. I like Canadianization too. But I think I can see the cost, and the cost is too high, and by the time the average laid-off worker realizes this, it's very late and much harm has been done" (Foster 1982).

Ouch. This was 37 years ago! And it is certainly not politically correct nowadays to describe voters as economic ignoramuses. But as detailed in upcoming chapters, the NEP would prove to be a disaster. The policy and its proponents did not last long. But botched policy happened again in 2007 when the Alberta government went after the oil industry, seeking "our fair share," a popular idea that won the Ed Stelmach PC administration a big majority government in 2008. In oil-producing Alberta no less! Stemach's New Royalty Framework didn't last long either, nor did its lead salesman.

Trying to explain the phenomenon of voters looking to governments for more solutions to more problems, Foster wrote about how government changed after the Depression. In a section titled "Growth by Public Demand" Foster wrote, "Apart from government's fundamental tasks of defense, the enforcement of law and order and provision of requisite public amenities, it has long been accepted that the state should intervene to regulate and control aspects of trade and environment excesses of unbridled industrial development, as well as legislate on a wide variety of areas of public interest, from safety on the job to speed limits on the highway" (Foster, The Sorcerer's Apprentices—Canada's Super-Bureaucrats and the Energy Mess 1982).

But after the Depression and World War II, the purpose of government had expanded.

> It was also enforced by the great post-war economic boom which led to the belief that the government had not only the responsibility but the ability to deliver growth, full employment and stable money. Thus, whenever a glitch appeared in the increasingly affluent lives of Canadians – and indeed the inhabitants of other industrial democracies – it became the norm to demand that the government "do something." The public seemed to see nothing contradictory in simultaneous complaints that the government was "too big" and that it was "not doing enough" (Foster 1982).

Due entirely to the market forces of supply and demand, the oil shortage disappeared, the price collapsed, and voters lost interest. When voters lose interest so do governments. Carbon politics was very peaceful for the rest of the twentieth century, mainly because the price was low and there was never a supply shortage. Governments continued to grow in every respect but focused their attention on other issues that attracted more attention and, coincidentally, votes. Many were social issues regarding sexual equality, gender parity, and multiculturalism. Public policy debates and election campaigns had little to do with coal, oil, or natural gas.

That all changed towards the end of the twentieth century when global warming caused by fossil fuel consumption began to appear in the news and the public consciousness. The United Nations "Earth Summit" in Rio de Janeiro in 1992 was the first major event receiving global publicity that talked about the link between fossil fuels and climate change. Building on the commitments from Rio, in 1997 the Kyoto Protocol was signed after a conference in Kyoto, Japan. Eventually signed by 192 countries, "The Kyoto Protocol implemented the objective of the UNFCCC (United Nations Framework Convention on Climate Change) to fight global warming by reducing greenhouse gas concentrations in the atmosphere to 'a level that would prevent dangerous anthropogenic interference with the climate system.'"[67]

Politically, the climate change issue has gained more attention for over a quarter century since Rio in 1992. There have been increasing demands by voters that governments "do something" and continuous commitments by politicians to respond to that demand. But as Peter Foster wrote in 1982, voters want governments to do more and less simultaneously. Readers should be reminded that in 2008 federal Liberal leader Stephan Dion campaigned on a "Green Shift" that included carbon taxes to deliver on Canada's emission reduction commitments. The Liberals garnered only 26% of the popular vote and the Conservatives won a majority government. Voters were indeed concerned but were not ready to pay for it themselves.

Canada was an active participant in Kyoto, signed the accord in 1997, and ratified it in the House of Commons in 2002 under Liberal Prime Minister Jean Chrétien. In 2012 Conservative Prime Minister Stephen Harper withdrew from the accord. In 2015 Justin Trudeau's Liberal government attended another climate change conference in Paris and committed Canada to lowering GHG emissions about 30% by 2030.

Since Kyoto, governments across Canada have introduced a myriad of programs and policies to reduce carbon emissions. In an update to its Paris commitments at the end of 2017, Minister of Environment and Climate Change Catherine McKenna revealed Canada revised its reduction targets downwards and actual results in 2030 would be one-third lower than promised.

In 2016, the last year for which data is available, the federal government reported greenhouse gas emissions of 704 megatonnes, one-sixth higher than in 1990 and about 2% more than 688 megatonnes when the country signed the Kyoto Protocol

---

67 "United Nations Framework Convention on Climate Change," *Wikipedia* (December 20, 2018). https://en.wikipedia.org/wiki/United_Nations_Framework_Convention_on_Climate_Change.

in 1997. Emissions peaked in 2007 at 745 megatonnes, 10% higher than 1997. The lowest level since 1997 occurred the 2009 recession, 682 megatonnes.[68]

Global carbon dioxide emissions in 2017 were 37 gigatonnes, up 54% from 24 gigatonnes in 1997 when Kyoto was signed. Since 1960 there has never been a year-over-year reduction in CO2 emissions.[69] Even before the year was over, initial reports indicate global emissions increased again in 2018.[70]

Unfortunately, these figures indicate that expecting your local politician to combat climate change using conventional politics is very unlikely to yield the desired outcome. The Canadian carbon tax debate is classic, as are the proposed policies. The latest rendition of federal policy is the promise to tax carbon but return the tax in the form of larger rebates than most will have to pay. This is what Foster noted 37 years ago, that voters want more and less government at the same time.

GHG emissions that change the climate are called "negative externalities" or "tragedy of the commons," where individual consumption of resources like fossil fuels for their own benefit creates problems for everybody else. This is a classic problem that only governments can solve. It is not unlike having government-created laws and a police force to protect everyone else from you.

The conclusion of this book will explain how governments can best achieve success.

---

68 "Greenhouse Gas Emissions," *Statistics Canada* (2018). https://www.canada.ca/en/environment-climate-change/services/environmental-indicators/greenhouse-gas-emissions.html.

69 *Carbon Brief: Clear on Climate* (2015). https://www.carbonbrief.org/analysis-global-co2-emissions-set-to-rise-2-percent-in-2017-following-three-year-plateau.

70 "'Brutal news': global carbon emissions jump to all-time high in 2018," *The Guardian* (December 5, 2018). https://www.theguardian.com/environment/2018/dec/05/brutal-news-global-carbon-emissions-jump-to-all-time-high-in-2018.

2
CARBON WEALTH CHANGES ALBERTA – LEDUC TO 1973

"What the new PC administration under Lougheed did was redefine the role of government, greatly expand its size and cost, and use carbon resources to pay for it. This was the beginning of new and ever-changing relationship between industry and government that continues to this day."

There's an old oilpatch saying that goes, "I'd rather be lucky than good." Like drilling for gas and striking oil. Nothing better describes the economic and political history of Alberta's carbon-resource industries than luck. No human being put the carbon resources in the ground under Alberta. For this we can thank the dead diatoms from 300 million years ago you met early in this book, along with trillions more dead animals and plants over many more millions of years.

Far too many Alberta politicians and businessmen have taken far too much credit for their own good fortune and that of the province, its communities, and their fellow Albertans. Another old saying in the 'patch goes, "Oil and gas is where you find it." Lots was found in Alberta. As the province was settled, people came looking for the oil sands and coal because they already knew it was here. Crude oil and natural gas were huge bonuses.

In retrospect, the federal government unknowingly gets a healthy portion of the credit for Alberta's spectacular success. It was, of course, an accident. Consider the federal/provincial carbon-revenue-and-control wars that followed the global crude price spikes of 1973 and 1979. Would Alberta be what it is if the federal government that created it in 1905 knew the true value of the province's massive carbon resource treasure? To heck with the British North American Act. Would Ottawa have ceded control of subsurface mineral rights to the province in 1930 if it had known?

OPEC and the world's growing thirst for fossil fuels should also be recognized for their role in making Alberta's carbon resources worth trillions more than anyone ever imagined. There was so much oil wealth piling up in what Central Canada still considered "one of the colonies," it inspired author Peter Foster to write a best-selling book in 1979 titled, *The Blue-Eyed Sheiks: The Canadian Oil Establishment.* It was one thing for oil and gas to be worth a lot of money. It was quite another for it not be in Ontario or Quebec. This is more fully explored later.

Too many of Alberta's famous and rich oil entrepreneurs have been reluctant to share the credit for their financial acumen and wealth creating genius. Much has been written about the players and their companies over the years. They were all certainly characters in their own way. But when the underlying asset value of a company or investment triples in six months like oil did in 1973—with all forecasts indicating there was more to come—even the most average appear exceedingly clever.

But none of the participants who have either taken credit or been given accolades for Alberta's incredible prosperity deserve it less than those fortunate politicians who were in office when the commodity price spiked skywards. This applies to oil *and* gas. History has been very kind to premiers who ran Alberta in good times, leaders like Manning, Lougheed, Klein. Those who had who had the misfortune to manage Alberta through a commodity-price downturn or economically flat period—Getty, Stelmach, Notley—are tarnished.

It is of course much more complicated than that. In the past 70 years several premiers and their administrations did their best to improve the situation during their time in office. Others did the absolutely wrong thing at the wrong time. But all were charged with managing Alberta as the province's economy reacted to events and circumstances beyond the control of the government of the day. As history would prove, along with good luck there is the other side of life's great coin toss: bad luck.

The premier of Alberta on February 13, 1947 was Ernest Manning. A religious man like Social Credit founding leader William Aberhart, Manning had been involved with the Socreds since their inception. He replaced Aberhart as party leader after Aberhart's death in 1943.

While reams have been written about the Social Credit dynasty that governed Alberta from 1935 to 1971, what the Socreds did best was to do little beyond the elected obligation of sound, predictable, and frugal public administration. They were typically more reactive than proactive. Like many governments in this period, they saw their job as providing the citizens the public services they needed at the lowest possible cost. Albertans and the private sector would shape the direction of the economy. The government's role was to do its part in areas where shared resources made the most sense—education, roads, management of Crown-owned resources, tax collection, legal administration (like marriage and driver's licenses), and essential public services like the police and fire departments, and later healthcare.

As was written in the introduction to this section, Alberta emerged from World War II as just another small struggling Canadian province. So one can imagine the good news the day Premier Manning learned there had been a big oil discovery just outside of the capital of Edmonton in early 1947. While it is unlikely Manning or anyone in his government had any idea exactly what the Leduc discovery meant for Alberta's economic future, the government's job was clear. With the memories of the struggles and challenges of the Great Depression and the Second World War fresh in everyone's memories, the Manning administration was determined to provide industry whatever it needed to grow this business and get as many people as possible working at better-paying jobs than what had been available in Alberta for the previous 20 years.

From a resource perspective, the impact of the Leduc discovery on the oil and gas industry is discussed in the previous section. In 1946 the industry spent $300,000 on "petroleum and natural reservations," or rights to explore for oil and gas. In 1956 the figure was $24.7 million. The total was $138.5 million for the 11-year period. After a discovery is made, part of the mineral rights reservation could be converted into a lease. In 1948 revenue from reservations and leases was $3.1 million. By 1956 this was up to $71.7 million and the total for the 10-year period was $315.7 million. On the percentage basis the growth in these figures was spectacular (Hanson 1958).

In 1946 total capital expenditures on agriculture, mining, construction, manufacturing, housing, utilities, "trade finance and commercial", and "institutions and government departments" was $122 million. This increased every year thereafter, rising to $909 million by 1956, 7.5 times higher. Total oil and gas investment in 1947 was $28 million. This also rose every year to reach $469 million in 1956, nearly 17 times greater. For the 10 years from 1947 to 1956, the total was $2.4 billion invested, unprecedented in the history of the province (Hanson 1958).

Under "The Income Benefits of Oil," Hanson wrote that of the $666 million in personal income collected by Albertans in 1946, only 2% came from those employed in the petroleum industry. By 1956 total personal income was up to $1.54 billion, 2.3 times greater. Of the total the non-petroleum component was

$849 million, about 28% higher than total personal income in 1946. The rest, 45% of $690 million, came from the petroleum industry. In only 10 years personal income from the oil and gas industry exceeded total personal income for all Albertans a decade earlier (Hanson 1958).

Alberta not only had more jobs, but it had better jobs. The total population was 803,000 in 1946, not much higher than 796,000 reported from the 1941 census. By 1956 the population had increased to 1.12 million, a 40% gain. As Calgary and Edmonton saw their populations double, every community of 1,000 people or larger grew by an average of 114%. The jobs were created where the wealth was created, a trend that continues today. Employment in the petroleum and related industries tripled from 11,100 in 1946 to 33,300 a decade later. Agriculture, the mainstay of the province before Leduc, declined from 40.2% of the jobs to 26.0% over the period. Manufacturing, utilities, construction, and finance all grew, primarily because of the petroleum industry (Hanson 1958).

Government revenues from oil and gas lease rentals, mineral lease sales, and production royalties enjoyed spectacular growth post-Leduc. From a total of only $1.3 million in 1947 the figure was $133.1 million in 1956. Alberta went from $118 million in net debt in 1947 to a $253 million surplus in 1956. At December 31, 1956 the province held $338 million in cash and investments, with the total provincial debt down to only $85 million. This was the same province that had defaulted on $3.2 million in debt 15 years earlier (Hanson 1958).

With the template set, the impact on the economy grew with production. More wells, more fields, more infrastructure, more pipelines, more support services, more taxes, more royalties, and more jobs. Conventional oil production averaged only 17,485 b/d in 1947. It was 393,650 b/d by 1956 and would peak at 1,431,199 in 1973. Natural gas output would be 55 times higher in 1973 than in 1947. Gas production would more than double again in the next 20 years. Sulphur extracted from natural gas would 872 times higher in 1973 than its first year of recorded production in 1952.[71]

After decades of government-funded research and support Alberta's massive but commercially recalcitrant oil sands finally began to contribute economically in a meaningful way when GCOS (today Suncor) came on stream in 1967. That

---

71   Canadian Association of Petroleum Producers, *Statistical Handbook for Canada's Upstream Oil Producers.* (Calgary, AB: CAPP, 2018). https://www.capp.ca/publications-and-statistics/publications/316778.

year production only averaged 1,965 b/d. With the plant working better with every passing year, by 1973 output was up to averaging 51,326 b/d. This was only 3.5% of conventional crude production for the same year, but it was certainly the start of something big, particularly 30 years later when output would exceed 1 million b/d.[72]

As the industry grew, Ernest Manning's Social Credit party was obviously not punished nor, by today's standards, even questioned. For the 24-year period from Leduc to when it was replaced in 1971, the Socreds won big in 1948, 1952, 1955, 1959, 1963, and 1967. The average plurality during the period was 85% of the seats. A bad year was when Manning won only 37 of 61 ridings, 61%. The Socreds were rewarded for a "job well done" by the voters by who elected 94% of the MLAs as Socreds in 1959 and 95% in 1963. The other years were all 85% of the seats or higher.

When the cash was flowing and the province was growing, Manning and the Social Credit party clearly proved Albertans loved their government.

In 1968 and after 25 years as premier, Manning, "… decided he had had enough," and abruptly resigned "entirely of his own volition." With a little help from Manning, later that year the Socreds chose Harry Strom, a farmer, as leader. In office since 1955, Strom served as minister of agriculture and later municipal affairs (MacGregor 1981).

Premier Harry Strom would never win an election as leader. Social Credit would lose the election in 1971 and, except for the name, gradually disappear off the political map in subsequent elections; the Socreds last elected an MLA in 1979. However it started, the party had abandoned its peculiar economic polices decades earlier a task made much easier when the oil money began to flow.

The post-Leduc Social Credit years were characterized by cautious, conservative, pro-business government, with every decision they made simplified by the ever-increasing flow of petro-dollars. It was the Socreds that created the APNGCB in 1938 in response to poor reservoir management and production practices at Turner Valley. The resources were a public asset, and that's what the Socreds believed government was all about—the careful, conscientious and prudent management of government's responsibilities, resources, and assets.

---

72 Ibid.

When the oil rush began to gather steam in 1948, the oil and gas industry had a structure and an efficiently run provincial regulator with which to deal. MacGregor wrote, "With an integrity all too rare, Ernest Manning ruled his cabinet with a rod of which was not sparing as soon as his keen perception detected any departure from the straight and narrow way. In their dealings with his government, the oil men were met with fairness and ability by the personnel of the APNGCB ... The Board made the rules as to spacing of wells, the interrelations of the companies and the rates at which they could produce from the fields. While the companies' individual interests often clashed with those of the Board, that body stood firm and came to one of the most highly respected regulatory authorities in Canada" (MacGregor 1981).

Preston Manning, Ernest's son who would go on to his own noteworthy political career as the founder and first leader of the Reform Party. He wrote the following about his father in 2002 in his book titled, *Think Big*.

> My father's philosophy combined fiscal conservatism with a compassion for people, born of Christian faith and the suffering of the Depression. While Aberhart was a crusader who could blast apart the status quo, my father was a builder and an administrator. He managed to reassemble Alberta's battered finances after the Depression and manage the economy through war and the post-war boom. He presided over the transformation of the province brought about by the discovery of oil at Leduc in 1947, his administration being one of the few in North America to remain in office throughout an oil boom without corrupting itself. He considered resource revenue to be money obtained from the sale of a depleting asset, which ought to be reinvested in other physical and social assets – roads, schools, hospitals, universities, libraries and performing arts centres (Manning 2002).

The Social Credit view of carbon policy was simple. Alberta was blessed with the fossil fuels the world needed. These resources created wealth and prosperity for everyone, from the workers to the government, which—through low individual and corporate taxes—created economic opportunity and expanded public services. Therefore, the purpose of the government was that of a regulator and facilitator. Such was the case with the rapid support for export pipelines once it became clear Alberta had reserves well in excess of domestic needs. When major initiatives to support the oil industry were required, they got done quickly and successfully. For example, to facilitate gas exports in 1954 the Alberta Gas Trunk Line natural-gas gathering system was created. The Socreds espoused a government and business environment that attracted capital and all the wonderful economic spinoffs growing profitable resource development can bring.

"In the eyes of the oil men and businessmen in general, the application of science to the regulation of the oil industry was only one of Alberta's attributes. The other was the stability of the province's long-time Social Credit government" (MacGregor 1981).

The Socreds were not without the occasional bout of fiscal generosity with public funds, something expected, even demanded, of all governments nowadays. In 1955, Alberta's 50th anniversary, the province built the Jubilee Auditoriums, first in Edmonton and two years later in Calgary.

And, of course, with the population growing, the government spent continuously on public works like schools, roads, and infrastructure. New oil discoveries in formerly remote places like Swan Hills and Rainbow Lake—made in the winter when the ground was frozen—required all-weather roads to complete development. People started to live nearby, requiring more communities and schools. As the oil and gas dollars flowed into the provincial treasury, they also went out the door just as quickly on infrastructure to keep up with rapid growth.

The population of Alberta of 796,000 in 1941 had more than doubled to 1,628,000 30 years later. As important were demographics or the age groups of the population. The "baby boom" is a term used to describe the 18 years after World War II during which the birth rate in North America surged, Canada included.

In 1970 the Canada Elections Act lowered the voting age to 18 from 21. By 1971 the age range of the "baby boomers" was from 16 to 26. Anyone born before August 29, 1953 could vote in the 1971 election. This comprised a good portion of the boomers. All they knew about the Great Depression and World War II is what they had heard from their parents, relatives, and older friends, or read about in books, magazines, and newspapers, or what they saw in the movies and on TV. Most had only experienced economic good times compared to the 1930s and first half of the 1940s.

In politics nothing lasts forever. Sensing Albertans were ready for change, Calgary lawyer Peter Lougheed announced in 1965 he would be seeking the leadership of the Alberta Progressive Conservative Party, or "PC" Party.[73] The Alberta PCs had won a handful of seats in the three elections prior to the 1963 election, when it didn't win any. Undaunted, Lougheed sought the leadership. In a 2011 article about the political rise of Peter Lougheed, author James Marsh wrote, "In 1965 the Conservatives had no leader, no seats, no money and no organization."

Lougheed sensed that the Social Credit dynasty was a one-man show—Manning—and it was only a matter of time before the much-younger PC leader's chance would come. Under Lougheed the PCs won six seats in the 1967 election (Marsh 2011).

By the 1971 election, the first campaign for 57-year-old farmer Harry Strom as Socred leader, the differences between the staid old Socreds and hip new PCs under the 43-year-old Lougheed couldn't be clearer, particularly on television. The election was more about change than policy. Nearing election day, most PC advertising carried a picture of Lougheed and his key team under the commitment-free slogan of "NOW." According to Marsh, when the PCs won 49 of 75 seats the media coverage included, "fresh breezes of change" and "euphoria of newness in the air" (Marsh 2011).

Looking back at the end of one political dynasty and what would prove to be the start of another, in 1993 Edward Bell from the University of Western Ontario wrote an essay for the *Canadian Journal of Political Science* titled, "The Rise of the Lougheed Conservatives and the Demise of Social Credit in Alberta: A Reconsideration". In his paper Bell quoted a passage from a prior work in 1979 by authors John Richards and Larry Pratt titled *Prairie Capitalism*:

> The new middle class was urban and secular in its outlook and impatient with Social Credit's blend of religious fundamentalism and the remnants of its agrarian populist past. Social Credit, the Lougheed Conservatives were fond of saying, had grown isolationist and was outside of the mainstream of North American culture, something which could never be said of Peter

---

73 As noted in the previous chapter, the federal conservatives had added the word "Progressive" to their name in 1942. In 1959 the Alberta Conservatives became the Progressive Conservative Association of Alberta (usually called the Progressive Conservative Party of Alberta).

Lougheed. While the pious Manning was still delivering his weekly radio Back to the Bible Hour sermons ... Lougheed was thumbing the pages of Theodore White's *The Making of the President 1960* and sharpening his television image in the Kennedy mould—the young, dynamic and safety right-of-centre heir to a family dynasty.[74]

Lougheed was a skilled politician, this being defined as an ability to create issues and debates at one point, then campaign differently to avoid distractions during the run-up to the election. While in opposition he talked about the importance of industrial diversification beyond oil and gas and how increasing the province's royalty take could be a way for the government to facilitate the process. Other issues included the ownership of companies like Home Oil and federal taxation. However, when the writ was dropped for the 1971 election, Lougheed skillfully played both sides of the party's name; progressive on the need for change and conservative when it came to the private sector. As is common in politics today, hard issues were replaced with generalities and platitudes.

Once in office, Lougheed had a very aggressive policy platform and brought forward new bills enshrining rights into law, more protection for those with mental health issues, and introduced daylight savings time (Marsh 2011).

To thank the boomers in a way they would surely remember, he also lowered the drinking age to 18. Now that's progressive! He made being an MLA a full-time job, by doubling their pay, and allowing them paid staffers. He ensured everyone understood he and PCs were much different than the Socreds by demonstrating that the government could be an active agent in furthering the common good, not just a manager.

If the role of a conservative government under the Socreds was to be the referee, in the great game of Alberta politics a progressive like Lougheed believed government should be the coach. "If this meant that government could do something better than business, then it should do it. His decisions were not always popular among many of his conservative followers" (Marsh 2011).

This was the start of big government in Alberta. Writing for website *Policy Options* in 2012, Lee Richardson, who worked for Progressive Conservative

---

74   Edward Bell, "The Rise of the Lougheed Conservatives and the Demise of Social Credit in Alberta: A Reconsideration," *Canadian Journal of Political Science*.

politicians in Ottawa and served as an MP on two occasions, wrote an article titled, "Lougheed: Building a Dynasty and Modern Alberta from the Ground Up." Lougheed's view of a "modern Alberta" included making the government a lot bigger. The cabinet was expanded from 15 to 22 and Lougheed extended the sitting of the legislature from two months to a year. The PC caucus met daily when the legislature was sitting. Richardson wrote, "Creating a modern Alberta in 1971 would require the implementation of efficient business systems, processes and controls where none existed ... Lougheed brought contemporary management systems, fiscal controls and oversight to Executive Council that would provide a basis for the transformation of the moribund administration he had inherited" (Richardson 2012).

Which is great, except Richardson was talking about the government, not the private sector.

This, of course, would require more money. Alberta's oil industry was doing very well by 1971. The Social Credit government had capped the maximum Crown royalty at 16-⅔% in 1948. This was great for attracting and servicing investment capital. But for a government determined to do more than simply supervise the game and support the private sector, more funds were required.

In 1972 Lougheed followed up on his commitment to increase the government's share from oil and gas production. This was highly controversial at the time. The industry's position was quite predictable. The general view, shared by companies, associations and the trade press, warned that investment capital would go elsewhere if returns were diminished. The basin was mature for oil and developing smaller prospects was less profitable thus requiring a higher return. Tearing up existing contracts, particularly the royalty rates under which the industry was developed, was a sign of bad faith.

Because the 16-⅔% maximum royalty was set by law, extracting up to an additional $90 million a year was achieved by superimposing a separate levy on the remaining reserves in the ground, the majority of which were, by law, owned by the Crown. The legislation was called a Natural Resource Revenue Plan. This effectively raised the royalty rate to about 23%.

This was the beginning of a new era in carbon politics for Alberta. To this point carbon resources—coal, oil, and gas—were a source of heat, fuel, employment, and tax revenue. By campaigning on an activist agenda to use the government to reshape Alberta and raising the province's share of production revenues to pay for it, carbon had moved beyond a resource or commodity to an agent of political change.

This was not entirely new. The creation of OPEC in 1961 was for this very reason—to ensure the jurisdictions that owned the oil could do more with their resources than stand by and watch the private sector produce their non-renewable resources at the highest profit and the lowest economic rent.

But in Canada, this hadn't happened before. The Crown's mineral-rights leasing and production royalty regime system that had taken this industry from nothing to valuable had served everyone very well. Because it was a developed jurisdiction in which local entrepreneurs and companies could participate in both the production and support industries as well as direct employment, all funds but corporate profits stayed in the province. The industry was growing, attracting new capital, and retaining existing capital. Profits were for the most part reinvested and supplemented by even more money through debt, equity, or inflows from parent companies. Many Albertans participated directly through employment, entrepreneurship, or investments.

What the new PC administration under Lougheed did was redefine the role of government, greatly expand its size and cost, and use carbon resources to pay for it. This was the beginning of new and ever-changing relationship between industry and government that continues to this day.

Total revenues from oil and gas had risen from $1.3 million in 1947 to $297 million in 1972, 228 times greater. For the 25-year period from 1947 to 1972, the total collected was $3.64 billion.[75]

But as would become increasingly common in the years ahead, the growing view on the oil and gas industry would be, "What have you done for us lately?"

As for the Progressive Conservative party, until it disappeared in 2017 it would struggle as to whether it was progressive, conservative or an ever-changing combination of both.

75   Government of Alberta, https://open.alberta.ca/dataset/d84937ce-8ca6-44fd-9de2-61b2e3de1205/resource/d010c813-40ea-4fe7-895f-03c61b6dd099/download/1947-1974-albertas-oil-and-gas-picture-page1-5.pdf.

3

OPEC CREATES THE FIRST BIG BOOM AND
THE FIRST BIG BUST - 1974 TO 1986

"Alberta would never be the same. Nor would the politics of carbon. Because what
immediately occurred was a polarizing split with owners and producers on one side
and consumers on the other about the definition of market price and the entirely
new concept of 'fairness.'"

Proving you can't beat luck in Alberta's oil business, the events of 1973 on the
other side of the world would change the province forever. Not a single living
Albertan, past or present, had anything to do with it.

The Yom Kippur War between Israel and a coalition of Arab countries was
launched October 6, 1973 by Egypt and Syria to recover territory seized by Israel
in 1967's Six-Day War. Egyptian President Anwar Sadat's stated objective was, "to
recover all Arab territory occupied by Israel following the 1967 war and to achieve
a just, peaceful solution to the Arab-Israeli conflict."

Israel fought back and by October 22 had inflicted enough damage in Syria and
Egypt that all three countries agreed to a ceasefire brokered by the United Nations.
What affected Alberta took place five days earlier on October 17. As it became clear
Egypt and Syria were losing the war, the Arab members of OPEC announced oil
exports would be reduced to the United States and other Western countries that
were supporting Israel. Supplies would decline by 5% every month until Israel left
the territories seized in 1967. In December 1973 this was expanded to a full oil
embargo against the United States, Portugal, the Netherlands, and South Africa.

Carbon resources had become a political weapon.

The price of West Texas Intermediate (WTI) crude almost tripled in four
months. WTI, which fetched US$3.60 a barrel in the first half of 1973, commanded
$US10.10 by January 1974. Although the embargo was ended in the spring of
1974, that year finished with WTI at US$11.20, about four times higher than the
average for the previous 20 years.

The year of 1973 was also when Alberta's conventional oil production had reached its all-time peak of 1.4 million b/d and oil sands production from GCOS had stabilized at 50,000 b/d. At that time one Canadian dollar was worth US$1.3. Using simple math, the one-year value of all this production pre-OPEC would have been $1.5 billion. After OPEC worked its magic the figure could have exploded to $4.6 billion, over three times higher and a $3.1 billion gain.

If, of course, Alberta and its oil producers had ever received the world price.

By doing nothing beyond being in the right business at the right time, Alberta's carbon-resource industry was worth billions more than it used to be. In his 1979 best-selling book *The Blue-Eyed-Sheiks—The Canadian Oil Establishment*, author Peter Foster wrote:

> There are no statues in Calgary to Sheik Yamani, but it is ultimately to the Saudi Arabian oil minister, the brains behind OPEC, that Alberta owes most of the fabulous increase in both wealth and power that the 1970s brought it. OPEC's actions, under Yamani's guidance, have been termed blackmail by a petroleum-hooked world that literally runs on oil. Yet Peter Lougheed, the iron-willed premier of Alberta, and the whole Canadian oil industry have clearly aligned themselves with Yamani (Foster 1979).

Alberta would never be the same. Nor would the politics of carbon. Because what immediately occurred was a polarizing split with owners and producers on one side and consumers on the other about the definition of market price and the entirely new concept of "fairness."

Having been subjected to the so-called "market" price forever, owners (like the Province of Alberta [aka The Crown]). and producers (the oil companies) believed the OPEC-created windfall was rightfully theirs. Except for reserving Ontario for western Canadian oil through the National Oil Policy in 1961 (explained in upcoming Chapter 6), there had been no interest among oil-consuming Canadians in paying more for oil or gas than necessary. The world was amply supplied. Sure, the price was going up. Yes, it was for a novel reason. But owners and producers had become accustomed to bearing 100% of the burden when prices declined. Should the opposite not be true?

But everybody who bought oil was outraged. It started with gasoline shortages and lineups in the US Fortunately, Canada was spared physical fuel shortages. America was not built on parked trucks and cars. That a handful of Third-World countries could hold the world hostage to their demands was immediately politicized as most in public office in the Western world who didn't directly own or control oil and gas reserves called OPEC's actions outrageous, immoral, illegal, vindictive, collusive, destructive, and just plain wrong.

Although Alberta was part of Canada—or at least it had been before October 1973—this form of "profiteering" on the part of Alberta was also seen by many as wrong. Ontario and Quebec felt it was okay for the new western "colonies" to be part of the country. But it was quite another for the four western Canadian oil producing provinces to enjoy illicit and obscene profits precipitated by a group of uncivilized dictators half a world away at the expense of their fellow Canadians.

From 1973 onwards, the word "oil" was rarely used without being accompanied by a pejorative adjective of some sort. This continues today.

The reaction in the respective capitals of Edmonton and Ottawa could not have been more different. Alberta Premier Peter Lougheed was a strong believer in provincial rights, including resource ownership, and since being elected in 1971 had communicated this message. Foster wrote about a speech Lougheed gave to the National Press Club in 1971 where he talked about Alberta's voice in trade exports of oil and gas, which had historically been managed by Ottawa. Lougheed said, "If Alberta poker chips are involved at the poker table, we will be at that table." Federal reaction at the time was that Alberta's new premier was out of line. As history would prove, Lougheed was just getting started (Foster 1979).

In the summer of 1973 Imperial Oil raised its oil price by $0.40 a barrel, bringing the total in the previous nine months to $0.95, or 32% more than it had cost. Ottawa was watching and worried but didn't know exactly what to do. In *The Politics of Energy* by G. Bruce Doern and Glen Toner, the authors wrote extensively about how Ottawa and the federal department of energy had little domain knowledge of global oil and gas markets and depended significantly upon producers for their information. In September Prime Minister Pierre Trudeau's administration started thinking about extending the pipeline for western oil from Sarnia to Montreal to provide a possible lower price option than Venezuela, Quebec's main supplier of imported oil. This was something the west and the industry had asked for but which had been rebuffed in the late 1950s (Doern and Toner 1985).

When Ottawa decided to raise the price of exports to the US by $0.40 barrel, it advised Alberta it would be keeping the difference. In addition, the federal government capped the Canadian crude price at current levels. Foster wrote of a speech Lougheed gave to the Calgary Chamber of Commerce the next day where the premier called the federal export levy "… the most discriminatory action ever taken by a federal government against a particular province in the entire history of Confederation … Jobs, both existing and in the future, are in jeopardy in Alberta today … We have to try to protect the Alberta public interest – not from the Canadian public interest – but from Central and Eastern Canadian domination of the west" (Foster 1979).

Albertans loved Peter Lougheed's fiery pro-Alberta, anti-Ottawa, anti-Ontario, and anti-Quebec rhetoric. This message had played well for years. The prior Social Credit government had an innate distrust of Ottawa ever since the federal government ruled its "funny money" illegal during the Depression. Unknown at the time was that Lougheed's tirades against federal government intrusion would continue for the next seven years.

The oil industry, still smarting from jacked-up Alberta resource levies a year earlier, was not as enthusiastic. Any time various levels of government start changing or disagreeing about the rules under which any industry conducts commerce, it introduces a level of uncertainty that inevitably affects the corporate bottom line. The people who found and produced all the oil and gas and figured out how to turn it into money would be, to a great degree, bystanders in the looming and protracted federal/provincial energy war.

Reacting to OPEC's unplanned but significant oil price shock, Ottawa concluded it had to do something. In December 1973 Prime Minister P. E. Trudeau announced the creation of a new national oil company that would be 100% government owned. It was eventually called Petro-Canada. Carbon was so important that the federal government had to own it on behalf of the citizens. The price protection western producers enjoyed in Ontario would end and an oil pipeline to Montreal for western crude would be constructed. Trudeau said, "We do not think it equitable or fair that surplus profits return solely to the provinces producing oil. In the government's opinion, the whole country should take benefit from any windfall profits" (Foster 1979).

In January 1974 Ottawa introduced more plans for oil prices and exports. While Canada had introduced a "made-in-Canada" price in 1973, this didn't apply to the ever-escalating cost of imported oil in Quebec and the Maritimes. To have a truly made-in-Canada price Ottawa now had to subsidize the difference between the legislated Canadian price and the world price for imports. Therefore, it needed more money. The export tax would help pay for that. Canadian prices would indeed rise so there would be additional funds to ensure Canadian producers had sufficient capital to continue to develop more oil supplies. But the increases would be gradual to prevent a major price shock to the Canadian economy. But as the price went up, who would get the rest of the money? Ottawa or Edmonton?

Furious about the heavy hand of Ottawa, the obvious solution for Alberta was to raise wellhead royalties, which it did. Instead of a flat rate of 16-⅔% plus the new tax on reserves, Alberta introduced new royalty rates for oil and gas, which were sensitive to price. Besides the base rate, 50% of the natural gas price from $0.36 to $0.72 per thousand cubic feet would now go to the Crown, and 65% of any gas price higher than that. For oil, 65% of any price above $4.41 would go the province (Doern and Toner 1985).

To collect more money yet spur continued investment, the province created "old oil" (defined as that found before 1974) and "new oil" (oil discovered there-after). Because old oil was theoretically bought and paid for, the logic was that it could withstand a higher royalty rate than new discoveries. The levies were applied accordingly.

The irony that 300-million-year-old oil discovered after 1973 would be called "new" went unnoticed at the time. New to the surface perhaps, but not to the earth.

This angered the federal government. Corporate federal income taxes were calculated on taxable income after royalties were deducted. This was because royalties were paid to the owner of the resource. Legally, the resource owner and corporate developer were partners through a lease. If royalties were privately held—freehold—they were treated as income for the owner and therefore still taxable (to the federal government). But in Alberta the province owned the bulk of carbon resources. One level of government was not permitted to tax another.

To prevent Alberta from grabbing more money than Ottawa deemed "equitable or fair," in the May 1974 budget the federal government decreed the recently increased Alberta provincial royalties would not be deductible from income for tax purposes by producers. Companies would have to pay the royalties on higher prices and taxes as if it were all income, and even though they didn't have the money. The cash was already in Edmonton. This of course further raised the temperature in the premier's office in Edmonton.

About this time the bumper sticker reading, "Oil Feeds My Family and Pays My Taxes" made its first appearance. Albertans were pissed and Lougheed was their saviour.

In 1973 slightly increased oil and gas prices, production volumes, and the new reservoir tax resulted in total revenues to the province of $520 million from the oil industry, almost as much as 1971 and 1972 combined. Thanks to OPEC, rising domestic prices, and the new royalty rates on higher prices, in 1974 Alberta collected $1.1 billion. The new Alberta—the one collecting hundreds of millions more carbon dollars than it needed for day-to-day operations—was born.

In 1974 the province created the Alberta Petroleum Marketing Commission, or APMC. Through APMC the Crown took its royalty share in oil, not money, and sold it on the market. Symbolically, one of its purposes was to highlight the issue of provincial ownership of its carbon resources. For the government, producing oil was not just a revenue collection and accounting exercise. In some ways the APMC's function mirrored that of Petro-Canada, Ottawa's new state oil company.

The battles between Edmonton and Ottawa turned into a series of skirmishes from 1974 to 1979. WTI crept up about US$1 per barrel a year and Canada in turn followed with routine increases in the made-in-Canada price. The IPL (now Enbridge) Line 9 pipeline extension from Sarnia to Montreal was completed in 1976, putting Quebec on domestic production for the first time. Despite ongoing disputes with Ottawa about the principles of provincial resource ownership, steady oil price increases and the new levies kept Edmonton in the chips. Total non-renewable resource income rose to $1.3 billion in the 12 months ending March 31, 1975. The figures for 1976, 1977, and 1978 were $1.7 billion, $2.1 billion, and $3.1 billion respectively.[76]

Premier Lougheed went back to the voters for a renewed mandate in 1975 and won 69 of 75 seats, 92% of the elected members. The former Social Credit dynasty won only four seats that year compared to 25 in 1971. That it had been raining provincial fiscal largesse through scores of new programs nobody had ever heard of until the PCs formed government in 1971 certainly didn't dampen Lougheed's popularity. Albertans were clearly happy with the economy and the

---

76 Government of Alberta, Ministry of Energy, https://open.alberta.ca/opendata/historical-royalty-revenue.

government's management of the carbon-ownership and revenue-sharing battles. Federally, Pierre Trudeau's slim majority government of 1972 was replaced by a solid majority mandate in 1974. Notably, the Liberals did not win many seats in the west in either election and none in Alberta.

With more carbon cash flowing in and a mandate to modernize Alberta, the Lougheed government got started with spending the money and building a larger and more expensive administration and cost structure. More air strips in rural Alberta. Grants to homeowners. Aid for mass transit. Funds for oil sands development, rural electrification, and municipalities (Marsh 2011).

Industrial diversification remained a key theme of the PC government. In the 1920s the Alberta government funded oil sands research and oil sands development to diversify away from agriculture. Fifty years later Lougheed wanted to diversify away from merely producing carbon resources. The government created the Agricultural Development Corporation and the Alberta Opportunity Company to provide government capital under attractive terms to farmers and businesses.

> Government spending soared from $929 million (from revenues of $958 million) in 1971 to $6.6 billion (from $9.4 billion in revenues) in 1980, and the number of civil servants increased from 18,595 to 33,607. Lougheed later admitted he may have overdone the new spending, but he argued that the Socreds had allowed the provincial infrastructure to deteriorate badly (Marsh 2011).

Growing government spending by a factor of six in under a decade required a lot of new ideas. Alberta bought control of Pacific Western Airlines, ostensibly to prevent it from moving its head office to Vancouver. Boasting of a $600 million budgetary surplus in 1976, spending promises were made for pay raises, mortgage subsidies, libraries, and research (Marsh 2011).

That year the government created the Alberta Heritage Savings and Trust Fund (Heritage Fund) with a contribution of $1.5 billion as seed capital. The commitment was that 30% of non-renewable resource revenue would flow into the Heritage Fund. Its purpose was to protect Alberta for a "rainy day" when oil and gas did not contribute as much as it could at that time. As was regularly repeated by Lougheed and his government, oil and gas were non-renewable resources. The supply was finite. The Heritage Fund would assist further industrial diversification and, at least on paper, relieve Alberta from the "boom-and-bust" cycle that historically characterized this industry.

Boom and bust? Boom, indeed. But where was the bust? By 1976 Alberta was on its 30th consecutive year of a post-Leduc oil boom of varying degrees of intensity, the last three of which were turbocharged thanks to OPEC. The Heritage Fund was a thoughtful and popular idea, which many Albertans regarded as a symbol of the province's prosperity and new role in the Canadian federation.

But whatever comfort it gave Albertans, the Heritage Fund would become a lightning rod for further federal political interference as Alberta enjoyed unparalleled prosperity while the rest of Canada—at least in the minds of many voters outside Alberta—paid for it.

With GCOS on production since 1967, the Alberta government was very interested in further oil sands development. The province put forward a number of different financial incentive packages, which finally resulted in the Syncrude consortium being formed in 1973. It didn't last long after a major partner withdrew in 1974. To see the project move forward, the governments of Canada, Ontario, and Alberta—unlikely bedfellows at the time—stepped up with $600 million to fund the missing 30% of the equity capital.

The three governments had been at loggerheads over oil pricing and other carbon-policy matters since 1973. But a big capital project to increase domestic production in the face of global supply uncertainty proved these governments didn't disagree on everything. Then, as now, politicians loved big projects with ribbon-cutting ceremonies that are ostensibly in the national interest. That it wouldn't produce oil for years was lost in the symbolism of thoughtful political leaders "doing something" about the oil crisis.

By the mid-1970s oil companies were about as popular as the bubonic plague. There were charges of a sellout by the governments to the multinational oil companies. University of Alberta political scientist and author Larry Pratt wrote a book in 1976 titled *The Tar Sands, Syncrude and the Politics of Oil*. In the preface Pratt wrote,

> The essential thesis is that Syncrude is likely to become the prototype for new energy ventures in Canada; and if this is correct, every Canadian should know something of Syncrude and the remarkable power of the oil lobby in our political system. If power is defined as the ability to realize one's will and to achieve

one's objectives, then the oil lobby necessarily must be reckoned as one of Canada's fundamental power blocs. (Pratt 1976)

The political thesis about oil producers, developed in the early 1970s and still in existence today, was that private oil companies wielded enormous power over governments and consumers. This was actually not the case but as the old saying goes in the news business, never let the facts ruin a good story. Major oil companies had seen its assets nationalized in whole or in part all over the world. The Lougheed government had tripled wellhead production royalties (depending on the price) since 1971. The federal government-controlled prices, taxation, and exports. Ottawa and Edmonton fought over what was once the oil industry's production dollars without consultation or a thought about compensation to the producers. But the level of distrust of private oil companies was so low that Ottawa felt compelled to fund a state-owned oil explorer and producer.

The publication of the book *The Seven Sisters—The Great Oil Companies and the World They Shaped* by Anthony Sampson in 1975 didn't help. Conspiracies abounded. The seven companies—Exxon, Mobil, Chevron, Texaco, Gulf, Shell, and BP—had at one point controlled 85% of the world's oil supplies. That all changed with nationalization and OPEC. But as one reviewer wrote a year later, "If ever there was an industry with profound international economic and political consequences, this is it."

Regardless, Pratt wrote:

> ... anyone who attempts to reconstruct Syncrude's lobbying activities in Alberta and Ottawa from 1972 to 1975 will be hard put to find a single issue of any substance where the owning interests in the consortium failed to win their basic objectives. This private power – and the absence of any countervailing power representing the public interest – is a phenomenon which most analysts of Canadian politics, who begin with the assumption that the state enjoys the last word in power, conveniently ignore (Pratt 1976).

Governments had lots of power. Good and bad. Petro-Canada was an extremely popular symbol but in the end was a financial disaster.

But distrust of anything to do with oil ran high. "The CBC aired a program based loosely on Pratt's book on September 12, 1977, portraying Lougheed as a dupe of the multinational oil companies. A furious Lougheed successfully sued the CBC for libel. He considered the Syncrude deal one of his five most significant accomplishments and felt that he had protected the province from any risk if the project failed" (Marsh 2011).

The notion Peter Lougheed and the oil and gas producers were the best of pals was dead wrong. "While Lougheed conferred often with the oilmen, his overall relations with the industry were frosty. The industry fretted over this interventionist mindset and the rising royalties he made them pay. Nevertheless, to them Ottawa was worse. 'He may be a son of a bitch' one executive told *The Globe and Mail* once assured his name would not be published, 'but he is our son of a bitch'" (Marsh 2011).

During this period the province also helped Alberta's petrochemical business expand significantly. Alberta Gas Trunk Line Company Ltd, (AGTL) was a government corporation established in 1954 to build, own and operate Alberta's natural gas gathering and transmission facilities. In the early '70s the it branched into petrochemicals and later changed its name to NOVA, An Alberta Corporation. To support AGTL, the province worked behind the scenes to create a fiscal regime to supply feedstock under attractive terms. Part of this was in response to the growth of the petrochemical industry in Sarnia, which used Alberta oil and natural gas as feedstock. The financial incentives were necessary to overcome the geographic obstacles of market proximity Alberta had been cursed with since creation. While it took some time to build, the Joffre petrochemical complex in central Alberta started producing in 1979. The facility is widely regarded as one of Lougheed's major achievements in moving the province forward as more than a "hewer of wood and drawer of water."

But then another event took place in 1979 that would shake up the carbon political landscape yet again. A revolution in Iran—a major OPEC oil producer. The existing monarchical administration was overthrown and replaced with a fundamentalist Islamic regime. The so-called Shah of Iran had been friendly to the West (sort of—Iran partially nationalized its oil industry in 1951). Iran's new regime was not. Grand Ayatollah Ruhollah Khomeini's revolution only reduced world oil supply by 4% but the "second oil crisis," as it was quickly coined, came only six years after the first. When would this end? This caused widespread panic in oil markets. In early 1979 WTI had crept up to US$15.90 a barrel. By year-end it was US$32.50. It peaked in mid-1980 at US$39.50 a barrel, more than ten times what it fetched only seven years earlier.

Besides the first and second oil price shocks, in the 1970s Western economies experienced the highest price inflation rates since the end of World War II, along with rising interest rates. Both hit double digits in 1974 and 1975. In 1975 the Trudeau administration introduced the Anti-Inflation Act, which introduced draconian, government-administered wage and price controls. The economies of Canada and the US were struggling. 1979's oil price spike put national governments all over the world in the position of having to be seen as doing something. Ottawa's response to global carbon geopolitics will be explored more fully in Chapter 6 of this section.

In the 1979 federal election the Liberals were replaced by a Progressive Conservative minority government. Due in part to the energy policies of Prime Minister Joe Clark's administration, the PCs were defeated on a non-confidence budgetary motion. Pierre Trudeau's Liberals were put back in power in early 1980 with another majority. On October 28, 1980, the government introduced the NEP. For 38 years analysts and commentators have not run out of negative superlatives to describe how intrusive, ill-conceived, and disastrous the NEP was. Most historians agree that outside of the ongoing threat of Quebec separation, no single policy has done more to split the country than the NEP.

After nine years of battling Ottawa for Alberta's control over its constitutionally granted resources, the NEP forced Peter Lougheed to notch up the rhetoric and response to previously unseen levels.

In 1979 Lougheed's PC government won another landslide victory claiming 74 of 79 seats, 94% of the total. At five MLAs, the opposition was the smallest in history. This was a powerful mandate for Lougheed to speak and act on behalf of Albertans.

In April 1980, prior to the NEP, Lougheed said publicly, "I frankly do not think it is possible for us – as we move into the 1980s – to develop a Canadianism that responds to the diversity of our nation without some major, major readjustments in the way we have operated as a nation in the past." After describing Alberta's huge financial sacrifice in selling its resources at half value since 1973, Lougheed continued, "Any unilateral action by the federal government – particularly one that's been rejected by western Canada – will be resisted by our citizens in the strongest and most determined ways" (Doern and Toner 1985).

Trudeau did it anyway. After reviewing the document, two days later Lougheed went on province-wide television and declared, "As far as I am concerned the NEP consists of proposals not programmes and they are not national, they are Ottawa," which, of course, indicated the NEP was more political than practical. He continued, "We will sit down and again try and negotiate ... However, no matter how stormy the weather, we will not capitulate. We will not give away the heritage of this province" (Doern and Toner 1985).

The issue was entirely political in Alberta's view. The larger provinces, where the Liberals had voter support, were ganging up politically on the smaller provinces and violating the constitution without taking the necessary legal steps to amend it. He would later say, "It is my belief that the Prime Minister's plan to unilaterally change Canada's constitution over the opposition of the majority of the provinces, is closely linked to resource development and western Canadian development ... If the country proceeds as intended by the Prime Minister, we will have a very different kind of Canada ... it will be primarily a unitary state with provinces other than Ontario and Quebec being in a second-class position ... the NEP's objectives are an attempt to take over the resource ownership rights of this province" (Doern and Toner 1985).

The problem for Lougheed was that the only tools at his disposal beyond righteous indignation were economic. In response to the NEP, Lougheed began to cut back oil production to Eastern Canada by up to 15% and withdrew essential provincial approval for two major oil sands development projects: Alsands Energy Ltd., which was a bitumen mining operation, and a significant expansion of Imperial Oil's Cold Lake subsurface thermal recovery project.

This time the bumper sticker read, "Let the Eastern Bastards Freeze in the Dark." The battle over carbon control and political fairness had reached the point that the threat of western separatism became real.

The problem for Edmonton and Ottawa was that in 1981 the world economy continued to deteriorate. Alberta's and Ottawa's respective positions during the federal/provincial carbon wars were popular with voters but financially devastating. In August of that year the bank prime lending rate hit 22.75%, the highest in history. Inflation averaged 12.5%. Worse yet, the price of oil had peaked and was declining. The pie that Alberta and Ottawa were fighting over was shrinking for the first time in eight years. The price of oil was down 10% from its peak in 1980. High prices were exerting their powerful forces in oil markets; demand was falling while new supplies from places like the North Sea and the north slope of Alaska were economic and flowing. Increased supplies from offshore made replacing reduced supplies from Alberta inconvenient and expensive, but not impossible.

On September 2, 1981, Alberta and Ottawa signed an agreement that effectively replaced the NEP with something Alberta could live with. Those who were around will never forget Lougheed and Trudeau smiling, shaking hands, and toasting with champagne on TV. What was that all about after a decade in which Edmonton and Ottawa were mostly at loggerheads over resources? Lougheed figured he won, sort of, saying, "The long-term consequence and benefits to Alberta are great. The Minister of Energy and Natural Resources [Merv Leitch] and I believe most Albertans will be pleased with the result" (Doern and Toner 1985).

The provincial resource ownership and control issue was reaffirmed and resolved. The federal export tax on natural gas was off the table. Most importantly, it provided indexed oil-price increases and a firm commitment to the amount to which each jurisdiction would be entitled.

The problem was that world oil markets declined to cooperate. The 1981 agreement was effectively stillborn because of falling oil prices and an extremely hostile macroeconomic environment. A year later oil prices were still falling, with WTI closing 1982 at US$31.70, 20% below the 1980 peak. Although the regulatory hurdles were removed, the proponents of Alsands and the Cold Lake expansion declined to proceed. The world price of oil was primarily supported by Saudi Arabia, which had shut-in millions of barrels of oil production to sustain the price.

Alberta's treasury wasn't impacted that badly, but the economy was. On the streets the oil boom was turning into an oil bust. Sky-high interest rates and the political fallout from the energy wars collapsed house prices. People walked on their mortgages by transferring title for $1. Alberta's total receipts from oil and gas royalties and lease rentals was holding at about $5 billion per year, ten times what it had been in the 1973/74 fiscal year. But as reinvestment dried up, the economy continued to deteriorate.

The crash spawned the third bumper sticker in a decade, which blared, "Please God, Give Us Another Oil Boom. I Promise Not To Piss It All Away This Time."

History would prove Albertans were just kidding.

In the five years after the agreement, oil prices would continue to fall, with WTI hitting bottom at US$11.60 in July 1986. Because the Canadian price was set below the world price, the Alberta/Ottawa agreement saw made-in-Canada prices continue to rise even as the world price declined This sustained some oilfield employment until 1985. But the Canadian and world price curves were destined to meet and then cross. Economic reality hit home. By 1985 things were so bad two Alberta banks—Canadian Commercial and Northland—went under. They were the first Canadian bank insolvencies since 1923.

This took its toll on the popularity of Alberta's hero, Peter Lougheed. In July 1984 the PCs released what some considered to be Lougheed's industrial diversification masterpiece, a white paper titled "Proposals for An Industrial and Science Strategy for Albertans—1985 to 1990". Citing the 1971 PC campaign platform of "New Directions for the Seventies", which included much of the government's spending and expansion plan, the preamble admitted that because of the political turmoil of recent years it required updating in 1983. It stated, "The forecast for 1984 is for a year of significant economic recovery from the downturn of 1983, but it is not expected there will be economic recovery to the peak levels of the 1979-81 period."

This was hopelessly optimistic.

The five-point plan included major capital spending described as "on a comparative basis is the largest of any Canadian provincial budget … to sustain capital expenditures during current economic conditions"; more money to stimulate the housing sector; "extensive, special job creation and manpower training programs"; more student assistance funding to expand skills and knowledge; and no new taxes to help for any of the foregoing. Oil and gas incentives, small business funding, and financial support for a heavy oil upgrader at Lloydminster were also included.

Lougheed's white paper and his vision for the province were poorly received. This might be the most comprehensive policy discussion paper his government ever produced and it was shelved almost immediately.

It got worse for the premier, according to the 1987 book *Running On Empty: Alberta After The Boom*. After a Canada Cup gold medal hockey game in Edmonton in September 1984 between Team Canada and Sweden, Lougheed was invited to centre ice after the game for the gold medal presentations. He was jeered and obscenities were hurled by the 10,000+ fans in attendance. The authors described Lougheed's reaction, which was televised, as visibly shaken by the crowd's hostility (Nikiforuk, Pratt and Wanagas 1987)

On November 1, 1985 Peter Lougheed resigned to return to private life. He was replaced by long-time colleague in football and politics, Don Getty. In the 1986 election the PCs won 61 of 83 seats, a solid victory by any normal measure. But as witnessed by the success of the Socreds and PCs when the oil money was flowing and the economy was growing, there was nothing normal about Alberta politics and broad support for their governments during economic good times. But in 1986 the economy was sinking and so was the popularity of the governing party.

By the end of 1985, oil producing giant Saudi Arabia had shut-in 7 million b/d to try to stabilize the world oil price. Its production had been 10 million b/d in 1980 and five years later it was down to three. Early in 1986 the Saudis, tired of absorbing major financial losses on behalf of every other oil producer in the world, abandoned the role of "swing producer" within OPEC. Oil prices went into freefall and would not rise in real terms until 2003.

Peter Lougheed passed away in 2012. He is revered by many as the founder of the "modern" Alberta. But as history would prove, spending a lot of one-time oil money to expand the role and infrastructure of government is infinitely simpler than reversing the process.

4

1986 TO 2007 – FROM DESPAIR TO ANOTHER BOOM

"What began in 2004 was Alberta's second great oil boom after the 1970s. One would hardly think this was a problem. But Albertans, back on their rightful perch of enormous wealth, made it one anyway."

History reveals the roller coaster was invented in the United States and first put into service at the Coney Island amusement park in New York in 1884. A small carriage on tracks, the device thrilled passengers by going up and down and from side to side at high speeds, subjecting the rider to physical forces to which everyone reacted differently. The experience ranged from thrilling to nauseating. Some became hooked. Others vowed to never ride one again. To make it more interesting, by 1898 the roller coaster of the day turned completely upside down.

This is an accurate description of the carbon politics of the Progressive Conservative governments of Alberta from 1986 to 2015, when the party was voted out of office. The party ultimately disappeared two years later. Think of the first line of the famous 1959 Australian country tune recorded by many artists including Johnny Cash: "I've been everywhere, man."

In the 29 years between the world oil price collapse of 1986 and the PC Party collapse in the election of 2015, the same party unsuccessfully spent billions of public dollars to get Alberta out of the oil and gas business and then periodically slashed royalties and taxes to expand it. When the industry's resuscitation from extended tough times proved more successful than anyone imagined—thanks again to external forces including skyrocketing commodity prices—Alberta's economy was declared "overheated" and the government reversed prior decisions on the appropriate economic rent from carbon resources in pursuit of "our fair share" on behalf of Albertans.

Under successive PC governments, Alberta amassed the highest public debt in the province's history, paid it all off faster than any political jurisdiction in Canada, if not the world, and then plunged back into red ink at the same time oil prices were at near-record levels. During this wild ride and under five different leaders, the PCs won eight more majority governments from 1986 to 2012.

World oil prices had begun to fall while the ink was still wet on the Alberta/Ottawa carbon-revenue sharing truce signed in September 1981, 11 months after the introduction of the ill-fated NEP of October 1980. While provincial revenues from oil and gas production royalties and mineral rights leases were still $5.4 billion in the fiscal year ended March 31, 1986, that figure would not be duplicated for 15 years. A year later they were cut by more than half. Under new Premier Don Getty, the PCs and all Albertans would learn the hard way that it was easier it was to manage Alberta with significant funds from carbon resources than without.

The Heritage Fund had been created a decade earlier to help protect Albertans from a "rainy day" when the oil and gas cash wasn't flowing from what previous Premier Peter Lougheed had repeatedly declared was a non-renewable resource. According to the 1990/91 Heritage Fund annual report, the inclement weather began in the 1982/83 fiscal year when about $900 million of investment income was transferred to "budgetary revenue" for general purposes. By 1983/84 the entire investment income of over $1.4 billion went to general government revenues. This was a period of continued double-digit interest rates; therefore, generating a positive investment return on nearly $12 billion in assets held by the Heritage Fund was not challenging. All that was required was a savings account.

During the mid-1980s and into the next decade the Alberta government continued to withdraw all the investment income from the Heritage Fund for use in general revenues, but still contributed non-renewable resource revenues. The value of the fund peaked at about $12.7 billion in 1987 and declined thereafter following the government's decision that Heritage Fund income was needed to pay today's bills, not tomorrow's. That year the government ran a $3-billion deficit with more to come. The debt began to pile up. The rain showers of 1983 had turned into a downpour four years later.

Because Premiers Getty and Lougheed had both played football for the Edmonton Eskimos (but not at the same time) and had worked together in politics since they were both elected as MLAs in 1967, it is no surprise they coached from the same playbook.

Since forming government in 1971, the PCs had preached industrial diversification away from the basic production and sale of carbon resources. Without passing judgement on the success of the various industrial diversification programs and initiatives the PCs had pursued in the previous 15 years, by 1986 it was clear that the non-carbon private sector was still insufficient to sustain the economy at historic levels, both in the private sector and in public administration.

What is interesting is that Alberta wanted to get out of the carbon-resource business 30 years ago. Just like David Suzuki wants the province to do today, but for an entirely different reason.

Getty concluded that if Alberta needed industrial diversification before, it really needed it on his watch. Since the private sector could not or would not achieve what the government desired, the Getty administration would lead the charge. What followed in the next five years was one of the country's greatest and most unsuccessful experiments in government-funded and directed economic engineering. And much of it was financed with borrowed money.

Ted Morton served as a PC MLA from 2004 to 2012. He held three portfolios as a cabinet minister, including energy and finance and ran for the PC Party leadership in 2006 and 2011. After he was defeated in the 2012 election, he joined the School of Public Policy at the University of Calgary.

In March of 2015 Morton and colleague Meredith McDonald co-authored a research paper titled *The Siren Song of Economic Diversification: Alberta's Legacy Loss*. While complimenting former PC Premier Lougheed for his valiant defense of Alberta's constitutional ownership of mineral resources against intrusion by the federal government during the period of high oil prices from 1973 to 1981, the authors were substantially less enthusiastic about Lougheed's vision of industrial diversification away from carbon resources. In the summary the authors wrote:

> But there is another aspect of the Lougheed legacy that is less remembered because it is less celebrated ... These were Lougheed's ambitious economic diversification projects. Between 1973 and 1993 the Lougheed-Getty 'forced growth' economic diversification projects are conservatively estimated to have cost Albertans $2.2 billion. While former Premier Don Getty got most of the blame for these losses, most of these programs began earlier (Morton and McDonald 2015).

The difference between Lougheed and Getty was that the Lougheed administration had enough oil money coming in the door to pay cash. Getty's did not. While not wanting to let either premier off the hook, it wasn't just the government of

Alberta that made bad investment decisions during the go-go days of the first great oil boom from 1973 to 1986. The oil business itself also made some colossal blunders during this period.

Morton and McDonald picked 12 initiatives the authors describe as, "the twelve most costly failed diversification projects undertaken by the Lougheed/Getty PCs." They were affectionately titled, "The Dirty Dozen."

In 1989 NOVA and Telus predecessor Alberta Government Telephones (AGT) figured Alberta could be a player in the emerging cellular telephone business. NOVA, which was mentioned in previous chapters, used to be a provincial pipeline company called AGTL, sold its interest to AGT for $42.5 million. During AGT's privatization share offering in 1990, it forecast a profit for NovAtel. A potential buyer discovered a loss. NovAtel was finally sold in 1992 with an estimated government loss to be between $544 and $614 million.[77]

Hazardous-waste disposal was an opportunity that looked promising. So in 1987 the Swan Hills Waste Treatment Plant opened to treat or incinerate the most awful stuff Alberta industry could create. It was owned 60% by Bovar Inc. and 40% by the province. But expensive processing and a remote location dampened utilization. Bovar's capital was treated like a utility with a guaranteed return on investment. Not only did Swan Hills not make money but to be profitable for the private sector partner, it didn't need to. The province put more money into the project until 2000 when it refused to advance any more funds. The loss to Alberta was estimated at $440 million but the plant still operates today.[78]

The Lloydminster Bi-Provincial Upgrader was a financial white elephant that remains in operation. Originally funded by Alberta and Saskatchewan with crude feedstock from Husky Oil, the upgrader was to take the region's heavy oil production and turn it into a lighter grade of oil easier to refine into gasoline, diesel fuel, and other products. Alberta would eventually invest $404 million and recover $32 million when it sold its interest. Today the upgrader is a profitable operation for Husky Oil and has been expanded.[79]

Forestry was a long way from oil and gas. Perfect. In 1987 the Getty government loaned Millar Western Pulp Ltd. $120 million to help finance construction of a

---

77 Ted Morton and Meredith McDonald, *The Siren Song of Economic Diversification: Alberta's Legacy Loss* (2015).

78 Ibid.

79 Ibid

new pulp mill. Ten years later the province had extended a total of $272 million in loans but wrote them off, only recovering about 10% of the total funds.[80]

The Gainers' meat-packing plant required government money for a different reason in 1986. Owned by not-yet-disgraced Edmonton entrepreneur and NHL Edmonton Oilers hockey team owner Peter Pocklington, there was a high-profile six-month strike following the new owner's decision to replace union workers with unemployed non-union workers. Morton and MacDonald concluded that the government just didn't want this mess in the news any longer, so it offered Pocklington a loan guarantee if the strike would end. Three years later Gainer defaulted on its debts, the province inherited the plant, and finally sold it for a total loss of $209 million.[81]

Magnesium smelting. Why not? In 1988 Getty's administration loaned the Magnesium Company of Canada $109 million for a new smelter located in the proven metal-processing hub of High River, Alberta. Just kidding. Morton and McDonald quoted Getty as calling this a "jewel" of Alberta's non-oil economic growth strategy. Magnesium prices tanked, the plant didn't work right, and soon the private sector sponsor abandoned the operation. Stuck with the facility and the operating costs, the province eventually sold the plant for under $5 million after losing $164 million.[82]

Other projects funded by the PCs before and during the Getty years involved a grain terminal in Prince Rupert called Ridley Grain, a giant new pulp mill processing aspen trees near Athabasca called Alberta Pacific Forest or Al-Pac, medical research under the name of Chembiomed, funds to support the ill-fated Canadian Commercial Bank (which failed in 1985), a canola-crushing plant in Sexsmith called Northern Lite Canola, and industrial lasers through a company called General Systems Research.[83]

The Ridley Grain debt of $161 million was still on the province's books in 2013. In 1995 the province wrote off $155 million in loan interest to Al-Pac if it paid back the original $250 million in loans. Alberta got out of Chembiomed in 1995 after losing $44 million. The Canadian Commercial Bank failure cost the

---

80 Ibid

81 Ibid.

82 Ibid.

83 Ibid.

province $56 million, the canola plant about $50 million, and the laser company went bankrupt in 1990, costing taxpayers nearly $31 million.[84]

The authors pointed out that not every investment ended up with red ink. Not surprisingly, the ones that were the biggest and best were in, or supported by, the carbon-resource business. Alberta invested in Syncrude to ensure that project was completed, created the locally owned independent oil and gas producer Alberta Energy Company (which later merged with PanCanadian Petroleums to create EnCana Corporation), established the Joffre natural gas/ethane based petrochemical plant at Joffre, and incorporated and Luscar Ltd., which mined coal.

Edmonton invested in the startup of the Bank of Alberta in 1985. It is now Canadian Western Bank and it is very successful. The purchase of Pacific Western Airlines worked out thanks to PWA's purchase of Canadian Pacific Airlines in 1986 to create Canadian Airlines, which was in turn purchased by Air Canada in 2000. These businesses are a little closer to Alberta's core carbon industry. In coal, oil, and gas they fly a lot. All three companies, their suppliers, and their employees borrow money and buy houses with mortgages.

The detailed recollection of these past government-led economic diversification failures is primarily to remind current and future Alberta politicians that using central planning and other people's money to retool the economy is fraught with peril. There is continued pressure from the public and pundits for Edmonton to diversify Alberta away from carbon resources, which is increasingly described as a sunset industry. The current mantra that Alberta can replace profitable and tax-generating oil and gas with subsidized renewables as a source of jobs and wealth creation deserves much deeper examination.

Back to the 1980s. It was not that the Getty administration was reckless. His government raised taxes and cut spending. But the economic headwinds for a formerly wealthy petro-province without oil and gas dollars were powerful. In 1987 privately-owned Principal Group, an Edmonton-based trust company, went broke. It was later learned it had been struggling for some time. Ensuring not too many people lost too much money cost the Alberta treasury another $85 million. And costs were indeed cut. In the early 1990s the Alberta government had one of the lowest rates of year-over-year spending increases among Canadian provinces. But the debt piled up.

---

84 Ibid.

Getty called another provincial election in 1989, only three years after the last one. The PCs won 59 of 83 seats, another majority government. The premier didn't win his own seat in Edmonton that year and had to run again in a by-election in PC-friendly Stettler to get into the legislature. Discontent within the party was growing. He resigned as PC leader in 1992.

Don Getty had to run Alberta without the oil money pouring in and a legacy-government cost structure that had been vastly expanded with carbon cash. For this reason his reputation was, and remains, tarnished while Lougheed's has become legend. Several events that took place during Getty's time in office would pave the way for a broad recovery in the 1990s. Two were provincial initiatives. He has never gotten much, if any, credit for either.

The deregulation of natural gas by the federal government in 1985 and the removal of the surplus test for exports in 1987 would forever change the gas business. The Foothills "prebuild" section of the Alaska Highway Natural Gas Pipeline that was never completed increased gas export volumes after it went on stream in 1981. Although conventional oil production had peaked in 1973 and has declined almost continuously for 45 years, by 1990 Alberta's gas production was 35% higher than in 1980. Three years later it would rise by another 35%. It was natural gas that powered Alberta's recovery for the remainder of the twentieth century.[85]

The impact of these events started to show up in provincial revenues during the Getty years. While gas production royalties had always been less than those of conventional oil, that changed in the 1987/88 fiscal year when they were about the same. Thereafter, gas royalties would be greater than conventional oil for the next 20 years. But total non-renewable resource income remained stuck at less than half of what it had been a decade earlier.

Another event that paved the way for the future was the conventional oil and gas royalty review of 1991. After five years of low oil prices and six years of moribund drilling and investment activity, Edmonton finally figured out that the royalty rates cobbled together by the Lougheed administration to protect provincial revenue from Ottawa's grasp were obstacles to investment in an era of low prices. Producers pressured the Alberta government to help the industry stay competitive as oil prices languished. Getty listened and that year Edmonton cut the maximum

---

85   Canadian Association of Petroleum Producers, *Statistical Handbook for Canada's Upstream Oil Producers* (Calgary, AB: CAPP, 2018). https://www.capp.ca/publications-and-statistics/publications/316778.

royalty rates for oil and gas significantly. By 1993 the number of gas wells drilled was twice that of 1991, and total drilling in the province jumped by 80% in the same period.

The Getty government also helped business stay afloat and grow in other ways. After two banks and one trust company went under and many businesses failed because they were over-leveraged, Alberta's economy needed equity, not more debt. One of the more innovative ideas, controversial at the time, was the introduction of the Junior Capital Pool (JCP) program by Alberta Stock Exchange and Alberta Securities commission to coax cash out of people's pockets.

Through JCPs, investors could buy shares in small and affordable amounts in what was called Blind Pool Companies, a publicly traded shell with some cash and founders and directors that hopefully weren't crooks. Selling shares in companies without assets was sneered at by Bay Street at the time. But some of the most successful companies in Alberta in future years started as JCPs. Many are still in operation today as elements of much larger companies trading on major stock exchanges in Toronto and the US.

This was accompanied by the Alberta Stock Savings Plan (ASSP) whereby Albertans could receive a 30% tax credit for investing in certain Alberta companies. There was no qualification as to what industries would qualify. There is a similar program today—the Alberta Investor Tax Credit (AITC)—however, the qualifying industries and sectors are determined by the government, not by entrepreneurs or investors.

The 1992 leadership race to replace Getty was widely contested. There were nine contenders on the first ballot. The second ballot was down to, in order of how they finished, Ralph Klein, Nancy Betkowski, and Rick Orman.

Ralph Klein's claim to fame was that he was a former TV reporter and the former mayor of Calgary, which had hosted the 1988 Olympic Winter Games. He was elected as a PC MLA in 1989 and served in the Getty government as Minister of Environment. After he won the leadership Klein became premier and kept that job for the next 14 years. Klein would eventually earn the nickname "King Ralph" because he won four consecutive majority governments in 1993, 1997, 2001, and 2004. His popularity rivalled that of Manning and Lougheed in 2001 when the PCs won 74 of 83 seats, 89% of the MLAs.

Klein is best known for his determination to cut costs and restore what was coined "The Alberta Advantage." It was the absolute opposite of the Lougheed and

Getty administrations. Klein's mission was to cut spending and get government out of the business of the private sector.

After winning the leadership and with the help of Treasurer Jim Dinning, Klein assembled an advisory group to have a cold, hard look at the province's finances. The 1980s had been brutal on the Alberta treasury. Frank Dabbs wrote a book in 1995 titled *Ralph Klein, A Maverick Life*. In a chapter titled "Miracle on the Prairies" Dabbs wrote, "When provincial treasurer Jim Dinning's independent Financial Review Commission reported, ahead of its March (1993) deadline, it warned the government that indebtedness was worse than expected. The provincial net assets of $12 billion had melted away to a net debt of $5 billion in only seven years. The per capita debt of more than $1,000 was the highest of any province and only slightly behind the federal government level of about $1,250. Alberta, the commission concluded, would inevitably hit a brick wall" (Dabbs 1993).

Klein and Dinning went to work. The premier ended provincial loan guarantees and froze funding to hospitals, municipalities, and school boards. MLA salaries were cut by 5% and their pensions terminated. The rest of the public sector took a now-infamous 5% rollback. In the spring budget Dinning cut spending by $700 million and assured the public that the operating deficit, $2.8 billion in the 1992/93 fiscal year, would be gone by 1996/97. Public service unions protested but Klein's support grew. Having the MLAs lead by example by taking pay cuts and ending their own pension plan made the process considerably more palatable within the government service. The public loved it. Many of the industrial-diversification initiatives of prior governments were terminated under Klein (Dabbs 1993).

Klein won the 1993 election with 51 of 83 seats. His main opponent was the Liberal Party under former Edmonton Mayor Laurence Decore. Not often remembered is that Decore's platform was to cut spending even more. When the Liberals won 32 seats that year it was their best showing since 1917 and a level of voter support that has not been repeated.

The public service spending cuts haunt Klein's legacy to this day. The growth in public spending that took place after Klein resigned in 2006 was justified as essential to catch up on the so-called "infrastructure deficit" of the 1990s under his administration. The renewed spending was due primarily to the spectacular economic growth caused by tens of thousands of people moving to the province. At that time in Alberta there were lots of jobs, taxes were the lowest in the country, and public spending was under control.

Only in Alberta in the middle of another oil boom would runaway prosperity ever be considered a problem. Alberta's current NDP government today still vilifies Klein as the enemy of fair and decent social services such as healthcare and education.

What is forgotten by those that vilify Klein is the positive impact the Alberta Advantage had on attracting business to Alberta. This was a form of industrial diversification through job growth and increased taxes that didn't cost the provincial treasury anything up front. Thanks to Klein and Dinning, major corporations like Imperial Oil, TransCanada Pipeline, and Canadian Pacific moved their head offices from Central Canada to Alberta. It proved the private-sector mantra that reducing corporate taxes would ultimately lead to higher corporate taxes as the corporate sector grew. How quickly people forget.

Despite the disparagement of Klein's legacy among the left-leaning, Alberta's modern fiscal conservatives, while still heaping high praise on Peter Lougheed, also idolize Ralph Klein. The legend goes that Klein's administration slayed the provincial debt while oil prices remained low. While this is true, people have forgotten about the situation with natural gas at the time. What is usually overlooked is that carbon resources worked their magic again, proving that Klein, like all of Alberta's most successful political leaders, was extraordinarily lucky with his timing.

In 1992 Alberta's marketed gas production was 9.6 billion cubic feet per day (bcf/D). By 2000 it would be up 44% to 13.8 bcf/D, the highest produced volume in the province's history. The price rose as well. In 1990 and 1991 the AECO natural gas reference price was $1 per million British Thermal Units (MMbtu). It was so low that the Alberta government introduced a minimum royalty on gas because after processing costs were deducted, the royalty was zero. One sage US oilman told a Calgary audience during a luncheon speech, "They call natural gas the fuel of the future. It must be because it has no present value."

That gradually improved for the rest of the decade. While not a straight line— commodity prices never are—by the end of 2000 the AECO reference price was over $3 per MMbtu and closed the year over $5.

The impact on Alberta's treasury was huge. While everybody remembers Klein's spending cuts and his colourful personality, over in the department of finance the cash was rolling in. Natural gas royalties rose from $1 billion in the 1992/93 fiscal year to $2.4 billion in 1999/2000. Total non-renewable resource income increased from $2.4 billion in 1992/93 to $4.8 billion in 1999/2000.[86]

Another major event took place in 1996 under Klein that would shape the economic future of the province as long as oil has value. That was the joint negotiation of the Generic Oil Sands Regime with Ottawa. Both governments were committed to developing a fiscal structure that would make the exploitation of this massive resource economically attractive under the existing flat oil prices. Alberta passed the new Oil Sands Royalty Regulation in 1997, which levied a 1% royalty until payout and 25% after capital costs were recovered. Ottawa chipped in with an Accelerated Capital Cost Allowance, which further enhanced the investment economics of multi-year, multi-billion-dollar oil sands investments.

The combination of the 1991 royalty changes, the new Generic Oil Sands royalties, corporate tax cuts, rising gas production, and the generally favourable business climate in Alberta began to work its magic. The industry's key barometer of future prospects is resource land sales or the leasing of mineral rights. These represent 100% profit for the government because the all they involve is a signed document assigning exploration or development rights. Land-sales revenue had declined from about $800 million per year in the mid-1980s to as low as $270 million 1992/93. The renaissance began in 1993/94 when it rose to $822 million. It burst through $1 billion for the first time since 1979/80 in 1993/94 and stayed at elevated levels for the rest of the decade.[87]

What operators do after buying mineral rights is drill. Alberta went from 3,627 wells drilled in 1992 to 13,543 in 2000. It blew through 10,000 new wells for the first time in history in 1996 after getting very close in 1994. In the best year of the prior boom, the total well count was only 7,044 in 1980. The biggest increase was in natural gas drilling. All this money being spent in the oil and gas fields trickled through every aspect of the economy in every region. This increased payroll taxes, corporate taxes, house prices, and housing starts. When Klein went to the polls in

---

86 Government of Alberta, Ministry of Energy, https://open.alberta.ca/opendata/historical-royalty-revenue.

87 Government of Alberta, Ministry of Energy, https://open.alberta.ca/opendata/historical-royalty-revenue.

2001 and ended election day with 89% of the seats in the legislature, King Ralph was firmly and rightly planted on the premier's throne.[88]

And it gets better. In the last six years of the Klein regime it is fair to say that the oil and gas industry went nuts again. Gas prices exploded in early 2001 reaching $9.80 per MMbtu in January. Gas royalties in the 2000/01 fiscal year reached $7.2 billion and total non-renewable resource revenues $10.7 billion. Land sales were $1.3 billion, the highest ever. In 2004 Ralph Klein proudly announced that Alberta had completely retired the enormous provincial debt his administration had inherited 11 years earlier. He was regarded as a fiscal, managerial, and political genius. Natural gas did not get anywhere near the credit it deserved, nor did Ralph or any of his close associates want to rain on his parade by mentioning it.[89]

In the late 1990s oil prices began to rise after 13 years in the dumpster. In 2000 WTI was in the US$30 range, double that of 1986. By 2004 it hit US$40 for the first time ever (corrected for inflation it was much higher in 1980). With the exception of the world economic crisis of the latter half of 2008 and 2009, oil would rise to and stay at unprecedented levels for a decade. Thanks to carbon and global forces over which it had no control whatsoever, Alberta was back in the chips again in a big way.

Like many governments before it, Klein's administration promoted the massive opportunities in the oil sands. An obstacle that one of the world's largest petroleum carbon deposits faced was a technical issue under which the US Securities and Exchange Commission (SEC) didn't recognize oil sands as "proven" reserves for accounting purposes. Assigned with the task of protecting investors, the SEC set the rules for what constituted long-term assets on an oil company's balance sheet. The US had 100 years of experience with conventional oil and almost none with oil sands mining. So determined were the American regulators to do the "right thing" to protect investors, in 2005 EnCana was required to reduce its stated reserves for public disclosure by 363 million barrels because of price fluctuations. The influential oil experts at Cambridge Energy Research Associates urged the

---

88 Canadian Association of Petroleum Producers, *Statistical Handbook for Canada's Upstream Oil Producers* (Calgary, AB: CAPP, 2018). https://www.capp.ca/publications-and-statistics/publications/316778.

89 Government of Alberta, Ministry of Energy, https://open.alberta.ca/opendata/historical-royalty-revenue.

SEC to review its policies so investors could compare the value of companies that produced petroleum from the oil sands with others with more conventional assets.

Under Klein, Alberta was if nothing if not unconventional in its approach to lobbying. In 2006 Energy Minster Murray Smith arranged to have one of the giant dump trucks used to haul bitumen ore parked on the National Mall in Washington DC to show policy-makers that oil sands were real. The truck was built in the US by Caterpillar. It was paid for by real oil sold for real money and turned into real products like gasoline and diesel fuel for the giant hauling machine.

By this time Alberta's largest oil sands mining operations, Suncor and Syncrude, were running great and were profitable. Imperial Oil at Cold Lake had been operating for almost 30 years. Major oil companies from all over the world renewed their interest in oil. Known offshore fields like the North Sea and the Gulf of Mexico were mature. There were very few places in the world where private-sector oil companies could develop long-term reserves and production that were not owned or controlled by state oil companies. The oil sands, which were heavily promoted by the Klein government and had a very simple and attractive fiscal regime in place, stood out. Only 20% of the world's oil reserves were not owned by governments or produced by state oil companies. And most of that oil was in Alberta.

In 2008 the SEC relented, changed its accounting policing policies, and everyone agreed the oil sands were real. At that time it was generally agreed that the oil sands contained 173 billion barrels of recoverable oil: the real thing at last for accounting purposes.

The oil companies had already figured this out. In the five fiscal years from 2004/04 to 2008/09, capital from all over the world invested an unprecedented $3.4 billion in oil sands leases, something that had never happened before and would never happen again. In the 2005/06 fiscal year, Klein's last full year as premier, gas royalties were $8.4 billion, land sales were $3.6 billion, and the total take was $14.4 billion from non-renewable resource revenues.[90]

This would be the high-water mark for a single year from carbon resources for the Alberta government. One cannot fathom the circumstances under which it will happen again. Perhaps oil at US$150 a barrel and gas at $10 per MMbtu.

With the deficit gone, the provincial economy booming and the treasury accumulating money faster than it could figure out how to spend it, the fiscal probity

---

90   Government of Alberta, Ministry of Energy. https://open.alberta.ca/opendata/ historical-royalty-revenue.

upon which Ralph Klein had built his political career disappeared. People were moving to Alberta from across Canada and all over the world. Several multi-billion-dollar oil sands projects began simultaneously, which placed significant stresses on the province's infrastructure and supply chain.

What began in 2004 was Alberta's second great oil boom after the 1970s. One would hardly think this was a problem. But Albertans, back on their rightful perch of enormous wealth, made it one anyway.

In 2005 the province declared a one-time dividend of $400 per citizen, a cash payout quickly dubbed "Ralph Bucks." This cost $1.4 billion but was a drop in the bucket compared to what was coming in the door. In the three fiscal years from 2004/05 to 2006/07 the Alberta treasury collected a staggering $36.2 billion. The Heritage Fund was growing again, and the Klein government set up a parallel Sustainability Fund of spare cash to help the province better manage times when the carbon cash quit flowing.[91]

In the 2006/07 fiscal year the province reported an $8.9 billion budgetary surplus, a contribution of $3.6 billion to the Sustainability Fund, raising the total to $7.7 billion; another $1.5 billion was put into the Heritage Fund raising it to $15 billion. Alberta reported net assets (cash and investment less all financial liabilities) of $34.1 billion. When Klein and Dinning had declared a state of fiscal emergency only 13 years earlier, the province was $5 billion underwater. The net improvement on the province's balance sheet was $39.1 billion. The Klein administration certainly gets the credit for making Alberta a great destination for oil and gas investment, but commodity prices certainly helped ... again.[92]

And what did the PC Party do to reward King Ralph for presiding over the what must surely be the second most remarkable financial turnaround in Canadian history? (The first being, of course, Alberta's great post-Leduc boom from 1947 to 1956.) They voted him off the island. At the required leadership review at the PC Party's annual convention on March 31, 2006, Klein received only 55% support from voting delegates, the political definition of "please go". There were too many politically ambitious people in the PC political machine to play second

---

91   Government of Alberta, Ministry of Treasury Board and Finance Annual Reports, https://www.alberta.ca/government-and-ministry-annual-reports.aspx#toc-0.

92   Government of Alberta, Ministry of Treasury Board and Finance Annual Reports, https://www.alberta.ca/government-and-ministry-annual-reports.aspx#toc-0.

fiddle to Ralph Klein any longer. Too much cash and too many opportunities to do something noteworthy with the money.

Klein, who had proven on multiple occasions he was no dummy politically, resigned as leader in September of 2006, sparking a new leadership race. He resigned as an MLA in early 2007 and passed away in 2010.

With the carbon-funded bank account still growing, the stage was set for what should have been the opportunity for Alberta to position itself financially forever. The bumper sticker praying for another oil boom so Albertans could prove they would better manage the money the second time around was ignored. Apparently, many of the senior people in the PC Party didn't know how to read

5

THE LAST EIGHT YEARS OF THE PC DYNASTY

"Whatever the plan was in early 2008 in Alberta, the rest of the world again refused to cooperate. Although not clear at the time, Alberta would soon be harshly reminded that it was not, nor would ever be, a master of its own economic destiny when it came to carbon resources."

By the time the PC leadership race began in the fall of 2006, history would prove a significant number of allegedly conservative Alberta politicians had lost their senses. Perhaps it was the fact that this boom was driven by the oil sands, a resource so vast Alberta's it was thought newly returned prosperity would last forever. But once again, there was much more focus on what the government did or should do, rather than how it was funded it or how it got there. Ralph Klein was clearly the "conservative" in PC. Time to roll out the "progressive."

The unprecedented prosperity that disappeared in the early 1980s had returned but a growing number of politicians were overcome by the seemingly uncontrollable hubris that accompanies politics when things are going well; prosperity must be due to the genius of government, not the private sector. Somehow the carbon resources that had made Alberta an economic powerhouse had become a liability instead of an asset. Despite the billions in carbon revenue coming in faster than the government could spend it, Alberta's politicians wanted more.

The boom of the 1970s, which ended so harshly and abruptly in the 1980s, was accompanied by record inflation, record-high interest rates, and the great carbon-wealth political war between the federal government and the provinces.

All this disappeared in the late 1980s and 1990s. In 1984 a new Progressive Conservative federal government tore up the NEP, deregulated natural gas, and signed the Free Trade Agreement with the US Ottawa was no longer the enemy even after the Liberals returned to office in 1993. The recession of the early 1980s tamed inflation, allowing central bankers to reduce interest rates. By 1992 the average inflation rate was down to 1.4% and would not rise above 2.9% for the next 20 years. While bank lending rates bounced around in the 5% range for most of the 1990s, this was a huge improvement from rates over 20% in the early 1980s.

This helped everything and everyone. The pace at which government debt escalated because of compound interest declined. Businesses could borrow at a lower cost. Property owners—corporate and individual—saw a larger portion of their monthly mortgage payments retiring the principal instead of servicing the interest.

Some of the aggressive government cost-cutting moves of the Klein administration remain controversial to this day. This is a belief perpetuated by proponents of big government who claim the cuts were all bad with no good. But the private sector and ordinary Albertans relaxed and got back to what had motivated them to move to the province—wealth creation and individual opportunity. In Alberta when you made a buck, you got to keep more of it. When you borrowed money, you could afford the payments. As evidenced in the previous chapter, thanks to natural gas the carbon-resource industry was growing again, drilling more wells and putting more people back to work across the province.

When the economy really started to rip early in the twentieth century, the lessons of the great crashes of the 1930s and 1980s were quickly forgotten. Perhaps the boom of the 1970s was still fresh in the minds of many. A return to prosperity was not a surprise. The world's population, the global economy, and demand for carbon energy were all growing. Carbon-warehouse Alberta had resumed its date with destiny.

The mantra, repeated over and over, went like this. China was emerging as a global economic force. It was industrializing at unprecedented rates. As the Chinese economy moved from rural to urban, from agricultural to industrial, its carbon energy needs grew in lockstep. In 2005 in the US there were 250 million vehicles on the road, nearly one for each of the 296 million Americans. Oil demand would grow forever. It was repeated over and over, "If only 10% of the Chinese bought a car…"

Population figures tell the story. In 1991 Statistics Canada reported Alberta's population at 2.5 million. In 1996, when things were clearly improving, it was 2.7 million. By 2001 it was three million and by 2006 3.3 million. In the 10 years prior to 2006, Alberta's population grew by 22% or 600,000 people. Big numbers. This was equivalent to 82% of the population of Edmonton in 2006 or 60% of the population of Calgary.

The problems associated with rapid growth became a public policy issue. Indeed, Klein had balanced the books, but complaints grew that it had been at the expense of public infrastructure. The Ralph Bucks of 2005 and growth in the Heritage and Sustainability funds proved the government had the money. Where were the public works programs? Klein warned the government should be cautious about competing with the private sector for materials and labour during this high-growth period. Nobody foresaw the rising commodity prices of the twenty-first century, nor that the oil sands would finally become a magnet for global investment capital.

Prior to the PC leadership race in the fall of 2006, royalties became a public issue. The auditor general had uncovered errors in the royalty-calculation-and-collection process. The long-standing haters of oil companies noted that oil sands developers only had to pay 1% royalties until payout. Record profits. Token royalties. Appalling. A negative perception of the environmental impact of the oil sands was emerging with lots of visceral images of boreal forests being replaced with giant, black holes in the ground. How could it be that the largest oil companies in the world were pillaging the environment while paying the people of Alberta peanuts for the right to do so?

When asked about royalties Premier Klein responded with one of his classic quips, "I don't give a tinker's damn about royalties." This was, as many of Klein's responses tended to be, controversial. And despite Alberta's resumption of carbon-fuelled prosperity, the distrust of oil companies and oilmen as a species remained ingrained with a great number of Albertans in and outside of the industry.

That oil sands were very capital intensive and that Albertans were getting billions in economic spinoffs during construction was overlooked. As was the fact that the Generic Oil Sands Royalty Regime of a decade earlier was one of the key reasons the cash was pouring in. By this time the economy was so frothy that people began to call it "overheated." Schools were full. Roads were crowded. The influx of people was pressuring an already stressed healthcare system.

The fact that having the public sector compete with the private sector for labour and materials during a boom would be inflationary was lost on the government's critics. Alberta had the money. Spend it. Having learned nothing from history, opposition politicians and pundits resumed the cry for the politicians to "do something."

Be careful what you wish for, because they did.

Of eight declared contestants, the three main contenders were former MLA and cabinet minister Jim Dinning, sitting MLA Ted Morton, and MLA and cabinet minister Ed Stelmach. Dinning and Morton were fiscal conservatives. Stelmach's economic views were, by comparison, not well understood. When the royalty issue was raised, none of the candidates were willing to publicly endorse Klein's position that this was a non-issue. A royalty review was added to the list of campaign promises. After all, conventional royalties had not been reviewed for 15 years and oil sands royalties for 10. Would it not be prudent to examine something this important?

To everyone's surprise, Ed Stelmach "came up the middle" after the second ballot as the second choice of the most voting party members in the PC preferential ballot process. He was declared premier after Klein formally stepped aside on December 14, 2006.

Stelmach, a farmer from the Vegreville area, would prove over time to be no friend of the oil and gas industry. He had a lot of support from Edmonton, as well as from northern and rural Alberta because he was not Calgary's leader, as were Lougheed and Klein. There existed a deep and not-well-understood resentment among many PC Party members against downtown Calgary and its wealthy and occasionally flamboyant multi-millionaires. It was these members who supported Ed Stelmach.

As promised, Stelmach began his royalty review in May 2007 with high-profile public hearings. This had never been done before. Historically, matters of economic rents and taxes were studied and negotiated behind closed doors by government, industry, and experts. This is what politicians are elected to do. The panel of six, led by former forestry executive Bill Hunter, travelled across the province over the next four months. The other five members of the panel were from business, academia, and finance.

At every hearing, industry representatives cautioned the government not to "kill the goose that laid the golden egg." The existing fiscal regime had been instrumental in Alberta's return to prosperity. It's a big world and capital is mobile and has options. Leave it alone. Like the 1950s and 1960s, Alberta was a great place to invest, attracting capital from all over the world. While all the presenters were polite in public, privately they could not understand why this province was having this conversation in this fashion. Where on earth did the government think all this money came from?

But in an open public forum with no qualifications for participation, everybody showed up. For every industry representative preaching caution, two others cited a long list of grievances, many historic, ranging from whatever was lacking in Alberta's infrastructure to healthcare, to classroom sizes, to the inherent amorality of oil companies. Remember, state oil company Petro-Canada was very popular when created in the 1970s. All wrongs could be righted or at least addressed if oil and gas producers gave the government more money.

Home-grown EnCana had reported a $6.4-billion profit in 2006, the highest in Canadian corporate history. Somehow this was perceived as a problem in oil-rich Alberta, even though it was clear evidence the last PC plan was working perfectly. What's going on here? It was classic socialist logic that if someone had more, it was only because others had less. But with the return to a carbon boom, didn't everybody in Alberta have more? Although the government had more money than it could spend intelligently, the popular reasoning outside the industry was more money spent by somebody other than capitalist oil companies would most certainly make things better.

The final report, released in September 2007, was a bombshell. The panel recommended the government collect what it called "Our Fair Share" and increase royalties substantially and immediately. In the oil company head offices in Calgary and abroad, and among the myriad of private and institutional investors the world over that had pouring money into Alberta for over a decade, the reaction was shock then anger. Who is Ed Stelmach? How did he end up as the premier of Alberta? Why was he wreaking havoc on an industry that was treating so many so well? That was contributing billions and billions to the provincial treasury? This was sheer lunacy.

It was an incredibly polarizing time, unprecedented in Alberta's political history. While the industry and government had never been in love, neither had they ever been at war. Through their associations or the CEOs of the larger corporations, the oil- and gas-producing companies had always been able to at least communicate with the premier of the day. They may not have agreed with what the government was doing, but at least they could talk.

The door to Stelmach's office was closed. Not only did the premier endorse the conclusions of his royalty review but he wasn't even prepared to discuss them with industry.

On October 25, 2007, Premier Stelmach announced what was called the New Royalty Framework. Effective January 1, 2009, royalties would increase about 20% across the board which, based on current prices and production volumes, would result in the province collecting another $1.4 billion per year. This was less than the $2 billion recommended by the panel. Existing regimes for oil sands projects that were producing or under construction would not be exempt. Rewriting the rules after the fact was compared to the hated NEP. To the benefit of producers, the royalty curves were adjusted to be more sensitive to price and production volume. Spokesmen for industry associations and anyone else in the industry who was asked expressed universal disappointment verging on despair.

A CBC News report that day stated, "Political observers believe the royalty decision is a defining issue for the premier, who has had the top job for just 10 months and is rumored to be gearing up for a fall election."

History would prove the "defining issue" comment correct. But for the wrong reason.

The public, however, liked it. "Big Oil" still had no friends. Nothing had changed. Stelmach started spending the money immediately. He ended individual healthcare premiums at a cost of about $1 billion annually. The province gave the teachers' union $2.2 billion to clear its unfunded pension liability. Responding to the growing climate change and carbon-emission issues, another $2 billion was set aside to co-fund carbon capture and storage projects to prove Alberta was a concerned and responsible developer of the fossil fuels. Oil sands had been vilified as the worst source of carbon emissions on the planet so the government demonstrated it was listening and would invest heavily in ways to reduce emissions.

Edmonton would stimulate heavy oil upgrading through BRIK, Bitumen Royalties In Kind. Alberta would take its share of oil sands production and use it as an economic tool to accelerate upgrading. Stelmach colourfully declared that selling raw bitumen without processing it "was like a farmer selling off his topsoil." However, the analogy made no sense because the supply of bitumen was virtually unlimited, capped only by price and market access. A farmer could only sell the topsoil once. But Albertans, who felt that upgrading and refining jobs should be in Alberta, loved the metaphor.

An election was called for March 2008 and Stelmach won a big majority government, 72 of 83 seats. For ordinary Albertans there was no risk. The economy was thought to be "overheated." A slowdown would be acceptable if not welcome. (If only other parts of the world could share Alberta's suffering!) Lots of government goodies were on the way, including more spending on infrastructure. The right guys, the uber-wealth and Big Oil companies, were howling so it must be good. What shocked many of the oil executives was how many of their own employees figured "Our Fair Share" was the right thing to do.

Whatever the plan was in early 2008 in Alberta, the rest of the world again refused to cooperate. Although not clear at the time, Alberta would soon be harshly reminded that it was not, nor would ever be, a master of its own economic destiny when it came to carbon resources.

In the United States, the rising price of natural gas, market forces, and human ingenuity combined to work their magic with new technologies to recover gas and later oil from enormous carbon deposits long reluctant to release their treasure using conventional recovery techniques. That tight shale reservoirs contained gas and oil was known. How to get it out was not. In the late 1990s the industry began experimenting with extended reach horizontal drilling in conjunction with multi-stage fracturing, where the reservoir was literally pulverized under high pressure to create a myriad of cracks and fractures where none existed before. This vastly increased the amount of rock exposed to the lower pressure in the wellbore and allowed the trapped hydrocarbons to be produced.

Around 2005 this was working very well in Texas in a reservoir called the Barnett Shale. By 2006 various American shale reservoirs had produced 1.1 trillion cubic feet (tcf) of what was coined "shale gas." This was just the beginning. Once the industry knew what to look for and how to do it, a shale gas revolution began that would radically change North American gas markets. What would ultimately be the most damaging for Alberta's natural gas was the discovery of huge quantities of shale gas in Pennsylvania and New York, the so-called Marcellus Shale. These reservoirs were much closer to the lucrative Ontario and Quebec gas markets that had been so valuable to Alberta since the late 1950s. Shale gas was cheap, and the transportation cost was much lower than shipping gas all the way from Alberta.

Then drillers applied the same technology to what would be called "light tight oil," shale rocks that contained petroleum. One of the reservoirs that responded immediately was the Bakken, a massive reservoir that straddled the North Dakota/ Saskatchewan border.

2008 began with great promise. That year WTI would hit its all-time high of US$147 a barrel. Gas was still valuable. The cash was flowing with exciting new places to spend the money. The Bakken in Saskatchewan looked promising. There were vast reserves of shale gas in northeastern BC in an area called the Horn River. Development opportunities were emerging all over the US.

But Alberta had hung a "CLOSED" sign in the store-front window.

With the New Royalty Framework and the Stelmach administration firmly in place, capital began to leave Alberta. Oil companies didn't have to travel to distant and lawless lands on other continents to make more money. They didn't even have to leave North America. As rigs, equipment, and capital fled Alberta, Stelmach's initial response was that it was geology, not him, that was the problem. Alberta didn't have a Bakken or a Horn River.

In fact, because so much of the WCSB was in Alberta, the province had more geological opportunities for the new technology than both BC and Saskatchewan combined.

Then the second shoe dropped: the world financial collapse. Problems had emerged in global credit markets in 2007. In Canada the Asset Backed Commercial Paper financial crisis had been underway since 2007 and required government intervention. In the spring of 2008, US investment banker Bear-Stearns folded. By the fall of 2008, a global economic crisis exploded with the unforeseen collapse of Lehman Brothers. Other major financial institutions were propped up by immense government-engineered bailouts in the US and Europe. Oil markets responded to what would surely be a huge drop in demand. WTI would fall by over $US100 a barrel by the end of the year.

But apparently nobody read the newspaper in Premier Stelmach's office. The New Royalty Framework was set to become law on January 1, 2009. It was pointed out by many that the considerable change in the global economic landscape had created a perfect opportunity to change direction and scrap or amend the policy. But like too many politicians before him, Ed Stelmach didn't want to look weak and admit that his new policies would not work in anything but the most robust economic environment. He chose instead to appear doctrinaire, isolated, reckless, and economically delinquent. Listening more to his supporters and himself than the rest of Alberta and the world, Stelmach proceeded with the 2007 plan anyway.

The 2007/08 provincial annual report showed Alberta's last big budget surplus. Estimated at $4 billion, it came in at $2.3. As promised, capital spending rose 46% or $2.2 billion. The operating budget increased by 10% or $2.5 billion. A chart showed how Alberta was spending more on healthcare per capita than any other province, almost one-third more than Quebec or Ontario. This was a badge of honour, not a management problem. The Heritage Fund grew a bit but there was no spare cash for the Sustainability Fund.[93]

A year later the wheels fell off. Spending was up another $2.7 billion but revenues fell by $2.6 billion. The first draw of $800 million was made on the Sustainability Fund to balance the books, resulting in a deficit of zero. Government operating expenses increased by 10% or $2.7 billion. Capital expenditures increased from $7 billion in the prior year to $7.6 billion.[94]

The year 2009 was a tough one for the global oil industry but it was worse in Alberta. Drilling in BC fell by only 16% because of its gas. Saskatchewan's dropped by 47% because the target was oil. In Alberta drilling plunged from 14,969 new wells in 2008 to only 6,894 in 2009, a 54% decline.[95] This was completely self-inflicted.

Politically, fiscal conservatives with financial support from a disgruntled oil industry flocked to the new Wildrose Party, which elected a new, articulate, and telegenic leader in Danielle Smith in October 2009. Wildrose won a Calgary by-election in the fall of that year. In early 2010 two PC MLAs crossed the floor to join Wildrose. Then a fourth MLA who had been ejected by the PCs joined Wildrose. The PC Party under Stelmach had lost its way. Internally, Stelmach was told if he stayed on this path the party had no future.

Bowing to pressure and the growing weight of evidence, early in 2010 a reassessment of the royalty review began, but not in public. The government met with industry organizations and conducted a "competitiveness review" based on the economics of oil and gas development in Alberta compared to other jurisdictions, including BC, Saskatchewan, North Dakota, Wyoming, and other US carbon-resource states. Stelmach's New Royalty Framework flunked a de-politicized and objective economic review.

The *new* New Royalty Framework (NRF) was put in place in June 2010 and Stelmach's three-year experiment was formally condemned as a complete disaster.

---

93 Government of Alberta, Ministry of Treasury Board and Finance Annual Reports, https://www.alberta.ca/government-and-ministry-annual-reports.aspx#toc-0.

94 Ibid.

95 Canadian Association of Petroleum Producers, *Statistical Handbook for Canada's Upstream Oil Producers* (Calgary, AB: CAPP, 2018). https://www.capp.ca/publications-and-statistics/publications/316778.

Like the NEP before it, the NRF proved once again that politically motivated Canadian carbon politics could not withstand international forces and world commodity prices in a global business. Like the Liberals' NEP of the 1980s, the New Royalty Framework was a blunder that took Stelmach's premiership with it. He resigned as premier early in 2011 and a new PC leadership race began.

With the fiscal regime repaired and competitive, a new drilling boom ensued as developers staked their claims to mineral rights in the Alberta shale basins that would respond to the new technologies. Land sales, which had been only $1.3 billion for three years, exploded in 2010/11 and 2011/12 as oil companies secured drilling and production rights in the Montney, Duvernay, Cardium, and Viking formations. Exploration and development spending resumed although the focus was more on oil and natural gas liquids. Gas prices continued to soften because of the continent-wide impact of the growing US shale gas phenomenon.[96]

The oil sands, however, were not so fortunate. Rising costs and the first glimmers that growing environmental opposition to new pipeline-takeaway capacity had oil sands developers rethinking capital commitments. In early 2009, Suncor walked away from billions of invested dollars and its partially built Voyageur oil sands upgrader north of Fort McMurray. The hulk sits by the highway today.

The exodus of development capital continued as more projects were cancelled or delayed indefinitely. But the ongoing long-term commitments on projects already underway would disguise the shift in investor sentiment for the next five years. Of the billions in heavy oil upgrading projects once planned for Fort Saskatchewan, only one would be built: the Sturgeon Refinery/Northwest Upgrader. It proceeded because of generous provincial support through the BRIK program.

Alison Redford was elected as the new leader of the PCs in early 2011. Pipeline opposition continued to grow. She made the rounds of the oil executives in Calgary assuring them things were different, that the conservative part of PC brand was back. In an attempt to convince Canadians to look more favourably on Alberta's growing oil production and buy into the idea that the oil sands were an asset and not a liability, Redford invented something called the National Energy Strategy. She crossed the country, reminding people how important oil and gas development was to Canada.

---

96 Government of Alberta, Ministry of Energy, https://open.alberta.ca/opendata/historical-royalty-revenue.

Back home the spending continued. In the 2010/11 budget revenue declined, spending grew, and the deficit increased to $3.4 billion. The Sustainability Fund, which had opened at $15 billion at the beginning of the fiscal year, was drawn down by $3.8 billion. Thanks to strong land sales and the recovery in oilpatch spending, revenue in the 2011/12 fiscal year improved but the government still reported a deficit. To fund the continued high levels of capital spending, the Sustainability Fund was drawn down by another $3.7 billion. Compared to two years prior, it was half gone even though the economy of Alberta was, by any measure, doing well.[97]

Total non-renewable resource revenues had been $21.8 billion in the previous two years, but this was no longer enough cash to cover the rising costs of the ever-expanding Alberta government. The population had grown to 3.6 million, another 300,000 since 1996, as more people flocked to the economic promised land.

Redford led the PCs to another majority government in 2012 after experiencing a big scare from the surging Wildrose Party. The PC seat count shrank by 16% and an avowed fiscal-conservative political party, Wildrose, formed the official opposition. In the 2012/13 fiscal year, revenue shrank and spending grew, resulting in a $2.8 billion deficit. Another $4.2 billion was withdrawn from the Sustainability Fund, leaving the balance at $3.3 billion. That year the name was changed to Contingency Fund, perhaps so when people went looking for the current balance it would be more difficult to figure out how much had disappeared in such a short period of time.[98]

By 2014 the Redford administration was in trouble. An aggressive Wildrose opposition and a series of scandals tainted the premier's performance. Under continued opposition attack, the PC Party internal brain-trust exerted sufficient pressure on Redford that she resigned March 19, 2014, and the third PC leadership race since 2006 was underway.

The province's finances for the 2013/14 fiscal year didn't look too bad; there was a slight deficit. It was a good year for total non-renewable resource income as oil sand royalties, assisted by increased production and high oil prices, exceeded $5 billion for the first time. Gas only contributed 20% of what it did in 2008/09 as shale gas continued to alter the North American gas market. Total spending for the year jumped $2.9 billion from the prior year. While total assets grew, by

---

97   Government of Alberta, Ministry of Treasury Board and Finance Annual Reports, https://www.alberta.ca/government-and-ministry-annual-reports.aspx#toc-0.

98   Ibid.

now the government was borrowing money again. Direct borrowing for capital prospects—$2.4 billion a year earlier—rose to $6.3 billion at March 31, 2014, 2.5 times higher.[99]

The PC leadership race began in mid-2014 and was concluded September 6. Former federal Conservative MP and cabinet minister Jim Prentice won by a landslide, defeating PC MLAs Rick McIvor and Thomas Lukaszuk. Oil prices were still strong, with WTI holding in the US$85 to US$100 a barrel range for the third straight year. While oil sands construction projects were not growing, activity was steady completing investments begun years earlier. Things were pretty good.

But by now pipelines had emerged as a major issue in Alberta's economic future. Redford's National Energy Strategy sounded encouraging and garnered a sympathetic ear in Ottawa. Prime Minister Harper had described Canada as an "energy superpower" and the Keystone XL pipeline as a "no brainer."

However, the approval process for Keystone XL in the US was being subjected to continuous delays, questions, and scrutiny. High-profile public protests were becoming common. Opponents came up with the message that oil sands pipelines were unsafe because the "sand" in oil sands was excessively corrosive. The oil sands output was declared to be "Dirty Oil." It was claimed that a big spill of blended bitumen in Michigan in 2010 only proved the point. The corrosiveness issue was fabricated but created another successful time and money consuming distraction and opponents achieved another delay.

The Northern Gateway pipeline to the west coast had been approved by the NEB and the Stephen Harper government in 2014 but was tied up in a court challenge by environmental non-government organizations (ENGOs) and Indigenous opponents. The anti-carbon, anti-oil sands movements had grown louder and stronger. Alberta's industry was finally recognizing the gravity of the problem and the challenges to the future of the province.

Jim Prentice looked like he had the ideal qualifications to be the next premier. He had been the federal Minister of Environment and Aboriginal Affairs, the portfolios that, at least in name, ought to have the most to do with the two main sources of pipeline opposition. As one senior oil executive told your writer, "Prentice will get us pipe."

---

99 Ibid.

But first he had to get elected. Because of three retiring MLAs and Prentice's need for a seat in the legislature, four by-elections were held on October 27, 2014. Three were in Calgary and one in Edmonton. The PCs won all four.

Not known to anyone but political insiders is that once it became clear Prentice was going to enter the PC leadership race, Wildrose began to falter from within. Having pushed two PC leaders out the door in three years—Stelmach then Redford—and leading in the opinion polls, in the spring of 2014 Wildrose began to view itself as the government in waiting. Wildrose had come close in 2012 but in the last week of the campaign urban support collapsed when Danielle Smith refused to fire a candidate over prior extreme social-policy positions.

However, some Wildrose MLAs viewed Prentice as unbeatable and concluded Wildrose would never form the government. The politically ambitious had joined Wildrose to replace the PCs, not just oppose them. After the loss of the four by-elections the Wildrose wheels started falling off, primarily due to internal strife and the failure of the leader to maintain control and confidence. Smith publicly calling for a review of her own leadership at an upcoming convention the very next day certainly didn't help. First two rural Wildrose MLAs crossed the floor to join the PCs in November. Then on December 17, 2014 leader Danielle Smith and eight other MLAs joined the PCs, bringing the total to 11. Five stayed behind.

A lot or promises were made by Prentice and his backroom operators about the role the floor-crossers would play. Their nominations in the next election would be secure, Wildrose core policy principles would be enshrined. Two defectors were promised cabinet positions. These were the terms of surrender the MLAs believed they had negotiated. The supporters of crossing the floor preferred to call it a merger, but this is a very tortured application of the true meaning of the word.

But when Prentice revealed his scheme to the PC caucus, the rest of his MLAs were outraged. All promises were withdrawn but the nine floor-crossers joined their two former colleagues in the PC Party anyway.

Anger does not adequately describe the view of the voting public. However, unless you've held the job or been close to the process, few in the public under-stand what a tortuous job it is to be the leader of a political party that feels destined to form government but has not yet tasted power. Everyone second-guesses the leader and eventually the leader may begin to second-guess him/herself. The knives are always out of the sheaths, just not necessarily in the leader's back. Yet.

To complicate matters even more, just before this bizarre event in Alberta politics had begun, OPEC announced it would no longer support the world oil price by restricting supply. It was just like 1986. Oil prices had been high enough for long enough that significant new supplies had entered the market from places like the US, Canada, and Brazil. High prices also had their predictable impact on demand growth, which had slowed.

Crude oil prices went into freefall. WTI, which had averaged US$103.59 a barrel as recently as July averaged only $US59.29 in December, a 43% reduction. A year later in December 2015, the average price would be US$39.16. In February 2016, when the price hit bottom, it averaged on US$30.32 that month, 71% below the price when Jim Prentice officially entered the PC leadership race as Alberta's next PC Party saviour.[100]

It was obvious to everyone that Alberta was again in another deep economic dive, a situation perhaps as serious as in 1986. In the first quarter of 2015, Prentice and his newly expanded team tried to figure out what to do next. While the Wildrose threat appeared to be neutralized with the capitulation of most of its elected members, what was the public thinking? So Prentice thought he would investigate.

In a public debate Prentice was asked why government spending was so high, why the Sustainability Fund was declining when times were still good, and why the Heritage Fund was, in real terms, only a fraction of what it was 30 years ago. Prentice responded that Albertans should "look in the mirror." The clear implication was that all Albertans enjoyed low taxes and opposed a provincial sales tax, but continued to demand ever-improved roads, schools, hospitals, and government services. The negative reaction was swift and hostile. Prentice was branded as arrogant and aloof. The public was reminded that his last job was that of a high-profile bank executive on Bay Street earning a salary most Albertans would only dream of.

Redford had passed a fixed election date law early in her term, which set the provincial ballot in the spring every four years. By terms of this law, the next election should have been 2016. But the PCs figured that with oil continuing to plunge in price, every economic problem Alberta had now would be worse in a year. The government announced Albertans would go the polls May 5, 2015, one

---

100 Energy Information Administration, "Cushing OK WTI Spot Price FOB," *Petroleum & Other Liquids* (2019). https://www.eia.gov/dnav/pet/hist/LeafHandler.ashx?n=PET&s=RWTC&f=M.

year early. Prentice explained that his government needed a mandate to execute the tough love Alberta required to weather the downturn.

Albertans didn't want austerity—at least from his government—or an early election. Heavy favourite Jim Prentice and the PCs were almost wiped off the political map. The NDP under Rachel Notley won 54 seats and a solid majority government. Wildrose, thought dead, came in second with 21 seats. The PCs won only 10 seats. The Alberta Party and the Liberals elected a single MLA each. Prentice was either so angry or so humiliated that he resigned from politics before the final election results were tallied. After 44 consecutive years of majority government rule—the longest of any political party in Canadian history—the PC Party was annihilated.

The NDP win was due to many factors. Prentice was seen as arrogant, new Wildrose leader Brian Jean was unknown, and Notley seemed, by comparison, nice, honest, and trustworthy. The backroom deal Prentice and his close advisers had cooked up to wipe out the Wildrose Party in late 2014 polarized a great number of Albertans, who weren't enamoured with politicians anyway.

Now what? The 2014/15 financial report explained for the new government Alberta's financial position after both oil and natural gas prices had collapsed. Total revenue for the year was down only slightly from 2013/14 at $45.2 billion. While operating expenses rose by $1.5 billion, not having to pay for huge floods like the one in 2013 saved the province $2.6 billion, resulting in a slight budget surplus.[5]

But the balance sheet continued to deteriorate. To maintain capital spending, the province had borrowed $3 billion. The province Premier Klein had declared debt free in 2004 now owed $12.4 billion in "unmatured debt." Including accounts payable, guaranteed debt to government agencies and municipalities, pension liabilities, and public/private partnerships, liabilities totalled $53 billion. When Ralph Klein left in 2006, the same figure was $20.6 billion.[101]

During some of the best years for Alberta's economy and after collecting $88.3 billion in non-renewable resource revenue—over 23% of all government revenue—the liability side of Alberta's balance sheet had grown by $32.4 billion.[102]

The business community was immediately concerned after the election. The NDP had campaigned on raising corporate taxes, raising income taxes on the

---

101 Government of Alberta, Ministry of Treasury Board and Finance Annual Reports, https://www.alberta.ca/government-and-ministry-annual-reports.aspx#toc-0.

102 Ibid.

wealthy, and conducting another royalty review. The party had been no friend of the oil industry or pipelines. The oilpatch was in a downturn and the only thing that was sure about the new government was that the cost of business would be going up. The NDP promised Government of Alberta a balanced budget by 2017. Nobody believed that.

What the NDP did and didn't do to Alberta once it assumed office will be dealt with in the next section, Carbon Future. Needless to say, as this book goes to press the last good year Alberta's carbon-resource industries enjoyed was 2014.

6

FEDERAL CARBON POLITICS – WHAT DO VOTERS WANT TO HEAR?

"The federal government has played an essential role in the development of Canada's carbon resources beginning before Canada became a country... But... has also been, from time to time, the carbon industry's worst nightmare. On occasion the federal government's policies have been so punitive and polarizing that the good will from the positive decisions and support is forgotten."

The federal government has played an essential role in the development of Canada's carbon resources beginning before Canada became a country. As noted previously, the Geological Survey of Canada was created in 1842 in the pre-Confederation Province of Canada to provide a public inventory of the new country's massive resource base.

But Ottawa, home of Canada's federal government has also been, from time to time, the carbon industry's worst nightmare. On occasion the federal government's policies have been so punitive and polarizing that the good will from the positive decisions and support is forgotten. If you went looking for a common thread amidst all the regulations, taxes, and policy decisions Ottawa has made in the 152 years since Canada became a country, look to politics. What they have in common is that they were either politically attractive to voters or politically neutral so they wouldn't cost those in power any votes.

The first major political decision that would impact future carbon-resource development was in 1905 when Alberta and Saskatchewan became provinces and Manitoba was expanded to its current size. In his book *Canada*, author Edgar McInnis wrote, "Now there was a strong desire for the admission of the settled region as a single province with full control over its natural resources" (McInnis 1969).

But Prime Minister Laurier didn't want one big province that would become too powerful and instead created three. "The federal government decided to create two provinces, Saskatchewan and Alberta, and to retain control of their resources, as had been done in the case of Manitoba. A generous federal subsidy helped to salve the disappointment" (McInnis 1969).

The Resource Transfer Act of 1930 turned the three prairie provinces into real provinces like the other six—New Brunswick, Nova Scotia, Prince Edward Island, Quebec, Ontario, and British Columbia—as was anticipated in the constitution in force at the time, the British North America Act.

That the transfer of resources to the new provinces would become a country-dividing constitutional issue 45 years later was, of course, unknown.

Through the 1940s and 1950s the general tenor of the federal government was supportive. The Pipe Lines Act of 1949 enshrined federal control over the interprovincial transportation of goods via pipelines. Ottawa took control over pipelines to ensure provincial jurisdictions could not interfere with the trans-provincial transportation of carbon resources, which was then regarded to be in the national interest. The legislation was passed in a hurry to expedite the completion of first main oil pipeline east and south from Edmonton through Saskatchewan, Manitoba, Minnesota, and Wisconsin in 1950. Three years later the Trans Mountain pipeline to Burnaby was completed.

Ottawa's Liberal government loved Alberta hydrocarbons so much that in the 1950s it intervened heavily, and at its own political peril, to get Ontario and Quebec connected to Alberta natural gas. The prime minister at the time was Quebecer Louis St. Laurent. He was supported by the now-legendary "Minister of Everything" C. D. Howe. Howe was regularly accused of placing "getting things done" before the process and politics of the House of Commons. In the 1950s the federal government helped Alberta change its mind about exporting gas. It promoted an all-Canadian route over the top of Lake Superior through Northern Ontario although it was longer and more expensive, helped find a way for governments to finance the Canadian portion (because at that time TransCanada Pipe Line was not Canadian-controlled), then set about jamming the whole thing through the House of Commons.

The political peril came from C. D. Howe's insistence that Ottawa advance TransCanada the funds to commence construction in 1956 and not lose another year in the seasonal construction schedule due to Canada's winter weather. Howe went so far as to try to persuade his Liberal colleagues to invoke closure in the House of Commons to force the funding through parliament. Closure had not been tried for anything since 1932. With an election coming within two years, the Conservative and CCF opposition called this massive political arrogance and accused the Liberals of being in the pockets of TransCanada's American shareholders. It was time-sensitive because TransCanada needed the money to pay for the steel for the pipe, which was in short supply.

The pipeline was not in itself as controversial as the charges of Liberal arrogance. The Liberals had been in power since 1935 and were accustomed to having their way. Eventually the bill passed, the money flowed, TransCanada got its pipe, and construction began. C. D. Howe wrote, "I should not like to face a general election at this moment. Fortunately, we do not have to."[103]

This may be the only time in Canadian history a federal government put its own political future on the line to assist Alberta's oil and gas industry. While Ottawa did lots of things to help the industry over the years, every other time the government intervened, it ensured the policy or plan was politically risk-free before proceeding.

With the pipeline debate lingering in their memories, voters were tired of the Liberals by the 1957 federal election. Led by Saskatchewan lawyer and noted orator John Diefenbaker, the Conservatives won a minority government in 1957 and a majority government in 1958. One the first acts of the new Conservative government was to appoint Toronto businessman Henry Borden to lead a Royal Commission on Energy. Terms of reference were not specific as the new Conservative government had no particular plan beyond distancing itself from the recent energy politics of the Liberal government it had just replaced. The public review would be known as the Borden Commission.

In the late 1950s western Canadian gas producers believed gas exports could be increased without jeopardizing secure supplies for Canadian consumers. Higher exports would lead to higher prices and higher cash flows from existing production which, in turn, would spur exploration. The WCSB did not appear to be fully explored or likely to run out of natural gas anytime soon. Existing Canadian customers, including Calgary and Edmonton, were of the view that proven supplies for cheap gas for Canadians should come before increasing exports. This would later manifest itself in the gas export "surplus test."

Producers also wanted to extend the oil pipeline from Ontario to Montreal, a large Canadian market supported entirely by crude oil imports, primarily Venezuela. In the late 1950s there was still more proven oil delivery capacity in

---

103 Robert Bothwell and William Kilbourn, *C.D. Howe: A Biography* (1979).

Alberta than existing pipelines could carry. About 50% of Alberta's productive capacity was shut-in. In 1957 the US introduced policies to protect its domestic oil and industry and froze oil imports into the country at 1954 levels. This included established suppliers Canada and Venezuela. Imports were only about 15% of US oil demand. Canada was restricted to existing pipelines to Superior, Wisconsin, and the extension of the Trans Mountain pipeline from Burnaby into the State of Washington. The only logical market in which to expand was Quebec.[104]

The politics of trying to get Quebec to buy Canadian oil were quite interesting. Canadian oil producers, led by independents Home Oil and Dome Petroleum, lobbied the Borden Commission hard to get some help expanding their market. In book sponsored by Interprovincial Pipeline, the author wrote:

> Canadian producers felt they deserved at least as much protection as their US counterparts … In their individual and joint presentations to Borden's public hearings, the independents did not pull their punches. They claimed that a new, efficient, large-diameter line from Edmonton to Montreal could deliver crude for 52 cents a barrel, 20 cents below IPL's (Interprovincial Pipeline) tariff then in effect to Toronto. They said this would make their oil competitive with imports. They claimed that even if Canadian oil cost 25 cents a barrel more than imported crude landed in Montreal, a pipeline could still be effective—with government help. The producers argued that the government should use tariffs, important controls and other means to guarantee this key market for them. They used the term 'dictatorship oil,' to describe crude from South America and the Middle East then being delivered to Eastern Canada."[105]

In *The Politics of Energy*, Doern and Toner wrote about the new Montreal pipeline carrying 200,000 b/d by 1960 and 320,000 b/d by 1965. This would supply all of Montreal's refineries with Canadian crude and save exporting $350 million a year to foreign oil producers. But the pipe would require long-term commitments from the refiners, which were primarily multinationals that had their own profitable

---

104 Interprovincial Pipeline Company, *Mileposts: The Story of the World's Longest Petroleum Pipeline* (Calgary: Interprovincial Pipeline Company, 1989).
105 Ibid.

supply sources from Venezuela and the Middle East. Another spin was that the multinational producers opposed a pipeline from Alberta not just because of the cost of Alberta oil or shipping, but because they made more money refining their own oil from offshore (Doern and Toner 1985).

Incredible. The same debate is going on 60 years later about Quebec and eastern Canada importing oil from all manner of nasty foreign producers while western Canadian producers are searching for markets beyond the US.

The Borden Commission finally arrived at some significant conclusions. It rejected the protectionist move of Canadian gas customers on the matter of increased exports. But it also rejected the Alberta-to-Montreal pipeline and the legislated designation of the Quebec market for western crude. Whether or not the independent and multinational producers had competing interests, offshore oil was cheaper than Alberta oil, so the commission was influenced by the view the North American market would be best served in the long run if it unfolded north/south, not east/west. This would reduce transportation costs and result in lower prices for consumers in both countries.

For western Canadian crude oil, what emerged from the Borden Commission in 1961 was the National Oil Policy (NOP), which created the Ottawa Valley line. Under the NOP, markets west of the Ottawa Valley, which included all of Ontario, would be reserved for Canadian crude and Quebec and Eastern Canada would continue to purchase imports. Concerns about slightly higher prices in Ontario were offset by increasing refining capacity and expansion of that province's petrochemical industry.

Another major element of the Borden Commission was the creation of the National Energy Board (NEB) in 1959. This also followed the recommendation of an earlier Royal Commission, called the Gordon Commission on Canada's Economic Prospects. It highlighted the oil and gas industry's growth prospects but noted high foreign ownership at about 80%. (The Gordon Commission had several suggestions about increasing Canadian participation and control, which would not influence public policy until the Liberals were in power in the 1970s). The NEB would take over administration of pipelines, energy exports including electricity, and setting tolls and tariffs on interprovincial regulated utilities like pipelines. The NEB website reads that the National Energy Board Act "... defined its jurisdiction and status as an independent court of record,"[106] an important new factor. This meant NEB decisions would be the law of the land after rendering a decision. While ultimately NEB decisions would make their way back to the federal cabinet, they did not have to be approved by parliament.

In many ways the formation of the NEB by the Diefenbaker government was intended to remove the energy decision-making process away from politics and the cabinet, as had been the case with the TransCanada pipeline in 1956. Noteworthy is the current restructuring of the NEB and the energy infrastructure application and approval process put forward by the Justin Trudeau Liberal government (Bill C-69) will enshrine final decision-making powers residing with the federal cabinet.

In the 1960s things got quiet on the federal carbon-policy front. The only movement on the Gordon Commission's concerns about foreign ownership was the Canadian Oil and Gas Lands Regulations, which applied to Canada Lands, Crown resources outside of the province in the north, or offshore. While exploration permits were available to any party, production permits could only be issued to Canadian-owned companies or foreign-controlled companies that made their shares available to Canadian investors.

The Liberals were returned to office in 1963 under Prime Minister Lester B. Pearson. Canadian ownership was still an issue, courtesy of the Gordon Commission, so some tax changes were introduced to make it more difficult for foreign companies to buy Canadian companies and easier for Canadians to buy shares in foreign-controlled companies. In 1966 the federal Department of Energy, Mines and Resources was created and the federal government purchased 45% of the equity in Panarctic Oils, which had a mandate to explore for oil in the Canadian Arctic. In 1969 the federal energy minister went to Washington, DC, to pitch a continental north/south oil policy to the Americans. By 1971 Ottawa was trying to convince the Americans to import more Canadian natural gas.

In 1968 Pierre Trudeau replaced Lester B. Pearson as Liberal leader and won a majority government after two Liberal minority governments in 1963 and 1965. Carbon resources and energy were not political issues and therefore were not the focus of major attention by Ottawa. The price of WTI had averaged US$3 a barrel from 1957 to mid-1967, US$3.10 through to 1969, and ended 1971 at US$3.60. Nothing politically interesting there.

---

106 "Our History," *National Energy Board* (2018). https://www.neb-one.gc.ca/bts/whwr/rhstry-eng.html.

The Trudeau Liberals were returned to office in 1972 but this time with a minority government. In mid-1973 Imperial Oil raised its price by $0.40 barrel but, as noted in Chapter 3 of this section, the total recent oil price increases had totalled $0.95. This was causing prices to rise at the gas pump. As history has proven, nothing is more public or political than the price of gasoline. This caused some concern in Ottawa but no major changes (Foster 1979).

The event that would change everything was the announcement by OPEC on October 17, 1973 that there would be an immediate cutback in oil shipments to countries like the United States, which supported Israel in the Yom Kippur War. Oil prices increased immediately and by early 1974 would be triple what consumers had paid for decades. This was a major "do-something" moment among Canadian oil consumers and Liberal supporters. The Pierre Trudeau administration quickly rose the challenge.

Ottawa was already intervening in Canadian oil markets. In March of that year Ottawa adopted the view that the US was getting preference over protected Canadian markets. In September, after several price increases, it froze domestic prices price at current levels. When OPEC dropped the price bomb, Doern and Toner wrote that it became apparent the federal government lacked independent, in-house knowledge of how the oil business worked and was dependent on multinational producers for information. In December 1973, Ottawa announced it was going to create a national oil company, later called Petro-Canada, to get a seat at the industry table. Petro-Canada was originally justified as a "window on the industry." Politically, with a minority government and the NDP holding the balance of power, a state oil company was a significant public-policy response to the emerging oil crisis. It was attractive to the NDP and, not coincidentally, congruent with the mission of any political party; to stay in office (Doern and Toner 1985).

As has been previously written, the combination of prices, export controls, and a state-oil company angered voters in the oil-producing regions. But western support for the Liberal Party was not material at the ballot box.

Security of supply and Canadian ownership of the industry became political issues. Ottawa felt it was imperative to complete an oil pipeline to Montreal, an idea rejected only 15 years earlier. This took longer than it should have given the apparent urgency of the issue. The delays were largely political since the typical proponents of oil pipelines—shippers and buyers—were not included in all the discussions and negotiations. The pipeline ran from Sarnia to Montreal and began delivering oil in June 1976.

With oil prices up everywhere in the world except Canada (and other oil producing countries with internal price controls), the battle with Alberta was on. The next big move in the 1974 spring budget was changes to the income tax act, making Alberta's rising oil royalties no longer deductible to corporations for tax purposes. This further angered the oil producing western provinces. However, high inflation rates were a major political issue and rising oil prices made the problem worse. The view of the Progressive Conservative (it had changed its name from Conservative in 1942) and NDP parties was that the budget didn't do enough to address inflation. The Liberal government was defeated in a non-confidence vote in May 1974. The Trudeau Liberals were returned to power with a majority government on July 8. One of the Liberals' main public messages was, "The Land is Strong."

In the rest of 1974 and through 1975, oil prices were higher but stable. In 1975 the Liberals passed the Petroleum Administration Act, which enshrined broad powers to set oil and gas prices. By this point Ottawa had implemented a made-in-Canada price, which was below world levels. This would protect Canadian consumers and non-oil industry from rising world prices, which were set and controlled by the evil OPEC cartel. The problem was imported oil. As Ottawa suppressed the price of domestic production below world prices, it simultaneously subsidized imports by covering the price spread between the domestic controlled price and the world market price. This created a new need for significant amounts of revenue. This would be offset by Ottawa collecting the difference between the domestic price and international price for oil shipped to and sold in the US By 1978 the market price of WTI was up to US$14.90 and the domestic price was increased regularly such that it was about 85% of the world price that year.

Security of supply and Canadian ownership were of great interest to the federal government during the 1970s. This manifested in several ways. Petro-Canada was founded as a crown corporation and made its first acquisition in 1976, the Canadian assets of Atlantic Richfield, an American company with nominal Canadian shareholders. In 1978 Petro-Canada bought Pacific Petroleums. Pacific had been founded in Canada but the search for capital in its expansion years led the company to partner with, and ultimately see its controlling shareholder become, US-based Phillips Petroleum, which sold refined product under the Phillips brand.

This put Pacific and later Petro-Canada into the retail fuel business. Thanks to Ottawa, Canadian ownership of the industry was growing and the "window on the industry" was getting larger.

The security of supply issue, and Ottawa's interest in exploring and producing from non-provincial or Canada Lands, resulted in a unique and short-lived tax policy called "super-depletion," formally called the Frontier Exploration Allowance (FEA). In his 1983 book *Other People's Money: The Banks, the Government and Dome*, author Peter Foster gives Dome Petroleum's CEO "Smiling" Jack Gallagher credit for convincing Ottawa to provide enhanced tax deductions for wells drilled on Canada Lands that cost over $5 million. It was a deduction of taxes of payable of 66⅔% of exploration costs above the threshold available at any time to the company or investors putting up the money. This was on top of the usual tax deductions for exploration and depletion (Foster 1983).

Foster cited the case of how an investor with an income exceeding $100,000 per year in Quebec could invest $1 in a frontier well qualifying for FEA and receive $1.38 back, regardless of whether or not the well ever generated production and revenue. Making Canada self-sufficient and Canadian-owned was more than important. It was also expensive (Foster 1983).

The 1979 federal election resulted in a surprise victory for the Progressive Conservatives, a minority government under Alberta-born leader Joe Clark. By now Canadian carbon politics were highly polarized. The energy wars with Alberta over resource control, pricing, exports, and taxation had been underway for six years. While neither the industry or the producing provinces were suffering thanks to oil and gas prices that kept increasing, the producers and owners were denied the full benefit of international prices.

For the first time Ottawa was interfering in the day-to-day operations of the western oil and gas industry in a big and not entirely supportive way. This was a highly political struggle, pitting Central Canada against the west, large provinces against small provinces, and Liberals against Progressive Conservatives. The election of Joe Clark on May 22, 1979 was seen in the oil- and gas-producing provinces as a big improvement. The PCs had campaigned on a platform to get rid of Petro-Canada as a failed socialist experiment. This was very popular in western Canada.

However, events in Iran would raise the temperature of the world's oil consumers even higher. The overthrow of the Shah of Iran in February 1979 and a national referendum to declare the country an Islamic Republic in April caused a disruption in Iranian oil output. While the total impact of Iranian oil on global supply was only 4%, the price of oil more than doubled, closing the year with WTI at US$32.50 a barrel. By the spring of 1980 it reached US$39.50, over ten times what it sold for only seven years previously.

To a world economy grappling with high inflation, high interest rates, and a recession, this was surely the definition of economic chaos. When the seven major industrialized nations met in Tokyo in 1979 for the annual G7 summit, oil dominated the agenda. All countries went home in agreement that inflation, exacerbated by skyrocketing oil prices, was a huge problem since the majority of G7 members had no significant domestic oil production. They all agreed to reduce demand and figure out how to reduce oil imports in 1985 to 1978 levels.

In the history of western Canadian alienation about the heavy hand of Ottawa during the OPEC years, forgotten is the behaviour of the Joe Clark's short-lived PC government. In its first and only budget in December 1979, Clark's PCs fell due to a non-confidence vote after only seven months in office. While the world oil crisis had been an issue in the spring election that year, it did not affect the outcome in the May 1979 election. Even though the Prime Minister was now an Albertan, the new Clark government tried but failed to negotiate a new federal/provincial oil pricing and revenue sharing agreement with Alberta that summer.

When Finance Minister John Crosbie read the budget, he described it as "short-term pain for long-term gain." Like carbon taxes today, this message has never been a vote-getter. He wanted to make Canada self-sufficient in oil by 1990 and increase oil and gas prices to 85% of international levels. The PC government also wanted more carbon cash. Tax measures would cap industry profits but ensure there were sufficient funds for exploration and development. This would send about half any of price increases exceeding $2 a barrel for oil and $0.30 per mcf of gas to the federal treasury to pay for conservation programs and to subsidize imports.

To meet its Tokyo commitments of reducing consumption, Ottawa would introduce an excise tax of $0.25 a gallon[107] on gasoline. With a $0.07 tax already in place, the total increase would be $0.18 a gallon for personal use. There would be rebates for fishing, agriculture, and public transit of $0.10 a gallon. The money would go back into the economy to encourage conservation, to alternative energy

---

107 Canada's conversion to the metric system began in the early 1970s by at the time of the Clark budget, gallons (not litres) were still used for measures of liquid volume.

sources, and to cushion the impact of higher prices. No details were provided. An energy tax credit of $80 per adult and $30 per child would help Canadians absorb higher prices. Super-depletion would be continued but reduced, and then replaced. The government would no longer permit tax rebates that exceeded the funds invested. Deductions for exploration and depletion would also be reduced.

The problem for the PCs was that by December 1979 the Clark government was growing increasingly unpopular as the energy crisis deepened. While Petro-Canada had not been an issue in the campaign leading up to the election in May, the government's determination to get rid of it when the oil world was once again in turmoil caused more concern among voters, especially in the oil-consuming-and-importing regions of Canada, the historic Liberal strongholds of Ontario and Quebec.

The Liberal Party's political strategists knew forcing an election only months after the last one was unpopular and would appear politically opportunistic. But polling indicated support for the Clark government was fading before the budget introduced the $0.18 a gallon gasoline tax. This would only enhance the chances for the Liberals.

Besides, being in opposition after another 16 years in power was boring. On November 21, 1979 Pierre Trudeau announced his intention to resign as Liberal leader. However, he was convinced to change his mind when his circle of political advisers indicated that the upcoming PC budget might provide the opportunity to force an election and return Trudeau and the Liberals to power. The battlefield of Canadian carbon politics was, after all, at its zenith (Foster 1982).

Two days after announcing the big jump in the federal gasoline excise tax, the Clark government faced a non-confidence motion in the House of Commons. However, more taxes on a country already struggling with high interest rates, high inflation and higher gasoline costs were very unpopular. In a non-confidence vote in the House of Commons over the budget on December 13, the Clark government was defeated. Three PC members were absent and five Social Credit members abstained. The NDP voted with the Liberals and the short-lived Joe Clark PC government fell by a vote of 139 to 133.

Consider the recent replacement of the Liberal government in Ontario by Doug Ford and the Progressive Conservative Party in June 2018, 39 years later. A Liberal government had completely reconstructed Ontario energy policy in the previous 15 years. It was another case of short-term pain for long-term gain.

The pain was skyrocketing electricity prices and carbon taxes as cheap coal-fired power was replaced with renewables. The gain—Ontarians were told—was the future of the world's climate.

Saving the world is a great idea. Paying for it is much more complicated.

The return of Pierre Trudeau's Liberal Party to power and its subsequent Canadian solution to the global oil crisis would polarize Albertans for a generation.

7

TWO TRUDEAUS – FROM THE NEP TO CLIMATE CHANGE

"Looking back to federal carbon-resource policies since Confederation, the success of the coal, oil, and natural gas industries has been inversely proportional to how often they became front page news. Judging by the frequency of media stories about carbon, challenging times were now a certainty."

The federal Liberals exploited the politics of carbon to win two majority governments, one in 1980 and another in 2015. Both prime ministers came from the same family. Pierre Elliot Trudeau was returned to office in 1980. His son Justin Pierre James Trudeau became Prime Minister in 2015.

That's it for similarities. The carbon-resource issues of 1980 were a perceived worldwide shortage of crude oil and skyrocketing prices. The threat was the economic well-being of Canadians due to the high price of carbon energy on top of a recession, high inflation, and high interest rates. Voters wanted price relief and assurances everything would be okay. The Liberals were listening and promised to deliver.

By 2015 carbon was still a problem but for completely different reasons. The world was awash with petroleum, the price was too low to curtail consumption, and the threat was the future of the planet because of climate change caused by fossil fuels. Voters were growing increasingly concerned about the global environmental threat. Oil sands and pipelines had become political flash points, and Canadians wanted a government that would deal with these issues and provide assurance that everything would be okay. The Liberals were listening and promised to deliver.

There were two books written about the NEP in the 1980s: *The Sorcerer's Apprentices—Canada's Super-Bureaucrats and the Energy Mess* by Peter Foster in 1982, and *The Politics of Energy—The Development and Implementation of the NEP* by G. Bruce Doern and Glen Toner in 1985. While most of the facts are the same the tone of, and approach to, each is quite different. As a business journalist writing for The *Financial Post*, Foster delved deeply into the politics and how governments operate. Doern and Toner were academics at Ontario universities; thus, their conclusions and observations reflected their background in government-funded academia, not the media.

After the defeat of the Joe Clark PC government in late 1979, Canadians went back to the polls February 18, 1980. The Liberals won a majority government, promising that no matter what happened in world oil markets, gasoline prices would not rise as much under the Liberals as the $0.18 a gallon the Tories had announced. Carbon-energy prices and security of supply were big issues. Not surprisingly after seven years of federal/provincial energy wars, the Liberals won no seats in BC, Alberta, or Saskatchewan, and won only two in Manitoba; however, they won 74 of 75 seats in Quebec and 52 of 95 in Ontario. The country was clearly split among provinces that produced carbon resources and those that consumed them.

Foster wrote that by 1980 energy was the place to be in Ottawa, the epicentre of political action because of major geopolitical events rarely unaccompanied by the adjective "crisis." (It was also a crisis in Alberta but for entirely different reasons). In his book Foster digs deep into the "Super-Bureaucrats" that would help the Liberals deliver on their election promises to save Canada from the global energy disaster (Foster 1982).

But whatever the politicians had said on the campaign trail, Foster wrote that the objectives of the government advisers were different. "The mandarin's principle concern was the flow of petroleum revenues in Canada, a three-pronged river of cash where the provincial and industry streams were at full flood while that which flowed into federal coffers was little more than a trickle." Polices such as super-depletion and subsidizing imports to hold the made-in-Canada price below world levels were costing an already-pressured federal treasury a bundle. The price of oil was going through the roof. All attempts to negotiate a "fair" revenue-sharing agreement with Alberta had failed, even under the PCs. It was time to do something radically different (Foster 1982).

Doern and Toner were more clinical in their analysis, preferring to set the stage and repeat what the government stated. Delivered as a budget on October 28, 1980, the goals of the NEP were, like any good public policy, all motherhood:

> To establish the basis for Canadians to seize control of their
> own energy future through *security of supply* and ultimate
> independence from the world market; To offer to Canadians,

all Canadians, the real *opportunity* to participate in the energy
industry in general and the petroleum industry in particular;
and To share in the benefits of industry expansion, to establish
a petroleum pricing and revenue sharing regime that recognizes
the requirement of *fairness* to all Canadians no matter where
they live (Doern and Toner 1985).

The italics were added by Doern and Toner in their reciting of the NEP document. They expanded that security of supply was to be achieved by 1990, the industry was to 50% Canadian-controlled by the same year thus creating an opportunity for Canadians to gain greater ownership. Fairness was defined as made-in-Canada prices; lower than world levels with the federal government getting a much larger share of the revenue.

The obvious implication is the amount of money going to the producing provinces—the constitutional owners—was unfair. The term coined to turn the oil industry from foreign to Canadian-controlled, using the heavy hand of government as necessary, was "Canadianization."

The NEP was a bombshell. To achieve the foregoing required a combination of taxation, subsidies, grants, crown corporations, nationalization, intergovernmental revenue redistribution, and lofty political rhetoric, all in a single political document. Foster's view that the policies themselves came out of the bureaucracy is reinforced by the fact not one single Liberal MP or cabinet minister had ever worked in the petroleum industry nor were the major oil- and gas-producing regions of the country represented in the government.

This is the fundamental flaw of politics as explained in the first chapter of this section; that ordinary people of varying backgrounds are, once elected to public office, expected to become experts in complex and often global economic and policy issues. They must rely on their external advisers, if they have any, and on public servants. The majority of the Liberal caucus of 1980 was not blessed with a network of oil and gas experts among their close friends and business associates.

To get more revenue the NEP introduced the Petroleum and Gas Revenue Tax, (PGRT) an 8% levy on wellhead production that was non-deductible for tax purposes. This gave Ottawa a whack of money at the wellhead, not the gas pump, as promised during the election. This looked just like a royalty which enraged producing provinces like Alberta.

To accelerate Canadian ownership, the role of Petro-Canada would be expanded. As Petro-Canada CEO Bill Hopper said at the time, "You and I know that generally speaking the Canadian public does not trust the oil industry. This has been proven many times by reliable public opinion polls." Give the voters what they want. Ottawa's interest in Syncrude and Panarctic would be transferred to Petro-Canada. To fund Petro-Canada, the NEP introduced the Canadian Ownership Account, a new levy at the gas pumps (Foster 1982).

Domestic prices would be controlled below world prices, increasing at a schedule set by Ottawa.

To encourage exploration, Ottawa would return funds to the industry depending on who spent what and where. Called the Petroleum Incentive Program (PIP), PIP grants depended upon the Canadian ownership and where the company drilled. On Provincial Lands the maximum PIP grant was 30%, a cash rebate of eligible expenses funded by the PGRT. On Canada Lands, where there was no provincial competition for the production revenue stream, the PIP grants would be as high as 80% if driller was 75% Canadian-controlled.

The two chosen instruments for exploration on Canada Lands were Petro-Canada and Dome Petroleum. Dome had engineered favourable tax treatment for drilling in the Beaufort Sea three years earlier. Although Dome was technically not 75% Canadian-owned at the time of the NEP, special-purpose entity Dome Canada, which appeared shortly thereafter, certainly was. In the spirit of Canadianization, Dome purchased Hudson's Bay Oil and Gas in 1981 in a transaction almost entirely debt financed. As interest rates rose and oil prices fell, this would prove to be the end of the company by 1988.

Possibly the most contentious element of the NEP awarded the federal government a 25% interest in all existing and future discoveries on Canada Lands. Without paying for it. This would be awarded to Petro-Canada or some other crown corporation and was effectively nationalization of private assets. Although their value was nominal because the only production of any consequence at that time was Imperial Oil's field at Norman Wells in the Northwest Territories. Hibernia had just been discovered offshore Newfoundland the prior year and looked very promising. The lead players were foreign companies Mobil Oil and Chevron Standard. Natural gas had been discovered around Sable Island, offshore Nova Scotia. So much gas had been found on the Mackenzie Delta a natural gas pipeline south had already been proposed. With the exception of Dome Petroleum in the Beaufort Sea, all this work had been done with foreign capital and good faith based upon the predictability and stability of Canadian law and energy policy.

This turned heads around the world, particularly in Washington, and the outrage in Calgary and Alberta ratcheted up yet another notch. The NEP was quickly branded as "confiscation without compensation." Trudeau's reputation as a socialist was sealed, as was the political future of the Liberal Party in western Canada.

The NEP also contained several other policy nuggets including the Natural Gas and Gas Liquids Tax (NGGLT) of $0.30 per thousand cubic feet, the phasing out of Earned Depletion Allowances on Provincial Lands, a minimum 50% Canadian ownership for production on Canada Lands, and a Progressive Incremental Royalty based on the profitability of new fields under federal jurisdiction. Ottawa's Foreign Investment Review Agency (FIRA), a result of the recommendations of the Gordon Commission, would review the activities of foreign-controlled companies if they diversified outside of the oil industry or acquired Canadian competitors (Foster 1982).

The reaction to the expanded role of Petro-Canada was expressed in Alberta by yet another bumper sticker. It declared, "I'd rather push this car a mile than fill up at Petrocan."

If foreign operators didn't like the NEP, Ottawa offered to buy them out. Foster quoted the NEP, which stated, "The Government of Canada recognizes the National Energy Program represents a fundamental departure ... from the current policy environment." If you didn't like it, "The Government of Canada is a willing buyer, at fair and reasonable prices." Petro-Canada would buy PetroFina in 1981 and the Canadian marketing arm of BP Canada in 1983 (Foster 1982).

World oil prices had peaked before the NEP was introduced and began to decline. The NEP update was released in early 1982 and the early stages of the complete undoing of the NEP began. To calm the waters Ottawa had already agreed to pay for its 25% "back-in" (ownership without payment) on Canada Lands. Foster wrote, "By the spring of 1982 it became obvious even to Ottawa that the National Energy Program had played a major part in the worst economic crisis in postwar Canada. Virtually all its projections had turned out to be disastrously wrong, as had the premises of the Federal-Alberta agreement of September 1981. The front running project, Alsands, had collapsed; the chosen instrument, Dome Petroleum, was on the point of bankruptcy; promises of moderate price increases had been totally abandoned in the grab for more revenue, but then these projected revenues … had disappeared" (Foster 1982).

What a mess. In 1984 when the Liberals went back to the polls, the party was decimated. Brian Mulroney's PC government won 211 of 282 seats. The Liberals plunged from 147 seats in 1980 to only 40. Carbon resources disappeared from the public's mind as oil prices continued to fall. The global oil shortage turned into a massive surplus and prices plummeted to less than one-third of 1980 levels by the summer of 1986.

The PCs began to systematically undo most of the Liberal energy policy initiatives of the previous decade. The entire PIP initiative was the first to go because of its cost. Natural gas was finally deregulated in 1986 and the PGRT eliminated. In Ottawa carbon resources returned to where they had been for decades with the market making decisions, not the cabinet. The pipeline from Sarnia to Montreal was depoliticized and reversed to bring imported oil in Ontario because it cost less. More oil was allowed to flow into the US because of lower transportation costs. US import protectionism had disappeared as demand grew and domestic production declined.

Funding to Petro-Canada was cut off and the company was left on its own. In 1990 it was partially privatized through a public offering of a portion of the government's interest. Its growth was severely dampened when it was forced to grow with its own cash, not Ottawa's. The final government interest was sold in 2004. Petro-Canada merged with Suncor in 2009.

The Mulroney government entered into the Free Trade Agreement with the US in 1988, which included a "proportionality" clause whereby Canada restricting exports to the US was no longer permitted. The definition of proportionality was Canada could not restrict oil and gas exports to the US to any greater degree than it was prepared to restrict supplies to Canadian consumers. Controversial at the time, the continental North American energy market was finally materializing as cost became the driving force, not nationality or security of supply.

The Liberals were returned to office in 1993 with a majority government under Jean Chrétien. The PC Party won only two seats in parliament because it had all but disappeared. In the west, PC votes migrated to the new Reform Party and in Quebec to the new Bloc Québécois. The Liberals resumed being very supportive of the carbon industry, cooperating with Alberta to develop the Generic Oil Sands Regime by contributing accelerated write-offs for oil sands capital investments.

It wasn't until the twenty-first century that carbon was back on Ottawa's political radar screen. As has always been the case, history would prove this would not be a positive development for carbon-resource industries.

The Liberals won the next two elections with majority governments in 1997 and 2000. They won again with a minority government under Paul Martin in 2004. This was not due to growing support for the Liberals, but to the end of fractured support for the other parties. The PCs and Canadian Alliance party (the new name for Reform) merged to create the Conservative Party of Canada (CPC) in late 2003. Support for the Bloc Québécois remained stagnant. In 2006 the CPC under leader Stephen Harper won a minority government, the first of three successive victories for Harper.

Carbon-resource policy under the CPC was generally supportive. As production from the oil sands grew, Harper referred to Canada as an "energy superpower." And it was. Canada was acknowledged as having the third-largest oil reserves in the world, and came in at number five for combined oil and gas production. The CPC supported all pipeline projects recommended by the NEB. As Keystone XL struggled with growing opposition in the US, Harper was publicly reminded he had once referred to the pipeline as a "no brainer."

There were a few non-conservative interventions from Ottawa that rankled carbon-resource developers. In 2006 Ottawa ended the so-called "income trust." An income trust was a legal structure through which corporations could avoid federal income taxes by distributing income to "unit holders." The income was then taxed in the hands of the unit holders, but not in the corporation. A lot of capital had been raised in the resource business under this structure and the abolition of the income trusts was not popular. That year the federal government also concluded the oil sands no longer required the enhanced capital investment write-offs introduced to spur investment a decade earlier.

Overall, pipeline got built as required, external capital flowed into the country and industry, and lots of taxes were paid at all levels. The Canadian dollar strengthened to the point that by 2012 it was at par or higher than its US counterpart. As the west's oilpatch boomed and Central Canadian manufacturing struggled because of currency exchange issues, the "Dutch disease" issue was raised. This term was coined about the Netherlands when that country discovered it sat upon huge natural gas reserves in 1959 then developed them for export. For the first time highly profitable natural resources became a big part of its economy. Holland's currency rose to the point that traditional, export-dependent domestic industries were negatively affected. By 2013 the oilpatch in western Canada was accused of creating the same phenomenon within the country. The Harper government, wisely, did nothing, preferring to let market forces sort things out. Which it did.

The carbon-as-menace issue emerged in the House of Commons early this century. Under Jean Chrétien, the Liberals attended the 1997 Kyoto conference and Canada was one of the first countries to sign in 1998. Under the Paul Martin government, Stéphane Dion was the Minister of Environment assigned with implementing the Kyoto agreement. Martin was defeated in the 2006 election by the CPC. Dion replaced Martin as Liberal leader after the election.

In 2007 Dion introduced a motion in the House of Commons to officially ratify Canada's Kyoto commitment. Although Conservative MPs voted against it, Harper did not hold a majority of the seats in parliament, so it was ratified. Because Dion's motion was not a confidence vote and largely symbolic, the minority Conservatives remained in government. In a prior debate Dion had asked Harper why he would not accept the science and help Canada solve the climate change problem. Harper noted that Dion had already admitted Canada would not meet its Kyoto commitments of a decade prior, and that political symbolism would not actually reduce global GHG emissions. This would be a consistent message from Harper until his party formed a majority government in 2010 and then formally withdrew Canada from the Kyoto Protocol in 2012.

However, the Liberals were not done with Kyoto or climate change. An avowed environmentalist, Dion believed there was enough voter support for the government to do something significant to combat global warming. In 2008 Dion introduced his new plan called "Green Shift," a national carbon tax plan with a series of rebates and incentives. Starting at $10 per tonne on GHGs, the levy would rise by $10 per tonne per year over four years. During this period the tax would generate about $15.4 billion annually. The tax would be revenue neutral as the government cut other taxes by $15.5 billion. The lowest income brackets would receive offsetting tax-rate reductions and total income taxes would be reduced by $11 billion. A new child tax benefit of $350 per child would be introduced, the guaranteed income supplement to seniors would rise by $600, and more benefits would flow to low-income Canadians. Average heating cost would increase by $200 to $250 per home, depending on whether the fuel was oil or natural gas.

Dion and the Liberals ran on the Green Shift in the 2008 election. No Canadian government had ever won an election on fuel-tax increases. Remember John Crosbie in 1979. This trend continued. The public was so unenthusiastic about saving the world with their own money that Liberal support was only 26% of the popular vote. Dion resigned and was replaced by Michael Ignatieff as Liberal leader. There was an attempted formation of a coalition government between the Liberals, Bloc Québécois, and NDP to replace the CPC in 2009. This failed and voters outside of Quebec didn't appreciate the Liberals teaming up with separatists to form power at any cost. In the 2011 election the Liberals won only 34 seats, the poorest showing in Canadian history. The CPC won its first and only majority government.

In the CPC's last term in office there was growing pressure from more voices for Canada to introduce carbon taxes. Harper's approach was not to oppose carbon taxes as a policy tool, but to oppose Canada intentionally handicapping itself economically by paying carbon taxes when its trading partners—particularly the US—did not. His argument was that global problems required global solutions. Canada alone could not change anything.

In the four years before the October 2015 federal election, the public debate had changed. Unfortunately for carbon-resource developers, fossil fuels, climate change, and pipelines were back on the front page of the newspapers. This is never good for the industry. A targeted campaign that started with significant US Environmental Non-Government Organization (ENGO) funding in 2008 successfully made the oil sands public enemy number one in the North American and European battle against climate change. Oil sands opponents focused on these countries because they had an open media and wealthy donors, certainly by international standards. The ENGOs, which had been legitimized as "friends of the earth" by the UN at the Stockholm conference in 1972, had come a long way.

The ENGOs realized that the oil sands resource base was, for all effective purposes, limitless in terms of reserves. But it was expensive and—as has been noted several times in this book—located far from markets in the landlocked interior of western Canada. The most obvious way to change the economics of the oil sands was to shut off the transportation routes. Therefore, the ENGOs opposed pipelines because without them the oil sands would be less profitable. Further, new pipelines involved public hearings. Not short of money, resources, lawyers, or ingenuity, the ENGOs were utterly shameless in the pursuit of publicity to attract and maintain support and raise more money. They figured out how to exploit the NEB process of public hearings for pipeline applications and approvals to gain significant publicity that was disproportionate to the number of people they represented.

There are several examples of success in spreading the message and keeping the funds flowing. Some failed, others worked brilliantly.

By 2012 the oil sands had caught the attention of the European Union to the point it was considering sanctions on imports of Alberta bitumen. The issue dated back to 2009 when Europe agreed to adopt the Fuel Quality Directive, which was intended to reduce GHG emissions from vehicles. More information was appearing that claimed oil sands had the highest GHG content of any petroleum on an extraction to tailpipe basis, also called "well-to-wheels." Canada immediately responded with an intense lobbying effort to set the record straight. But pressed by the ENGO lobbyists and sympathetic politicians, by 2011 the EU executive commission opted to ignore all the information from Canada and declare oil sands crude dirtier than other oil supplies. A vote was finally held in February 2012 but never garnered a clear majority. Securing that outcome was expensive and time consuming for Canada.

An illustration of how quickly the image of the oil sands had deteriorated in Ontario was demonstrated when pipeline operator Enbridge applied to reverse the flow of the once-strategic Sarnia-to-Montreal oil pipeline to again carry crude to Quebec. The TQM pipeline had been considered critical to the national interest in 1973 when the Trudeau government ordered it constructed. In the 1990s the line had been reversed to carry imported oil from Quebec to southern Ontario as foreign oil was cheaper than that from western Canada. Led by Suncor, which owned a refinery in Montreal, the industry wanted to return the pipeline to its original configuration. Enbridge began the application process in 2012.

The ENGOs rallied and turned what became known as the "Line 9 Reversal" into a political battlefield. There were high-profile disruptions of the NEB hearings, which made national news. Protesters chained themselves to facilities and vowed the line would never carry dirty bitumen across southern Ontario. Indigenous opponents were mobilized. Finally, two years later the NEB approved the reversal and eastward oil flow resumed in early 2015.

Two other flashpoints were TransCanada's Keystone XL (KXL) pipeline to carry more oil sands product to the US, and the Enbridge Northern Gateway pipeline from Edmonton to the west coast to export oil sands to Asia. Although both were approved by the NEB, both were the subject of ferocious opposition by ENGOs.

NEB pipeline hearings had been for decades some of the most formal and boring expenditures of public money in history. No longer.

The NEB hearings into Northern Gateway in 2012 was an example of how ENGOs were utilizing the internet and social media to intervene in the process. The NEB website reports ForestEthics Advocacy alone submitted "approximately 4,000" Letters of Comment in 2012 on behalf of individual members of the public. The cover letter from ForestEthics indicated they were solicited and collected online but were diverse and from separate concerned individuals and thus each should be reviewed separately. This was meant to create the impression of significant public concern, even though it was a form letter.

The increasing interference in NEB pipeline hearings by ENGOs caused the Harper government to amend the National Energy Board Act in 2013 to restrict who could intervene and appear at NEB hearings. This was immediately described by the ENGOs as suppression of free speech. Minister of Natural Resources Joe Oliver disagreed because the process still "permits submissions from individuals impacted by the project. The democratic right to express a public opinion is honoured Canada. Focusing submissions ensures the [NEB] review is informed by the facts material to the scope of the hearing and protects it from being used as a tool to delay decision. This concern arose in the context of the Northern Gateway hearing when over 4,000 people registered to be heard, but only 1,179 actually showed up at the hearings."[108]

A year before Oliver had publicly claimed that those opposed to the pipeline were supported by "foreign-funded radicals."

Northern Gateway was approved by the NEB and later the Harper government. That decision was challenged in the Federal Court of Appeal and overturned in 2016.

In the US, opposition to oil sands expansion by blocking Keystone XL was not required to focused on just one pipeline. US ENGOs were standing up for the

---

108 David P. Ball, "Advocates Take Harper to Court over 'Pattern of Muzzling' Energy Critics," *The Tyee* (August 14, 2013). https://thetyee.ca/Blogs/TheHook/2013/08/14/NEB-Restrictions-Lawsuit/.

future of the world. In his 2017 book *Dysfunction: Canada after Keystone XL,* former TransCanada executive Dennis McConaghy explained the type of pressure that was put on US President Barack Obama to persuade him to use his powers to block the pipeline. McConaghy wrote about the activities of Bill McKibben, founder of 350.org, a vocal oil sands and pipeline opponent. According to McConaghy, McKibben had previously written that the pipeline was a "… fifteen hundred mile fuse to the biggest carbon bomb on the continent, a way to make it faster and easier to trigger the final overheating of our planet" (McConaghy 2017).

McConaghy continued, "In October (2012) McKibben called on the major American ENGOs to sign a letter to President Obama, calling for his decisive action to block the pipeline. 'We expect nothing less,' they wrote. They described the pipeline battle as 'perhaps the biggest climate change test you face between now and the election (November 2012)' adding that denying the permit (to allow oil to cross the Canada/US border) would trigger a 'surge of enthusiasm from the green base that supported you so strongly in the last election.' Most significant was the clear threat that failure to block KXL would cause the ENGO movement and its donors to reconsider their support for President Obama's re-election" (McConaghy 2017).

Eventually US President Barack Obama, under intense pressure from American environmentalists, killed Keystone XL in late 2015 by refusing to issue the necessary White House approval for the crude to enter the US.

This is what the Canadian political landscape looked like leading into the 2015 federal election. The incumbent, Stephen Harper, was a Kyoto-rejecting, oil-sands-loving, Canadian-energy-superpower-promoting, bitumen-pipeline enthusiast.

The contender was the young, hip, attractive, and telegenic new Liberal leader Justin Trudeau. He had replaced Michael Ignatieff as Liberal leader in 2013. Regarded even in his own party as somewhat of an intellectual and economic lightweight, he was handsome and demographically attractive. And, of course, his last name was Trudeau.

Trudeau's campaign, both scripted and off the cuff, was not friendly to the oil sands or western Canada's carbon-resource industries. The official Liberal campaign platform was titled "A New Plan For A Strong Middle Class." Under *A Clean Environment and Strong Economy,* the pledge read, "We will provide national leadership and join with the provinces and territories to take action on climate change, put a price on carbon, and reduce carbon pollution."[109]

A very critical issue was summarized in the statement, "We will make environmental assessments credible again." Following multiple acrimonious public hearings into pipelines and the changes the Harper government made to the NEB, the Liberals figured the way to make the issue go away was label the existing process flawed and commit to retooling it. The problem was simply that Harper's government wasn't doing it right. The policy read, "Canadians must be able to trust that government will engage in appropriate regulatory oversight, including credible environmental assessments, and that it will respect the rights of those most affected, such as Indigenous communities. While governments grant permits for resource development, only communities can grant permission."[110]

But it was what Trudeau said that was not in the policy manual that raised eyebrows. On September 10, 2015, while campaigning in Vancouver and while the Federal Court of Appeals was studying an appeal of the NEB and federal government decisions on Northern Gateway, Trudeau promised an oil-tanker ban off BC's northern coast. *Canadian Press* reported, "… the Liberals would place a moratorium on oil tanker traffic along the northern coast of British Columbia. The move would effectively kill any pipeline projects through the area — including the controversial Northern Gateway pipeline that Liberal Leader Justin Trudeau opposes."

The Justin Trudeau Liberals won a majority government in October 2015. CPC leader Stephen Harper announced his intention to step down as party leader shortly after and resigned as an MP in 2016. Trudeau's inner circle was heavily populated by career environmentalists and social activists. As Peter Foster said about the Pierre Trudeau government in 1980, the carbon file in Ottawa was once again the place to be, but for entirely different reasons.

Trudeau chose as chief of staff his long-time friend Gerald Butts. Butts was formerly the head of the Canadian arm of the World Wildlife Fund and had been active in the Ontario Liberal government's green-energy-policy agenda in the prior decade. Catherine McKenna, appointed Minster of Environment and Climate Change, was formerly a lawyer specializing in human rights and social

---

109 Liberal Party of Canada, *A New Plan for a Strong Middle Class* (2015). https://www.liberal.ca/
  · wp-content/uploads/2015/10/New-plan-for-a-strong-middle-class.pdf.
110 Ibid.

justice. Professional environmentalists who had worked for the Pembina Institute and the Sierra Club assumed key positions in the ministries of environment and energy, mines, and resources.

Fifteen months later Trudeau made another public comment that revealed his views on the future of Canada's carbon-resource industries. He told an audience in Ontario in early 2017, "You can't make a choice between what's good for the environment and what's good for the economy. We can't shut down the oil sands tomorrow. We need to phase them out." Immediately attacked, he later "clarified" his remarks. In late 2016 Trudeau announced that Northern Gateway was officially terminated, and the oil tanker ban on BC's northern coast would become federal law.

For Alberta's carbon-resource industries, the election of Justin Trudeau's government was bad news. The Liberal agenda to embrace climate change initiatives, cancel pipelines, make oil tankers illegal in certain waters, and change the major-project-review-and-approval process was the third body blow in less than a year. The first was the collapse of oil prices in late 2014. The second was the election of an NDP majority government in Alberta in May 2015.

Looking back to federal carbon-resource policies since Confederation, the success of the coal, oil, and natural gas industries has been inversely proportional to how often they became front page news. Judging by the frequency of media reports about carbon, challenging times were now a certainty.

As for the country under a Liberal Prime Minister named Trudeau, Albertans with long memories were reminded that the interests of Quebec would always be more important than those of Alberta. Nobody can recall either stating the need for any industry in Quebec to be "phased out" in the national or international interest.

8

CARBON CONCERNS – THE EVOLUTION
OF MAN-MADE CLIMATE CHANGE

"The environment was a perfect cause for the UN to embrace because it was emerging as a threat to the citizens of member countries and would maintain financial and public support for the UN. The environment vastly expanded the UN's mandate and influence from its original purpose. The founding charter was about the people of the earth. The environment was about the earth itself."

You don't have to be a scientist to figure out the relationship between the sun, the atmosphere, the atmosphere's composition, the climate, and the weather. Everybody understands the following.

The primary source of heat on Earth is the sun. Heat generated by living creatures makes a very modest contribution, probably immeasurable. Volcanoes and forest fires can change things periodically and temporarily, but not consistently.

What makes the earth habitable is the atmosphere. What makes the moon uninhabitable, at least to carbon-based life forms like us and plants and animals, is the lack of atmosphere. The layer of gases surrounding the earth deflects and diffuses the intense heat of the sun during the day and retains heat at night.

The earth's temperature variation between night and day differs from place to place but in intergalactic terms is very modest. The average temperature of the earth is either 14°C or 15°C, or somewhere in-between depending on the data source. This is an annual figure, which includes all regions and seasons. Because the earth rotates on its axis every 24 hours, the temperature rises when exposed to the sun during the day and declines at night when it is dark. The heating or cooling cycles end on average every 12 hours, one-half of one full rotation. The difference between the daytime high and nighttime low changes by season and location. In Canada it is in the range of 15°C.

Because the earth is titled at a 23° angle to the sun and it takes a year (365 days) to complete an orbit around it, the angle of the sun at any given point on the planet continuously changes. We have this neatly divided into equal quarters called seasons. The days get longer or shorter and temperatures rise or fall depending upon the angle of the surface to the sun and the number of hours of sunlight. This varies a little, a phenomenon called polar wobble. Major changes in the location of the north and south poles—either end of the earth's rotational axis—are believed to have caused major climate change events in Canada. Examples are when the high Arctic islands of the Northwest Territories contained tropical forests or when Alberta was buried under two kilometres of ice.

That early scientists would study the atmosphere should surprise no one. People quickly figured out clouds were part of the atmosphere. So was vaporized water, most often called humidity. Instruments were not required to determine how the composition and behaviour of the atmosphere affected the weather. When it was cloudy it got cooler. When it was cloudless it got hotter. When the air contains a lot of water—high humidity—it retains more heat than when the air is very dry. That's why the temperature drops sharply at night in the desert and does not in the jungle. When the air moves at very high velocity, it has enough force to make things dangerous by hurting people and wrecking things. When the atmosphere releases a lot of moisture in a hurry (rain or snow) it is often inconvenient and all too frequently deadly.

As science progressed it was learned that the atmosphere contained essential components like oxygen and carbon dioxide. Life cannot be sustained without either. Shortly after the invention of fire, man became acquainted with soot and smoke. Over time people learned the hard way that other compounds like carbon monoxide were dangerous and often lethal. Human ingenuity invented poisonous airborne compounds such as pesticides to fight insects, or weapons of war to fight each other.

The atmosphere was clearly important to the continuation of the human species. It should surprise no one that people study what has happened to the atmosphere before, what is happening now, what might happen next, or what the main drivers are that can change it. An easily measured example of the impact of the atmosphere caused by a major change in its composition occurred in 1991 with the eruption of volcano Mount Pinatubo in the Philippines. It was classified as the second-largest volcanic eruption in the twentieth century after an event in Alaska in 1912. Pinatubo is estimated to have released 20 million tonnes of sulphur dioxide and particulate (tiny particles) into the upper atmosphere or stratosphere, the largest quantity since the eruption of volcano Krakatoa in Indonesia in 1883. By blocking the sun, the global temperature fell by an average of 0.5°C for two years thereafter.

Natural sources such as the sun and volcanoes notwithstanding, the largest new source (in geological terms) of heat into the atmosphere has been man's invention of fire and the subsequent combustion of anything that will burn to create heat such as wood, peat, coal, oil, and natural gas. This has grown for centuries and accelerated with the Industrial Revolution. When steam-powered, coal-fuelled mechanical devices—primarily trains and boats— were invented to move things around faster, farther, and in larger quantities, the transportation and consumption of carbon fuels was greatly assisted. The growth in fossil fuel production and consumption is well documented.

The issue with fossil fuels and the atmosphere is not so much the extra heat contributed by their use, but the gases associated with production and consumption, primarily carbon dioxide. As more scientists studied the impact of increasing CO_2 concentrations on the atmosphere, they thought it could cause atmospheric warming by trapping more heat. Hence the term greenhouse gases, or GHGs, was coined. A greenhouse traps heat but still lets the sun shine through so plants can grow. The theory was that these gases had a greenhouse effect on the earth.

Today other GHGs include methane, nitrous oxide, ozone, chlorofluorocarbons (CFCs), hydrofluorocarbons, sulphur hexafluoride, and nitrogen trifluoride. The last four are by-products of industrial processes or the manufacture of other industrial chemicals.

The concern that carbon emissions would change the climate is not new, nor did the idea gain broad acceptance quickly. Documented research dates back 120 years. It is only in the last 30 years that man-made GHGs have emerged as a real and serious factor affecting the climate and ultimately the weather.

Before proceeding, it should be pointed out that the words *weather* and *climate* may appear interchangeable but technically they are not. "The weather is the condition of the atmosphere in one area at a particular time, for example if it is raining, hot, or windy. The climate of a place is the general weather conditions that are typical of it."[111]

This means that the weather is what you see out your window or experience outside. The climate is the overall type and range of weather expected in a specific

---

111 "Weather," *Collins Dictionary* (2018). https://www.collinsdictionary.com/dictionary/english/weather

region. That is why climate change is used to describe what is happening to the whole world. When the climate changes it affects the weather, but this varies greatly depending upon location, time of the year, and other regional variables.

Studying the atmosphere started out as pure science. An organization that published a useful document titled *The Discovery Of Global Warming* is the American Institute of Physics. The history tab on its website reads, "The American Institute of Physics (AIP) was founded in 1931 in response to funding problems brought on by the Great Depression. At the urging of the Chemical Foundation, which provided initial funding, leaders of American physics formed a corporation for the 'advancement and diffusion of knowledge of the science of physics and its application to human welfare.'" The AIP was created to foster research and disseminate information. The AIP states, "The mission of the American Institute of Physics is to advance, promote, and serve the physical sciences for the benefit of humanity."[112]

The AIP opens with atmospheric measurement of carbon dioxide in the period 1800–1870 as about 290 parts per million (ppm). The average temperature of the earth was 13.7°C, about one degree cooler than today, depending upon the data source. That the world would be much colder without an atmosphere is credited to research by French mathematician Joseph Fourier in 1824. In 1859 scientist John Tyndal identified what would be called the Tyndal Phenomenon, which states that some gases block infrared radiation and therefore changes in the concentration of gases could bring climate change.[113]

The first calculation of global warming from human CO_2 emissions was credited in 1896 to Swedish chemist Svante Arrhenius, nicknamed "the father of climate change." He experimented with how the water and CO_2 content of the atmosphere affected the temperature. An article in *The Guardian* in 2005 stated, "Between 10,000 and 100,000 calculations later, Arrhenius had some rough, but useful, results that he published in 1896. If CO_2 levels halved, he concluded, the earth's surface temperature would fall by 4 to 5°C. There was a flipside to his calculations: doubling CO_2 levels would trigger a rise of about 5 to 6°C."[114]

---

112 Spencer Weart, "The Discovery of Global Warming," *American Institute of Physics* (2018). https://history.aip.org/climate/index.htm.

113 "The History of Climate Change," *LiveScience.com* (January 31, 2007). https://www.livescience.com/1292-history-climate-change-science.html.

114 Ian Sample, "The Father of Climate Change," *The Guardian* (June 30, 2015). https://www.theguardian.com/environment/2005/jun/30/climatechange.climatechangeenvironment2.

Most of the first half of the twentieth century was preoccupied with World War I, the Great Depression, and World War II. However, AIP writes that in 1945 the US Office of Naval Research began significant funding of scientific research, some of which found its way into climate science. The first computer model for analyzing the behaviour of the climate was developed in 1955 by American Norman Phillips, a "theoretical meteorologist" (an appropriate title for someone studying weather that had not yet happened). By 1960 it was reported atmospheric CO_2 concentrations were up to 315 ppm thanks to the Industrial Revolution and the emergence of large and cheap supplies of crude oil in the last century.

In 1963 calculations determined a connection between water vapour in the air and CO_2 levels. That higher humidity retains heat had been known for centuries. A conference was held in 1965 in Boulder, Colorado, to discuss climate change and the potential for significant shifts. A linkage between the collapse of the Antarctic ice cap and ocean levels was released in 1968. Concerned about the environment in general, the first "Earth Day" was organized in 1970 to create awareness of the relationship between the activities of mankind and the growing pollution and exploitation of air, land, and water.

In 1971 a conference was organized in Stockholm and it was attended by 30 scientists from 14 countries. It was sponsored by the Massachusetts Institute of Technology (MIT), the Royal Swedish Academy of Sciences, and the Royal Swedish Academy of Engineering Sciences. Concerns were raised about the impact of rising CO_2 emissions, but little evidence existed at that time that it was having any impact on a global scale.

What emerged from this conference was a report titled *Inadvertent Climate Modification: Report of the Study of Man's Impact on Climate*. A review of this document in a synopsis of events titled, "Chronology of Climate Change Alarmism in Climate Science" read, "Although the findings were mostly ambivalent on the question of any global or regional climatic effects, and although it had minimal impact on the Stockholm conference, this study is important for introducing anthropogenic climate change to the sphere of the global environment movement."[115]

It was in the late 1960s that the future of the climate was no longer the exclusive purview of scientists. At the time global cooling was thought to be as big a risk

---

115 Bernie Lewin, "Chronology of Climate Change Alarmism in Climate Science," *Enthusiasm, Skepticism and Science* (n.d.). https://enthusiasmscepticismscience.wordpress.com/chronology-of-climate-change-science/.

as global warming. In 1969 President Richard Nixon received briefings on the possibility and risk of both. Justifiably confused, he did nothing.[116]

In the early 1970s there were several headlines about the pending climate-cooling disaster, which one newspaper called "The Big Freeze." In 1975, *Newsweek* magazine would carry a feature story that declared, "There are ominous signs that the earth's weather patterns have begun to change dramatically and that these changes may portend a drastic decline in food production—with serious political implications."[117]

During the 1960s the environment emerged as a public issue. By 1962, the economy had improved to the point that a growing number of people had the time and money to worry about more than themselves. They had been preoccupied for centuries with the full-time job of providing food, clothing, and shelter to themselves and their families. A pivotal event was the publication in 1962 of the best-selling environmental book titled *Silent Spring* by American author Rachel Carson. *Silent Spring* brought attention to the negative impact on the environment caused by the growing and broad use of chemicals, particularly pesticides. This was something the average person could relate to. By 1972, the US had banned DDT, the world's most popular insecticide.

The DDT story is worth reviewing because, like carbon energy, it is another miracle-to-menace story. Technically called dichlorodiphenyltrichloroethane, the chemical was certainly pleading for an acronym. Although it was invented in 1874, DDT wasn't until 1939 that it was determined by Swiss chemist Paul Herman Müller that it was lethal to insects. During the latter half of World War II, DDT was used extensively to control malaria and typhus by killing the bugs that carried the diseases. This benefited both troops and citizens. In was made available for public use in the US in 1945 and promoted by government and industry as a household and agricultural pesticide. Müller was awarded a Nobel Prize for his contribution to humanity in 1948.

DDT did indeed save many lives and solve a lot of problems but like carbon, it was later identified as a threat. Eighty years later no other insecticide has been developed that is as effective as DDT. To control malaria, DDT has been

---

116 Ibid.

117 "The Cooling World," *Newsweek* (April 28, 1975) https://www.washingtontimes.com/news/2006/apr/2/20060402-112828-5298r/.

reintroduced in parts of Africa after governments determined the benefits of its use vastly exceed the cost of collateral damage.

Environmental pollution as a political issue was growing fast. The Love Canal, located in Niagara Falls NY, was a buried chemical dump upon which housing and schools were constructed. It started leaking to the surface in the 1960s. Acid rain became an issue in the lakes of Eastern Canada and the US. It was caused by the combination of industrial smokestack emissions—particularly sulphur dioxide—with water in the atmosphere. When it rained the chemically altered water contaminated lakes downwind of industrial emissions by increasing their acidity.

Significant pollution in the Great Lakes was becoming glaringly obvious. The first closing of a public beach in Toronto on Lake Ontario occurred in 1959. because it was no longer safe to swim in the water. Another sewage plant to dump more waste in Lake Ontario from Toronto was scheduled to open in 1960. When your writer first attended the University of Alberta in Edmonton in 1970, a movement called STOP—Save Tomorrow Oppose Pollution—was well organized on campus with booths, placards, and petitions.

By 1972 the environment became an issue of sufficient gravity that it attracted the attention of the United Nations. Founded in 1945, the first line of the UN's original charter read, "to save succeeding generations from the scourge of war, which twice in our lifetime has brought untold sorrow to mankind."[118] The UN also pledged to protect human rights and equality, to promote justice and the rule of law, and to promote social progress and freedom.

The environment was a perfect cause for the UN to embrace because it was emerging as a threat to the citizens of member countries and would maintain financial and public support for the organization. The environment vastly expanded the UN's mandate and influence from its original purpose. The founding charter was about the people of the earth. The environment was about the earth itself.

There are two interesting elements of the UN's move to embrace the environment as one of its crusades. As noted, one is that it was never mentioned in the founding charter of 1945. The other is that the UN was had begun to focus more on the indirect negative impacts of human activity—protection of the air, land,

---

118 "Preamble," *Charter of the United Nations* (1945), http://www.un.org/en/sections/un-charter/preamble/index.html.

and water—and less on the direct negative impacts such as wars, persecution and injustice.

At the invitation of the government of Sweden, in 1972 the UN sponsored the first of many official multi-country gatherings to discuss the future of the world. It was called the United National Conference on the Human Environment. Forty years later the BBC would call it, "Stockholm: Birth of the Green Generation." Its mandate was "stimulating and providing guidelines for action by national government and international organizations."[119]

This was the first public and official recognition of the important role of ENGOs that have become household names at the forefront of the climate change movement today. Greenpeace, for example, was founded in 1971 to protest underground nuclear weapons testing on a small island called Amchitka, at the end of Alaska's Aleutian Island string in the middle of the North Pacific Ocean.

The Stockholm Conference concluded with the Stockholm Declaration of 26 principles and 109 recommendations. These included protecting human rights, natural resources, and the environment and combatting excessive population growth. An issue that emerged that continues today is that of income and economic inequality among developed and non-developed countries. Developed countries had demonstrated that economic growth helped combat pollution and therefore developing countries needed financial assistance. India's Prime Minister Indira Gandhi highlighted a direct linkage between environmental protection and alleviating poverty.

Of interest is that global warming or climate change did not rate mention. Other commitments were to end weapons of mass destruction, expand environmental education, and stop the pollution of oceans. International ENGOs would help improve the environment by spreading the word directly among the people.

It was also mutually agreed in 1972 that environment policy should not hamper a country's economic development. Forty-seven years later this should be noteworthy in Alberta.

The next biggest step for the UN took place in 1990 with the first report of the Intergovernmental Panel on Climate Change (IPCC), created in 1988 as a joint initiative of the United Nations Environment Program and the World Meteorological Organization. The IPCC website states its purpose is to, "… prepare, based on

---

119 "United Nations Conference on the Human Environment," *Wikipedia* (2018). https://en.wikipedia.org/wiki/United_Nations_Conference_on_the_Human_Environment.

available scientific information, assessments on all aspects of climate change and its impacts, with a view of formulating realistic response strategies." The first task was, "… a comprehensive review with recommendations with respect to the state of knowledge of the science of climate change, the social and economic impact of climate change, and possible response strategies for inclusion in a possible future international convention on climate."[120]

The IPCC's conclusions were that the greenhouse effect was real and that CO_2, methane, CFCs, and nitrous oxide were indeed GHGs. These would increase the concentration of water vapour in the atmosphere and cause global warming. The IPCC was confident that half the greenhouse effect was caused by CO_2 alone and an immediate reduction in CO_2 emissions was essential. It contained an analysis of known temperature increases since the invention of the thermometer and the current increase of CO_2 in the atmosphere was greater than anything the earth had experienced in 10,000 years. While the IPCC admitted its forecast models for future temperature increases were not precise, temperatures and ocean levels would rise.

At the time the IPCC conceded that observed temperature increases were, "… of the same magnitude as natural climate variability. Thus, the observed increase could be largely due to this natural variability … The unequivocal detection of the enhanced greenhouse effect is not likely for a decade or more."[121] This would not matter to many who read the IPCC report.

The next big UN environment conference was held in 1992 in Rio de Janeiro. Officially called the United Nations Conference on Environment and Development, it was later coined the "Earth Summit." This was well attended, and the focus was on global warming and the IPCC report of two years earlier. No need to wait a decade as the 1990 IPCC report had cautioned. The conference was chaired by Canadian Maurice Strong, who had officially been named the event's secretary general. Strong had done many things in his life. He had been the first CEO of Petro-Canada and had played a major role on behalf of the Canadian government in helping organize the Stockholm conference in 1972. The Earth Summit was attended by representatives of 178 countries and 30,000 delegates.

---

120 "History of the IPCC," *IPCC* (n.d.). https://www.ipcc.ch/about/history/.
121 Ibid.

The Earth Summit ended with near-unanimous consensus among attendees that climate change was an issue. The website *UN Chronicle*, in an article titled "From Stockholm to Kyoto, A Brief History of Climate Change," described the outcome in the Rio Declaration which, "… reflected a global consensus on development and environmental cooperation. Chapter 8 of Agenda 21 dealt with the protection of the atmosphere, establishing the link between science, sustainable development, energy development and consumption, transportation, industrial development, stratospheric ozone depletion and transboundary atmospheric pollution." It went on, "As the most important action thus far on climate change, the Convention was to stabilize atmospheric concentrations of 'greenhouse gases' at a level that would be prevent dangerous anthropogenic interference with the climate system."[122]

The term "near consensus" was used in the conference documentation because of the position of the United States. That country only agreed to the consensus report once actual targets were removed and the result was a more of a motherhood statement. A review published by one participating delegate noted that complete agreement was likely impossible. It used the example of Saudi Arabia with its entire economy highly dependent on producing oil.[123]

Of interest is the Canadian version of what happened at Rio, contained in a public report on the Government of Canada website written in 1992. Under the heading *Framework Convention Climate Change* it stated, "The main principles of this convention state that the developed world must take the lead in combatting climate change and its adverse effects. Canada could be a leader in achieving the commitments of the convention by reducing greenhouse gas emissions to the 1990 level by the year 2000. It should be noted the convention principles state that 'any policies and measures to deal with climate change should be cost-effective so as to ensure global benefits at the lowest possible cost.'"[124]

Put this "cost-effective" comment in your memory banks for future consideration.

---

122 Peter Jackson, "From Stockholm to Kyoto: A Brief History of Climate Change," *UN Chronicle* (June 2007) https://unchronicle.un.org/article/stockholm-kyoto-brief-history-climate-change.

123 Geoffrey Palmer, "The Earth Summit: What Went Wrong at Rio?" *Washington University Law Review* (1992). https://openscholarship.wustl.edu/cgi/viewcontent.cgi?article=1867&context=law_lawreview.

124 Stephanie Meakin, *The Rio Earth Summit: Summary of the United Nations Conference on Environment and Development*, (Ottawa: Government of Canada, November 1992). http://publications.gc.ca/Collection-R/LoPBdP/BP/bp317-e.htm.

Rio '92 introduced the word "anthropogenic" into the future vocabulary of the carbon-resource industry. It means "caused or influenced by humans." And for readers who have been preoccupied with other things for the past 27 years, climate change became a real international public policy issue over a quarter century ago.

Stockholm and Rio had both ended with homework assignments and commitments for attendees. Measurable steps to ensure progress were to be implemented at certain dates. In Rio it was decided that starting in 1994 conferences were to be held annually to further consider the climate change challenge. This led to the first Conference of the Parties (COP) in 1995.

COP 1 was held in Berlin. The delegates agreed to the Berlin Mandate, which reinforced the commitment to combat climate change but also expanded on how rich nations had caused the problem. It stated, "The fact that the largest share of historical and current global emissions of greenhouse gases has originated in developed countries, that the per capita emissions in developing countries are still relatively low and that the share of global emissions originating in developing countries will grow to meet their social and development needs..." was the reasoning behind the unfairness of the climate change challenge among development and undeveloped countries.[125]

Climate change was now more than an environmental challenge. It was also a social issue. "Fairness," first introduced in Canadian carbon politics during the oil price spike of the 1970s, had gone global. Reducing GHG emissions should be disproportionately financed by the developed countries accused of creating the problem so the developing countries could release as much GHGs as necessary until they caught up economically. The direct linkage between carbon and prosperity was clear.

The next big gathering was held in 1997 in Kyoto, Japan which resulted in what would be known as the Kyoto Protocol, the first formal international structure by which countries would cooperate to reduce GHG emissions. Following COP 1 in Berlin, in 1995 Kyoto was, "The cornerstone of the climate change action … the adoption in Japan in December of 1997 of the Kyoto Protocol to the UNFCCC, the influential climate change action so far taken. It aimed to reduce the industrialized countries' overall emissions of carbon dioxide and other greenhouse gases by at least 5% below the 1990 levels in the commitment period of 2008 to 2012.

---

125 "Framework Convention on Climate Change," *Conference of the Parties* (June 1995). https://unfccc.int/resource/docs/cop1/07a01.pdf.

The Protocol, which opened for signature in March 1998, came into force on February 19, 2005, seven years after it was negotiated by over 160 nations."[126] One of the key proposals to reduce carbon was the concept of an international emissions-trading system to measure, monetize, and penalize increasing emissions and reward reducing them.

Canada attended the Kyoto conference and agreed to everything, signing the protocol in 1998.

In 2001 the IPCC released an updated report in which it answered some of its own questions from 1990. By this time the IPCC stated global warming as theorized was indeed occurring, and carefully defined the pending impacts, all bad. Following the release of this report the phrase "the science is settled" entered the lexicon of concerned politicians and the growing number of ENGOs that had increasingly embraced the climate change issue since the Rio Earth Summit in 1992.

The UN empowered ENGOs in 1972 as a legitimate, non-government forces on environmental matters. While the Sierra Club dates back to 1892, more than a century later its purpose is no longer about how to enjoy and protect the mountains. Today it exists to the save the world. The World Wildlife Fund (WWF) was created in 1961 in Switzerland. Today's WWF operates in 100 countries and its mandate is to "protect the future of nature," which obviously includes the atmosphere. The Pembina Institute, which was founded in Drayton Valley in 1982 after its citizens were subjected to high concentrations of hydrogen sulphide gas from the second sour gas blowout in five years, grew considerably and morphed into climate change activism. ForestEthics, now Stand.earth, started on Vancouver Island in 1979 to protest clear-cut logging. It now opposes oil sands development and pipelines because of climate change. America's National Resources Defense Council was founded in 1970. It has moved to the forefront of the climate change movement and opposes oil sands directly and indirectly with opposition to the Keystone XL pipeline. Greenpeace has evolved from opposing nuclear power and saving whales to Canada's most visible facilitator of anti-oil sands publicity stunts.

These groups have not only embraced all IPCC reports and UN initiatives, but they have increasingly exerted significant influence on politicians and governments.

---

126 Peter Jackson, "From Stockholm to Kyoto: A Brief History of Climate Change," *UN Chronicle* (June 2007) https://unchronicle.un.org/article/stockholm-kyoto-brief-history-climate-change.

There would be many more UN-sponsored climate change conferences and commitments. At COP 15 in Copenhagen in 2009, The Copenhagen Agreement was signed and seen as a binding successor agreement to the Kyoto Protocol of 1997. The Copenhagen Agreement recognized the importance of holding the increase in global temperatures below 2°C, stating "deep cuts in global emissions are required according to science," and had a schedule for emission reduction commitments by country. Copenhagen was regarded by those following the subject as ground-breaking and significant.

In 2015 the Harper government was defeated by the Liberal Party under Justin Trudeau. The Liberals ran on a very activist pro-environment and anti-climate change agenda.

COP 21 was held in Paris in November 2015 and had a material impact on Canadian public policy. There had been 19 other COP gatherings since Berlin in 1995. Paris was big for Canada because two newly elected governments—the NDP in Alberta and the Liberals in Ottawa—vowed not only to attend COP 21 in force, but to commit their respective governments to significant future reductions in GHGs. Instead of sending representatives, Premier Rachel Notley went to in Paris on behalf of Alberta and Prime Minister Justin Trudeau represented Canada.

Paris attacked climate change with surgical precision. This gathering would yield measurable results. The objective was to ensure that by 2100 the earth would warm by no more than 2°C from pre-industrial levels, and ideally only 1.5°C. It was agreed that a 1°C average temperature increase had already taken place. At its conclusion, 192 countries agreed to specific steps. Canada committed to cut its emissions by nearly 30% by 2030. The signatories agreed to set up a $100 billion fund to help developing nations participate.

However, the history of UN climate change conferences had been a lot of talk resulting in little action. As noted in Chapter 1 of this section, despite the commitments at Kyoto in 1997 over 20 years ago, there has been no reduction in total GHG emissions. The appearance of Barack Obama in Paris in 2015 to commit the US to the final decision made Paris more real. On a much smaller scale, but not insignificant now that Albertans are paying a carbon tax, was Premier Notley's arrival in Paris with the ink not yet dry on the NDP's Climate Leadership Plan, which included phasing out coal electricity generation, carbon taxes, and a cap on emissions from the much-denigrated oil sands.

Regardless of all that has been written and done about fossil fuels and their impact on the atmosphere and the climate, there still exist significant pockets of resistance among voters and industry leaders. Those who question the science behind or the phenomenon of climate change as it has unfolded in the past 25 years are called "deniers" and are branded pariahs. Because "the science is settled," climate change deniers are publicly criticized by the growing number of people who know better.

When Alberta's NDP introduced its Climate Leadership Plan, the official opposition of the day—the Wildrose Party—asked questions and were quickly branded as "deniers" by the government. What else need be said? Opponents of the government's Climate Leadership Plan could only be Neanderthal, knuckle-dragging, anti-science lunatics, clear menaces to the future of the world.

COP 21 got its first major setback after only a year when in 2016 newly elected US President Donald Trump said he would withdraw the US from its Paris commitments. The wording of the agreement was such that he couldn't actually do that because participation remained voluntary. No enforcement mechanism on sovereign nations exists. But the US president became the world's most high-profile climate change denier. This has not legitimized other climate change deniers. Trump has recently softened his position somewhat but is unlikely to change his views while serving as president.

Of interest is that in the US, by 2017 the continued replacement of coal by natural gas for electricity generation, the growth of renewable electricity sources, and increased vehicle fuel economy had put America in first place in the world in the reduction of GHG emissions. According to the BP Statistical Review of World Energy, in 2017 US GHG output declined by 0.5%, more than any other developed country that signed COP 21.

The most recent UN gathering was COP 24 in Katowice, Poland in December 2018. A key purpose of the gathering was to finalize the rules under which the COP 21 emission cuts would be implemented. Part of the Paris agreements was a $100 billion fund to help developed countries meet their targets. At COP 24 it was revealed the funding commitments were only about half complete. Countries like India, China, Brazil, and South Africa noted that developed countries were, "far from realizing their climate finance commitments of mobilizing $100 billion per year by 2020."[127]

Part of the discussion focused around October 2018's IPCC pronouncement that the Paris emission reduction targets were not adequate and should be accelerated.

COP 24 of course ended in a public agreement. They always do. But on December 17, 2018 in *The Globe and Mail* Shawn McCarthy noted that while the details of the Paris implementation were agreed upon by attendees, final action was "delayed for another nine months new commitments for more aggressive action to reduce greenhouse gas emissions."[128]

The article also reported, "Canada has yet to indicate how it will meet its current commitment to reduce GHGs 30 per cent below 2005 levels by 2030."[129]

---

127 John Light, "HomeClimate: What is on the Agenda for the COP24 Climate Conference in Katowice, Poland," UN Dispatch (November 30, 2018). https://www.undispatch.com/what-is-on-the-agenda-for-the-cop24-climate-conference-in-katowice-poland/.

128 Shawn McCarthy, Negotiators Set Paris Treaty Rules, Delay New Targets, *The Globe and Mail* (December 16, 2018). https://www.theglobeandmail.com/world/article-negotiators-agree-on-rules-for-paris-climate-treaty-implementation/.

129 Ibid.

9

MODERN MEDIA AND CLIMATE CHANGE – RISE AND ROLE OF THE INTERNET

"To understand how carbon got to where it is without examining the role of the internet would be to miss the whole story. Because the internet is the story. Without the internet there is no possible way events surrounding the fossil fuel industry or the climate change issue would have unfolded the way they have."

O n August 30, 2018, the Federal Court of Appeal announced it had reviewed and overturned earlier decisions by the Government of Canada and the National Energy Board to approve the expansion of the Trans Mountain oil and refined product pipeline from Edmonton to Burnaby. As was the case of the approval of the Northern Gateway pipeline in 2012, Trans Mountain's opponents were so aggrieved that every possible avenue to stop it was pursued. They challenged the legality of the decision and won. For the second time in two years.

This century pipelines have become regular headline news. The Trans Mountain court decision was no exception. When the story broke, every interested person who could type or talk had something to say. In recent years millions of words have been spoken or written about pipelines, fossil fuels, and global warming by tens, possibly hundreds of thousands of people. If the only means of disseminating the mega-mountain of messages remained the printed word on paper, more trees would have been required than went up in flames during the terrible forest fires of 2018.

But little paper was required. Communication of almost everything is now electronic, courtesy of the internet. People increasingly don't even phone each other, preferring email or text messaging. To understand how carbon got to where it is without examining the role of the internet would be to miss the whole story. Because the internet is the story. Without the internet there is no possible way events surrounding the fossil fuel industry or the climate change issue would have unfolded the way they have.

Of all the stakeholders directly affected by the emergence of fossil fuel climate change as *the* menace to the future of the world, the carbon-resource industry—the developers of coal, oil, and natural gas—comes in dead last in figuring out the root of the problem. Perhaps the providers of the fuels and products that power civilization and the modern world, resource industry executives, can be forgiven for only paying attention to the behaviour of their customers. After all, people buy as much carbon product and energy as industry can produce. Normal methods of measuring consumer satisfaction—high sales volumes—would indicate people love coal, oil, and gas.

But the resource industry failed to understand that consumers love and hate fossil fuel the same time. When the price of gasoline goes up, they hate it more. A growing number of customers—even fossil fuel's most determined opponents—demand the end of the same products they depend upon.

In 2010 a book was published about how the oil industry lives in a world of its own. Written by John Hofmeister, former President of the US division of Royal Dutch Shell, it was titled *Why We Hate The Oil Companies—Straight Talk From An Energy Insider*. Hofmeister tried to explain the disconnect between what oil companies do and what the world thinks of them. In the introduction he wrote about the amazing things Shell was doing with technology around the world to continue to supply global customers with essential fuel. "Yet it is a company in one of the most hated industries in the United States, an industry that consistently ranks at the bottom in reputation polls" (Hofmeister 2010).

This obvious contradiction of being despised yet consistently supported at the cash register initially caused great confusion among industry leaders. As more customers protested the product, the initial reaction from the industry was ridicule of its customers. How could opponents maintain a straight face as they flew on an airplane or drove their cars to an oil sands or pipeline protest? Wasn't that smart phone on which they got their information and directions made of plastic? Wasn't the electricity powering their homes and computers often created by coal-fired electricity generation?

This was so blatantly hypocritical it couldn't possibly last.

Even as more headlines appeared about the perils of fossil fuels, and as more commentators wrote about the growing threat of climate change, the response by industry was to ignore the problem or shoot the messenger. The newspapers were biased, the journalists were lefties, and the media in all its forms were suckers for a socialist plot.

Hofmeister blamed it on oil company indifference or even arrogance.

> Instead of being accessible to the media, many energy companies choose to buy advertising space to tell a guarded version of the truth. Instead of educating consumers on the real risks and real costs of energy, they choose to sponsor cultural events and television programs. Instead of being on-site to respond to a crisis, they send the lawyers … I actually heard one executive from another company say, "Why waste money to communicate to people who don't like us? If they don't want our products, there are plenty of others who do." (Hofmeister 2010)

But because of the internet, the contradictions and challenges grew. The incongruences are best explained by how the world now communicates. As media content changed, how much the *medium* had changed went unnoticed. Carbon developers regularly complained about what they read in the newspaper. What newspaper? That shrinking combination of paper and ink that was published every day? It has taken years, but the newspaper most industry leaders grew up reading no longer existed. The once powerful and profitable print media had been decimated; the daily paper a fraction of what it once was.

Your local newspaper once had an entire section of classified ads where everyone looked for jobs or people to hire and bought or sold everything that mattered, from a house to a car to golf clubs. It is all gone. The advertising revenue that supported the staff and infrastructure to turn out "all the news that is fit to print" every 24 hours has been crushed by the internet and the proliferation of online sites using tools that supply more information and from more sources at a fraction of the cost.

In Alberta, *The Globe and Mail, National Post, Calgary Herald/Edmonton Journal,* and *Calgary Sun/Edmonton Sun* were once separate companies competing for advertising, readers, reporters, and the next big story. Today all but *The Globe and Mail* are owned by one company, PostMedia Network Inc.

As the papers shrank, so did the number of reporters and columnists. Read all three PostMedia newspapers in Calgary or Edmonton today and you'll find the same articles written by the same people. While individual editors try to publish a reasonable cross section of commentary, opinion, and summary of what happened yesterday, the available space has shrunk significantly. Less ads, less pages, less news, less reporters, less columnists, less research, less content. It's that simple.

The electronic media—radio and television—has fared somewhat better, perhaps because nobody remembers a TV or radio station going broke. They all started out free via the airwaves, with advertising paying for everything. Now TV stations obtain some revenue from cable and satellite TV, and a few radio stations have figured out how to collect money from listeners.

Magazines have been clobbered. Weekly news magazines once specialized in the "deep dive," investigative research into the story behind the story. Now they have gone broke, changed hands, or are struggling to find a formula that works. After 70 years, *Newsweek* published its last print edition in 2012. *US News and World Report*, founded in 1933 and distributed as a resource about the world to students for decades, abandoned its print edition in 2010.

In September 2017, *The New York Times* published an article titled, "The Not-So-Glossy Future of Magazines." It reads, "Magazines have sputtered for years, their monopoly on readers and advertising erased by Facebook, Google, and more nimble online competitors. As publishers grasp for new revenue streams, a 'try anything' approach has taken hold. Time Inc. has a new streaming TV show, 'Paws & Claws,' that features viral videos of animals."[130]

How the mighty have fallen. Corporate flagship *Time* magazine was best known for years for naming "The Man of the Year," the most important person in the world in the previous 365 days. Now Time Inc. has gone to the dogs, pun totally intended.

More than the advertising and content is shrinking. Getting the facts right used to be a badge of honour. Every self-respecting media participant had trained journalists, experienced editors, researchers, and "fact checkers," people whose sole purpose was to assure accuracy. If a reporter quoted somebody, for example, the source would be contacted to verify the interview had taken place and that the quote was correct.

The most expensive part of any traditional media outlet is the one that generates no direct revenue: the editorial staff and its support team. A 2018 study by the Pew Research Centre, which analysed data from the US Bureau of Labor Statistics and Occupational Employment Statistics, revealed print newsroom staff had shrunk

---

130 Sydney Ember and Michael M. Grynbaum, "The Not-So-Glossy Future of Magazines," *The New York Times* (September 23, 2017). https://www.nytimes.com/2017/09/23/business/media/the-not-so-glossy-future-of-magazines.html.

from 71,000 in 2008 to 39,000. Digital outlet staff had almost doubled from 7,000 to 13,000. Cable television and television broadcasting were flat, while radio had shrunk from 5,000 to 3,000.

This diminishing staff means the media is increasingly dependent upon third-party sources for the news, networks like *Reuters, Bloomberg,* and *Canadian Press,* which create and distribute content. This provides economies of scale whereby a single reporter can feed stories to dozens, if not hundreds, of media outlets. Redistributors of bureau-generated news didn't have to phone anybody, do any research, or even leave their desks.

Others quickly figured out how much information could be spread far and wide at a low cost. This included ENGOs, political parties, governments, research organizations, non-profits, trade associations, and charities. Once received, there were a few criteria for distribution by the many online media outlets. The first, and most important, was that the information be interesting and topical. The second was that the source be an organization that sounded credible, but if something were incorrect, the redistributing online outlet would bear the blame.

The third, and sadly the last, was that the information actually be true. This criterion was ignored because the redistributing media outlet lacked the resources to verify the content.

An analysis in 2017 by the *Reuters Institute for the Study of Journalism* examined where people got their news in the age of the internet. Of those aged 18 to 24, 64% got it online, including from social media. Thirty-three percent of this age group got its news exclusively from social media. Only 5% read the printed word in newspapers or magazines, while 24% watched television.

For those 55 and older it was the exact opposite. Only 28% got their news online, 7% from social media, while print news was more than double that of the youngest generation at 11%. Of this group, 51% watched their news on TV.

What is most meaningful is that twenty years ago, nobody got their news online. Broadly speaking, online media through the internet didn't exist. Although people still receive conventional news stories through social media, unlike the newspapers of old they can easily avoid anything and everything that doesn't interest them or disagrees with their preconceived views.

As the media shrank and restructured, the news business has become more competitive with massive growth of online sources from blogs to networks. No giant printing presses to create the product; no paper; no distribution network such as trucks or a door-to-door delivery. All replaced by a stream of electrons. This was a fantastic financial advantage.

The online media was also completely unregulated. The "new news" could write and distribute whatever it wanted. While fraudulent or defamatory information remained illegal, the only recourse for the maligned was through the courts. With print media you could punish the publisher by cancelling a subscription or refusing to buy the product at a newsstand. If enough people cancelled their newspaper, it would reduce advertising rates. A credible news organization would have its reputation tarnished by a lawsuit. This kept news outlets on their toes.

Except for litigation, it is almost impossible to financially penalize a media outlet that distributes its product for free.

As attacks on the fossil fuel industry grew, the intense opposition finally created measurable financial damage as pipelines to export oil were delayed or obstructed. Only then did the carbon industries begin to accept there was growing public support for the climate change issue, however irrational it may be. While demands to end the carbon era intensify, nobody can imagine a world without cars, trucks, trains, ships, boats, airplanes, and the myriad of consumer products like plastics and chemicals.

Confused? Not your fault. You should be.

The carbon industry eventually figured it out. Through polling, research, and watching the behaviour of their children, the industry realized the millennials were not listening to the message the industry was sending out. The Millennials had grown up seeing and hearing climate change information acquired elsewhere. They depended more on social media and their cell phones to find out what was happening than the mediums via which the industry communicated its message. If it communicated at all.

If only the fossil fuel producers could get through to the Millennials on social media, surely fewer people would hate the industry and adopt a more rational approach to the future of carbon.

One of the more illuminating presentations of the impact of the internet on the new media came courtesy of a pollster who was guest speaker at the banquet following the annual general meeting of the Petroleum Services Association of Canada in Calgary in 2013. He said it wasn't that young people were reading less or didn't read. In fact, they were reading more. Thanks to mobile devices, there was more information available at their fingertips than ever before. But they got to choose what they wanted to view, listen to, or read.

Pre-internet, media content was set by owners, publishers, and editors of newspapers, magazines, TV, or radio stations. Traditional media ensured the news presented a cross section of information that included politics, government, business, lifestyle, politics, and economics. Readers or viewers got a bit of everything, whether they wanted it or not.

With the internet and the equivalent of functional computer in everybody's pocket, the individual became the sole judge of what he or she wanted to know. People could pick and choose among literally thousands of information streams ranging from sports scores, to obscure music, to social trends, to the activities of their close friends and acquaintances.

Be assured Kim Kardashian did not go from obscurity to fame and fortune because of talent, intelligence, or some worthy contribution to the future of civilization. She is a product of the internet and has millions of followers.

As an increasing number of people changed the channel, the industry's response was often a website that supported carbon resources by explaining all the wonderful things that are made of plastic. Or how much it paid in taxes, gargantuan numbers in the billions that are incomprehensible to most.

An interesting book that delves deeply into history to explain how powerful the internet has become in shaping the world was written in 2017 by Niall Ferguson and is titled *The Square and the Tower—Networks and Power from the Freemasons to Facebook*. The author's definition of a "network" is the size and structure of a group of people with common interests, common causes, and the ability to communicate with each other either directly or by sharing information. Networks are vehicles through which leaders and members influence the government or the economy, or what people read, heard, thought, or believed.

Ferguson traces the history of networks back to the fabled Illuminati, the medieval organization that, according to urban legend, still controls the world by influencing politicians, banks, and big business. The narrative then progresses through the historical development of networks and defines their characteristics. Ferguson explains that people are social animals and seek out each other's company, particularly those with whom they share common roots or interests, or who think and behave the way they do. Some of the strongest early networks were family-based, such as the Medici family of Italy during the Renaissance, or the Rothschilds in the 1800s (N. Ferguson 2017).

The author devotes a chapter to Johannes Gutenberg's invention of the printing press in Germany about 1450. These were in broad use across Europe by the end of that century. A hundred years later there were fifteen "printing establishments" in England, which turned out 115 books. By 1695, "some seventy printing establishments produced 2,092 titles." This technological advancement in Europe caused massive disruption, particularly in religious institutions, which no longer controlled the message (N. Ferguson 2017).

Early in the book, Ferguson writes,

> [Social] networks include patterns of settlement, migration and miscegenation that have distributed our species across the world's surface, as well as the myriad cults and crazes we periodically produced with minimal premeditation and leadership. As we shall see, social networks come in all shapes and sizes, from exclusive secret societies to open-source movements. Some have a spontaneous, self-organizing character; others are more systematic and structured. All that has happened – beginning with the invention of written language – is that new technologies have facilitated our innate, ancient urge to network. (N. Ferguson 2017)

The last section of the book is dedicated to the biggest and most powerful network in human history, the internet. Its reach and power make the printing press insignificant by comparison. Ferguson focuses on Facebook, the social media website that has become one of the most valuable companies in the world and has recently been accused/credited/identified as helping shape the outcome of the 2016 election of Donald Trump and the so-called Brexit vote in Britain, the referendum in which a majority voted to leave the European Economic Union.

Today's Facebook has 2.27 billion active users, nearly one-third of the world's population. Other social media outlets include Twitter, YouTube, Snapchat, LinkedIn, Reddit, and WhatsApp. The Pew Research Centre reports 45% of Americans get their news from Facebook, with lower figures for the other platforms. Twitter—founded in March 2006—has unprecedented influence.

While oil companies and their executives have long been accused of collusion, price-fixing, profiteering, and shameless exploitation of the common man, the real money is in the modern digital industry. Six of the 12 richest people in the world are modern practitioners of the internet economy or digital-information business with products like Facebook, Microsoft, Amazon, Google, Bloomberg News, and Oracle.

How the mighty have fallen. The long-despised "oil barons" are nobody, replaced in the global economy by the people who have created the tools with which the anti-carbon movement is stripping them of their remaining wealth, power, and influence. If they ever actually had any power or influence in the first place.

Michael Bloomberg, founder of the massive media organization that bears his name, is a professed believer in fossil fuel-driven climate change and a promoter of cleantech alternative energy sources, companies and opportunities through a separate media entity called *Bloomberg New Energy Finance*. As will be discussed in the next section, Bloomberg is also a driving force behind more public company disclosure by carbon-resource developers of future commercial influences caused by climate change and the anti-carbon movement.

The use of Twitter by politicians to communicate their views, thoughts, and commentary has reached unprecedented levels. US President Donald Trump appears to be running the most powerful country in the world, 280 characters at a time, using only his cell phone. But a point Trump has made since being elected is beginning to resonate; that is the public proliferation of "fake news." Anybody can write anything, put it on social media, and somebody will read it. If it reflects the views of the reader—the essence of social networks as defined by Ferguson—its sticks and is accepted as factual.

Only now is the mainstream media conducting surveys into the public's understanding of how much of the news is fake. The public is catching on but people don't know what to do about it. On June 19, 2018, The *Toronto Star* carried a story titled, "7 in 10 Canadians say the government should regulate fake news, polls say." It explained a Nanos Research study commissioned by *Canadian Journalists for Free Expression* "… found a strong majority of Canadians – more than eight in 10 – say false information that looks legitimate is making it harder to find out what's real. To help prevent the spread of fake news, seven out of 10 Canadians think the government should step in."[131]

Oh dear. The means governments can and should regulate the media but obviously are not themselves the source of any fake news. Think again of the famous quote from George Orwell, repeated earlier in this book, from his 1946 essay, "Politics and the English Language". Political language, Orwell wrote, "is designed to make lies sound truthful and murder respectable, and to give an appearance of solidity to pure wind."

Further research into fake news was revealed in a *Global News* article from September 5, 2018 about an Ipsos poll. It opened with, "US President Donald Trump's relentless attacks on the media have made 'fake news' a household term, but many Canadians still don't know as much as they think they do about the concept. ... Nearly six in ten Canadians (58%) and Americans (62%) defined 'fake news' as a story in which the facts were wrong. However, 46% also applied the term to politicians and news reports that only present one side of an argument."[132]

Social media and instant communications are increasingly used to organize and expedite major political upheaval. In a Google article Eric Schmidt and Jared Cohen wrote about how people can organize mass demonstrations using nothing but their smartphones. Eric Schmidt is the executive chairman of Google. Jared Cohen is the founder and director of Google Ideas.

At a conference at Tufts University in Boston in October 2017, Google's Schmidt and Cohen discussed *The New Digital Age: Transforming Nations, Businesses and Our Lives*, which they co-authored. News coverage about the presentation read, "The new technology has created 'a massive shift in power to individuals' Schmidt said. He predicted that 3 billion more people getting smartphones over the next five years 'is a one-time way of changing the power structure, with enormous implications, most of which are positive – but not all.'"

> Technology will change world politics, the Google executives argued. "In the future, revolutions will be easier to start and happen faster, but will be much harder to finish," said Cohen.

---

131 Canadian Journalists for Free Expression, *Cision* (June 19, 2018). https://www.newswire.ca/news-releases/canadians-concerned-about-fake-news-and-want-government-to-take-action-685929162.html.

132 Josh K. Elliott, "Canadians can't agree on what 'fake news' really is: Ipsos poll," *Global News* (September 5, 2018). https://globalnews.ca/news/4428483/canadians-fake-news-poll/.

"What we've seen from the Arab Spring, the Ukraine and various other examples is that it's very easy for people to organize in virtual town squares around the common idea of 'We don't like this particular dictator. Let's get him out of power.' But that's the only thing people agree on, and after the dictator is unseated, the expectation that change and transformation will happen just as quickly is not met."[133]

This article explains the impact the internet has had on everything to do with climate change, pipelines, and oil sands. The internet was integral in advancing the concerns about carbon in the atmosphere from scientific research to a growing global political movement. And, as Schmidt and Cohen from Google pointed out about "the massive shift in power to individuals," it isn't necessarily rational because it is so much easier to complain about what is wrong than explain how to fix it. Governments have indeed been overthrown but often replaced with chaos, not a new or improved government.

On September 8, 2018, *The Globe and Mail* reported, "More than 18,000 people marched Saturday in Paris as part of an international mobilization to show popular support for urgent message to combat climate change in advance of a San Francisco summit…they called on politicians to spearhead a transition to 100 per cent renewable energy. The front page of France's daily *Liberation* newspaper featured a call from 700 French scientists for the government to "move from incantations to acts to move toward a carbon-free society."[134]

If successful, of course, none of them would be attending the conference in San Francisco unless politicians got with the program and ordered the immediate invention and implementation of the solar-powered airplane.

But that's the internet. It is history's most pervasive and easily accessible supplier of information, most of it probably accurate. Maybe. It allows people who think alike to find each other and share ideas. But the internet cannot answer questions that have not been answered. If information doesn't exist—for example, about a new and qualified government is ready to replace the one the people just forced

---

133 "Google's Eric Schmidt and Jared Cohen on 'The New Digital Age'," *Tufts* (October 30, 2017). https://sites.tufts.edu/hitachi/googles-eric-schmidt-and-jared-cohen-on-the-new-digital-age/.

134 The Associated Press, "Global Marches Seek Urgent Action on Climate Change," *The Globe and Mail* (September 8, 2018). https://www.theglobeandmail.com/world/article-global-marches-seek-urgent-action-on-climate-change/.

out of office, or the discovery of a replacement fuel for airplanes—the internet can't find it.

Now that the mainstream press knows what to look for, stories about the proliferation of fake news are appearing everywhere. On December 5, 2018, *Bloomberg Businessweek* carried a feature story about the latest election in Brazil titled, "Democracy in the Age of Disinformation." Apparently not all the claims made by those seeking public office were true. The article read, "Brazilians are among the world's top users of social media, leaving them especially exposed to fake news and political influence campaigns online."[135]

CBC chimed in during the US mid-term elections in November 2018 with an article titled, "Fake News, Even Fake Fact-checkers Found in Runup to US Midterms." Canada's "Mother Corp" wrote, "All the major social news networks have made attempts to clamp down on fake news, but the trickery has only grown more insidious and pervasive, with new derivatives of fake news, such as fake fact-checkers." This is apparently where one side takes the opponent's spin and re-spins it, calling the rebuttal "fact checking."[136]

The proliferation of fake news is even having an impact on mainstream media organizations that take pride in getting the facts straight. The website *Voice of Europe*, which advertises "uncensored news" (whatever that is), carried an article on November 23, 2018, titled, "Brits No Longer Trust Mainstream Media and its Presenters, Report Shows." It highlights a survey that showed public trust in what viewers saw on TV as declining 5% from 67% in 2017 to 62% a year later. The article stated, "Journalists are also revealed to be among the five least-trusted professions, coming above only ad executives, politicians and government ministers."[137]

The news nugget that best described the bleak future of mankind because of mobile devices and the internet was written by the Associated Press, January 8, 2019. It was titled, "Russian Church Head: Smartphone Could Precede the

---

135 Shannon Sims, "WhatsApp Groups and Misinformation Are a Threat to Fragile Democracies," *Bloomberg Businessweek* (November 2, 2018). https://www.bloomberg.com/news/features/2018-11-01/whatsapp-groups-and-misinformation-are-a-threat-to-fragile-democracies.

136 Ramona Pringle, "Fake News, Even Fake Fact-Checkers, Found in Run-Up to USU.S. Midterms," *CBC News* (November 6, 2018). https://www.cbc.ca/news/technology/fake-news-midterm-elections-1.4892305.

137 Politicalite, "Brits No Longer Trust the Mainstream Media and its Presenters, Report Shows," *Voice of Europe* (November 23, 2018). https://voiceofeurope.com/2018/11/brits-no-longer-trust-the-mainstream-media-and-its-presenters-report-shows/.

Antichrist." It was a warning by the head of the Russian Orthodox Church who said in an interview on state TV that anybody that can figure out what people want to hear and what they are afraid of, then disseminate it They can then use this information to gain "centralized control of the world. ... The Antichrist is the person who will be at the head of the world wide web that controls the entire human race."[138]

Alberta's carbon-resource industries are certainly accustomed to their share of fake news. The rise in the earth's temperature is undeniably true unless all the world's thermometers are wrong.

But the anti-oil sands movement is rife with information that is not correct. In February 2012 a group of American ENGOs—Natural Resources Defense Council, National Wildlife Federation, Pipeline Safety Trust, and the Sierra Club—released an important looking document titled, *Tar Sands Pipelines Safety Risk*. It opens with, "Tar sands crude oil pipeline companies may be putting America's public safety at risk. Increasingly, pipeline supporting tar sands crude oil in the United States are carrying diluted bitumen or 'DilBit' – a highly corrosive, acidic and potentially unstable blend of thick raw bitumen and volatile natural gas liquid condensate – raising risks of spills and damage to communities along their paths."[139]

After reciting the usual problems of boreal forest destruction and "immense lakes of toxic waste," the sponsors allege, "There are many indications that DilBit is significantly more corrosive to pipeline systems than conventional crude. For example, the Alberta pipeline system has had approximately sixteen times as many spills due to internal corrosion as the US system."[140]

This allegation was dutifully disseminated by multiple news outlets. This resulted in the industry having to perform extensive research to refute the report. This subject has not been mentioned since and was not a significant issue at the public hearings into either Northern Gateway or Trans Mountain. Another factor not mentioned in the article was Alberta's very high standards for reporting pipeline

---

138 "Russian Church Head: Smartphones Could Precede Antichrist," *AP News* (January 8, 2019). https://www.apnews.com/310ae6ecfaa1414c8f3717d62870784c.

139 Anthony Swift & Susan Casey-Lefkowitz, "Tar Sands Pipelines Safety Risks," *Natural Resources Defense Council* (March 15, 2011). https://www.nrdc.org/resources/tar-sands-pipelines-safety-risks.

140 Ibid.

leaks of any consequence. Many non-Canadian jurisdictions with pipelines have less stringent or non-existent requirements on reporting pipeline spills. Also not mentioned is that when pipelines have leaked in Alberta in recent years, the factor was generally age or shifting terrain, not the contents.

Another accusation that was raised in the Trans Mountain expansion hearings involved the frequency of oil-tanker spills. In 2012 a Vancouver-based online media outlet[141] released an article titled *The Cost of an Oil Spill in Burrard Inlet: $40 Billion for Starters*. It took every major offshore oil spill in the world from Exxon Valdez to the BP Macondo blowout in the Gulf of Mexico and created some economic assumptions. None had cost that much to clean up, but of course they had not spilled bitumen. "The Aframax tankers now using Vancouver Harbor carry up to 700,000 barrels of bitumen, the deadliest crude on Earth. ... Consider a 500,000-barrel bitumen oil spill in Burrard Inlet, 70% of an Aframax tanker. Globally, there has been a spill of this size about every 18 months worldwide for the last 40 years."[142]

There is no evidence anywhere that oil spills of this magnitude have happened even once in the past 16 years let alone 18 months. According to data from the International Tanker Owners Pollution Federation, the worst year for oil spills in recent times was in 2002, and it was about 60,000 tonnes, or 410,000 barrels. That statistic included every event in the world, not just a single spill. From 2003 to 2016, the worst year was about 10,000 tonnes in 2007, which saw 68,000 barrels spilled. The other years were lower. The online article's allegation was absolute rubbish but obviously if your intentions are noble, among your social network, facts don't matter.

Another "fact" regarding bitumen is that it sinks in water and therefore cannot be cleaned up using traditional oil-spill containment methods on the surface. A *Maclean's* article titled "Does Spilled Pipeline Bitumen Sink or Float?" was published January 7, 2017. The subtitle read, "A scientist seeks to answer this pressing environmental question amid a national pipeline debate." The article involved the work of Heather Dettman who "has spent nearly two decades taking

---

141 Rex Weyler, "The Cost of an Oil Spill in Burrard Inlet: $40 Billion...For Starters," *The Commonsense Canadian* (May 10, 2012). http://commonsensecanadian.ca/cost-of-oil-spill-burrard-inlet-40-billion-kinder-morgan-rex-weyler/.

142 Ibid.

her lab coat on and off at a federal research complex ... in the Alberta town of Devon," an oil community built after the Leduc boom of 1947.[143]

Dettman was tasked by the Stephen Harper administration's World Class Tanker Safety initiative and the Trudeau government's Ocean Protection Plan to study the subject and report back. The article stated, "Most of her findings thus far side with the pipeline (Trans Mountain) and oil companies" because bitumen does indeed float. The issue was picked up by the anti-oil sands movement after the disastrous Enbridge pipeline spill in Michigan in 2010. At that time the US National Academy of Sciences concluded bitumen didn't float as well as conventional crude, making it harder to clean up.

But there are a variety of factors that influence the buoyancy of bitumen, including temperature and type of water. Merv Fingas, an Environment Canada oil-spill expert who co-authored the US report, agreed with Dettman that bitumen does not sink quickly, telling *Maclean's*, "Under typical Canadian climactic conditions, you've got up to three weeks to clean it up before you've got any major sinking."

When the Trans Mountain decision was overturned it had nothing to do with bitumen corroding pipelines, tankers spilling oil, or the difficulty of cleaning up "the deadliest crude on Earth." The issues were aboriginal consultation and the impact of tanker traffic on killer whales.

Oil companies had an image problem long before the invention of the internet or the emergence of fossil fuels as an enemy of the planet. Originally it was high prices and an engineered oil shortage. Today it is low prices and unlimited availability. The internet has allowed climate change alarmists to accelerate and broadly disseminate an enormous amount of information about what is wrong with carbon resources, not all of it true. The worldwide web and social media have certainly contributed to the opposition to pipelines and helped organize opposition on multiple fronts.

There has been significant criticism of carbon industry executives for allowing the misinformation to go this far; that the fossil fuel sector has done a poor job of spreading its message and the essential nature of its products.

But if the internet and smartphones can help organized opponents overthrow a government, a pipeline is, by comparison, helpless.

---

143 Jason Markusoff, "Does Spilled Pipeline Bitumen Sink or Float?" *Macleans* (January 7, 2017). https://www.macleans.ca/society/does-spilled-pipeline-bitumen-sink-or-float/.

This trial by fire has forced carbon-resource developers to better understand the communications challenge and how it grew so quickly and so broadly. But history has proven, and the executives of Google have admitted, that the internet revolution is infinitely better at identifying problems than solving them. In this respect it validates industry confusion about how people can simultaneously vilify and depend upon fossil fuels.

Some progress has been made by pro-industry organizations using social media as their primary form of communication. Recent public demonstrations in favour of pipelines and the oil industry in western Canada have been entirely organized using the same tools that helped stop them in the first place.

But the carbon-resource industries are years behind and their messages must reach beyond those who are directly affected because they work in the business. At least the captains of industry appear to have finally figured out how to do it. The next section further explores the enormity of the challenge.

10
MANKIND'S LONG HISTORY OF
PREDICTING SELF-DESTRUCTION

"In 1988 in New York City, climate change was real and the cause was known. The predictions of impending doom, including dates and disasters, were on the public record. A senior US political body had sought and received the information from leading experts, and other elected American politicians were on the track of forwarding the changes necessary to prevent or mitigate the upcoming climate catastrophe."

The very fact you're reading this chapter proves previous predictions about the end of world or civilization have not yet materialized. That doesn't mean it won't, or that mankind has ceased its frequently described, determined, and inevitable date with disaster. However, the lack of confidence so many of us have in the behaviour and future of our own species has a major impact on how modern society thinks and behaves.

Past predictions of disaster have been for the most part blamed on shortages. Technology and market forces have been able to solve these problems. Climate change is different because it is an environmental phenomenon based on too much of something, not too little. Let us all hope the patterns of market forces, new technology, and human ingenuity will continue to make the future of the world better than it appears today.

Past predictions of disaster have been for the most part blamed on shortages. Technology and market forces have been able to solve these problems. Climate change is different because it is an environmental phenomenon based on too much of something, not too little. Let us all hope the patterns of market forces, new technology, and human ingenuity will continue to make the future of the world better than it appears today.

Going back centuries, there have been dozens of recorded predictions of the end of mankind, civilization, or the world itself. Most had their roots in religion, often involving the arrival of a messiah or an evil entity. Man has always been more comfortable understanding things that start at a certain time and end at another. Why would human civilization be any different? In a world where everything is measured—weight, height, distance, time, temperature, wealth—the idea that the universe has always existed, always will exist, and has no physical end measured by time, is incomprehensible.

The theory of human evolution has always troubled a lot of people if for no other reason than man is so different than any other living things, current or past. While science says we evolved from creatures that crawled out of primordial seas millions of years ago, nobody sees the characteristics of an ancient amphibian, reptile, or mammal in the mirror. In 1959 in Tanzania's Olduvai Gorge British paleo-anthropologists Louis and Mary Leakey found the skull of *Australopithecus africanus*, the earliest modern human ancestor. Affectionately named Lucy, she was estimated to be 1.9 million years old. This species would lead to *Homo erectus* and *Homo sapiens*. This time frame is impossible to comprehend for a society that measures the performance of its athletes in to the nearest one one-hundredth of a second.

In modern times virtually all the pessimistic predictions about the future of the world have been based upon the activities and behaviour of man. Clearly at the top of the evolutionary ladder and food chain, human beings were surely responsible if something went wrong. Even during the darkest days of the Cold War and the nuclear arms race, if "mutually assured destruction" came to pass it would be because man created and then launched the bombs. And as the most intellectually successful creature to ever inhabit the earth, *Homo sapiens* were the only living creatures with the capability and time to even think about such things.

The first forecast most relevant to this book came from Thomas Robert Malthus who lived in England from 1766 to 1834. These were the early days of the carbon-fuelled Industrial Revolution and the general feeling at the time was that life and civil society was getting better for many.

But, like today, at least one person concluded it was too good to be true. Something had to go wrong.

Malthus believed that if living conditions continued to improve at the prevailing rate, it was inevitable that families would have more children. This would put more pressure on the food, housing, and clothing supply, thereby eliminating the recent progress. He reasoned that because two people could have many children, the population would grow geometrically, while food production would grow arithmetically, or much closer to linear. It was only a matter of time before people would be starving again.

The Malthusian theory included "preventative checks" whereby people would recognize the problem and have fewer children. This was achieved through voluntary actions such as marrying later and abstaining from procreation until parents were economically capable of supporting their children. (One of Malthus' counsels was to exercise moral restraint.) Also included in the theory were "positive checks;" factors that would shorten life span. These included disease, war, famine, or terrible working and living conditions. Such conditions would control the population to match the food supply, thus preventing what was later coined the "Malthusian catastrophe."

History would prove Malthus wrong. The population did not grow as rapidly as anticipated and advances in food supply for the most part kept pace.

However, there were two noteworthy elements of the work of Malthus that would shape all future predictions of man-made catastrophe.

The first was the 15th century invention by Johannes Gutenberg of the printing press, which ensured broad distribution of cataclysmic forecasts based on the beliefs and fears of the time, in this case 300 years later in the England of Thomas Robert Malthus. It had a big impact. Spread the word.

Malthus also sparked the recognition that the pursuit of your needs and wants could impact the future of the world within a comprehensible time frame: your lifetime. These needs were easy to grasp; food, family, and shelter. But while preying on the powerful emotion of fear, the Malthusian Theory also contained a message of hope. If we acknowledged the problem and took the appropriate remedial actions immediately, the future could be different.

Leaping ahead 200 years but on the same subject, in 1968 Paul Ehrlich published a book that garnered a great deal of interest. It was titled *The Population Bomb: Population Control or Race To Oblivion*. To ensure it grabbed the reader's attention the book's cover stated, "While you are reading these words four people will have died from starvation, most of them children."

Needless to say, everyone took notice. The problems associated with the rising birth rate in the 1950s and 1960s coined the increasingly more popular term "population explosion." The earliest editions read, "The battle to feed all of humanity is over. In the 1970s hundreds of millions of people will starve to death in spite of any crash programs embarked upon now. At this late date nothing can prevent a substantial increase in the world death rate" (Ehrlich, The Population Bomb—Population Control or Race To Oblivion 1968).

To ensure there was no confusion, the author wrote, "We must rapidly bring the world population under control, reducing the growth rate to zero or making it negative. Conscious regulation of human numbers must be achieved." Because Americans ate so much, the US should be a leader. If America did it first, the world would follow. Understanding the nature of human beings better than Malthus, Ehrlich didn't expect he could get people to voluntarily refrain from having sex or children.

Ehrlich proposed, then rejected, that the government add sterility-causing chemicals to drinking water and the food supply. His only confessed objection to the idea was that insufficient biomedical research had been conducted. He then proposed taxing additional children above the set quota, levying luxury taxes on childcare, and offering tax incentives to men to undergo voluntary sterilization.

In Ehrlich's vision a government "Department of Population and Environment" would enforce the law and conduct further research into improved contraceptives and drugs for mass sterilization. The department would also take steps to create more men. Ehrlich concluded couples would have more children if their first child is a girl. If they had a boy first or sooner, the result would be fewer women and thus a lower birth rate.

Over two million copies of The Population Bomb were sold. It was regarded by most as "out there" in terms of its radical proposals to solve the problem. But in the growing environmental movement it was considered by some to be not draconian enough, and by others as a useful, if impractical, contribution to public awareness of the seriousness of too many people competing for what was regarded as the finite resources of the earth.

The book was written half a century ago when the world's population was half what it is today. While the problem of starvation has never been eradicated, the percentage of the population going hungry continues to decline because of major technological and process advancements in the food supply.

In 1972 a book appeared titled The Limits to Growth, which expanded on Ehrlich's theme. The cover of the first edition subtitle read, "A Report for THE CLUB OF ROME's Project on the Predicament of Mankind". Including updates as recent as 2004, over 30 million copies have been published and distributed in 30 languages.

The Club of Rome started in 1968 when about 30 European academics, economists, scientists, and businessmen met in Rome to discuss global issues. Besides agreeing on a name, nothing happened. By 1970, with a structure formalized, an expanded group met in Bern, Switzerland. The term "man's predicament" was part of the group's stated purpose, and a scientific analysis rooted in mathematics would be required in a hurry. Unable to agree how to do this, the members contracted researchers from MIT to develop what was called the "World3 Model" and report back. In 1971 a report was tabled at gatherings in Moscow and Rio de Janeiro. A book followed a year later.

The future of mankind could be, as the sponsors feared, awful. Key variables driving the "predicament" were the growth in population, food production, industrialization, pollution, and demand for non-renewable resources. Algorithms assumed these would grow exponentially but that technology to provide solutions would only grow linearly. The report admitted the results were only predictions and based on trends and human behaviour. Therefore, the outcomes ranged from the collapse of known civilization by the mid- to late-twenty-first century, to a growing but stable global economy and society as problems were solved.

A key subject in *The Limits to Growth* was non-renewable resources. Thanks to OPEC, the book's methodology was reinforced the year after publication by artificially engineered crude oil shortages. The MIT model took the known sources of key resources such as iron, chromium, gold, and petroleum and divided known reserves by the predicted demand growth. This result was what the oil industry calls a "reserve-life index" based on static demand or exponential demand. Either way the world would eventually run out of everything that came out of the ground and therefore life would be disrupted.

What was important about *The Limits to Growth* was the methodology, not just the message. By hiring MIT, one of world's most respected research universities, human-induced catastrophe was more believable than the popular cartoon image of a bearded guy in robes walking around downtown carrying a big sign on a stick that read, "THE END IS NEAR." This added much more credibility to the warnings than any single person or group of concerned individuals would have achieved.

As the oil crisis unfolded and the price exploded, believers would say "I told you so." But in 1981, Julian L. Simon wrote a book titled *The Ultimate Resource*, which questioned much of the thinking of the 1970s, particularly considering the impact of the oil price shocks of 1973 and 1979 on the world's economy. That the world was running out of oil and other resources was clearly factual based on the world oil shortage and sky-high prices. But the subtitles on Simon's book read, "Natural resources and energy are getting less scarce," "Pollution in the US has been decreasing," "The world's food supply is increasing," and "Population growth has long term benefits" (Simon 1981).

About the time *The Ultimate Resource* was released, the high price of oil was working its magic on supply and demand. Combined with a global recession, demand was flat or contracting. High prices and the urgent need for new hydrocarbon sources had resulted in completely new supplies of oil on stream from remote and hostile places like the North Slope of Alaska and Europe's North Sea. As noted in Chapter 7 of this section, the 1981 energy-pricing and revenue-sharing agreement between the federal government and Alberta was signed after the price of oil had already peaked. The fundamental perception of the energy shortages of the 1970s and the justification for the NEP was ancient history by 1983.

But Simon wrote about more than petroleum. On food he wrote, "Contrary to popular impression, the per capita food situation has been improving for the three decades since World War II." "Agricultural land is not a fixed resource, as Malthus and many since Malthus have thought. ... Paradoxically, in the countries that are best supplied with food, such as the US, the quantity of land under cultivation has been decreasing because it is more economical to raise larger yields on less land than to increase the total amount of farmland" (Simon 1981).

After the oil price shock of the 1970s, Simon's book was controversial. On natural resources he wrote, "Hold your hat – our supplies of natural resources are not finite in any economic sense. Nor does past experience give reason to expect natural resources to become increasingly scarce. Rather, if the past is any guide, natural resources will progressively become less scarce, and less costly" (Simon, The Ultimate Resource 1981).

For the growing population, Simon's main issue was the accuracy of the forecasts. "Population forecasts are publicized with confidence and fanfare, but the record of even the official forecasts made by US government agencies and by the UN is little (if any) better than that of the most naïve predictions. For example, experts in the 1930s foresaw the US population as declining. ... And official U.N. forecasts made in the year 1970 for the year 2000, a mere thirty years in advance, were five years later revised downward by almost 2 billion people, from 7.5 billion to 5.6 billion" (Simon, The Ultimate Resource 1981).

Simon went on to write about how many of the things people worried about were wrong. Pollution was declining, immigration had a positive effect on the US, and poor people did not "breed like animals" and did indeed consider the consequences of more children before they had them.

As he was writing the book, in 1980 Simon made a very public wager with The Population Bomb author Paul Ehrlich about his own economic view of the future versus that of Ehrlich, the biologist and environmentalist. This was apparently in response to an earlier claim by Ehrlich, who had written, "If I were a gambler, I would take even money that England will not exist in the year 2000" (Ehrlich 1968). Finally, the pair wagered US$1,000 on the inflation-adjusted value of five commodities chosen by Ehrlich—chromium, copper, tungsten, nickel, and tin—in the subsequent decade. In 1990, Simon won. Two of the five were lower in current dollars and the other three were lower when corrected for inflation. Ehrlich paid Simon US$576.07, meaning the five commodities had declined in real terms by 43%.

Other than Simon's unabashed optimism, which was not widely publicized at the time, the environment was bumped off the front page during most of the rest of the 1970s and the first half of the 1980s because of the world oil crisis, sky-high inflation, and record-high interest rates. By the mid-1980s the recession was over, oil prices were down, inflation and interest rates had stabilized, and the economy was growing again. This gave more people, particularly in the US, more time to think of things other than themselves.

In 1988 the US Senate Energy and Natural Resource Committee met in New York City to discuss, among other things, global warming and the greenhouse gas effect. It was a very hot day in in June and the meeting room apparently did not have air conditioning. One of the testimonies came from Dr. James E. Hansen, described by The New York Times on June 24, 1988 as an expert who was the director of the NASA (North American Space Administration) Institute for Space Studies in New York.

The title of the article was, "Global Warming Has Begun, Expert Tells Senate,"[144] and it stated, "The earth has been warmer in the first five months of this year than in any comparable period since measurements began 130 years ago, and the higher temperatures can now be attributed to a long-expected global warming trend linked to pollution, a space scientist reported today." The newspaper wrote that the linkages between GHGs and the climate had been cautiously suggested by scientists. Hansen was not the least bit cautious telling the gathering he was 99 per cent certain this was not a natural variation but the effect of the continuous buildup of GHGs. Several US senators in attendance considered Hansen's presentation to be a call for immediate political action."[145]

The article continued that by 2025 to 2050 if the current buildup of GHGs continues, "the effect is likely to be a warming of 3 to 9 degrees Fahrenheit (1.7°C to 5°C)." It wouldn't happen at the same rate in various places in the world, but the result would be major increase in ocean levels as the ice caps melted. Inland, droughts would cause water levels in the Great Lakes to decline."[146]

In 1988 the pending climate change disaster was all but assured and the only courses of action defined. The testifying experts concluded immediate action was required. This included employing more nuclear power, ending or slowing deforestation, and reducing the use of CFCs, which were thought to contribute to the greenhouse effect.

Thirty-one years ago in New York City, climate change was real, the cause was known, the predictions of impending doom including dates and disasters were on the public record, a senior US political body had sought and received the information from leading experts, and other elected American politicians were on the track of forwarding the changes necessary to prevent or mitigate the upcoming climate catastrophe.

The IPCC and UN would take up the climate crusade in the early 1990s, particularly after the so-called "Earth Summit" at Rio de Janeiro in 1992. While the first IPCC report in 1990 had urged caution until more quantifiable evidence was obtained, the delegates to Rio in 1992 sounded a lot like James Hansen in 1988.

---

144 Philip Shabecoff, "Global Warming Has Begun, Expert Tells Senate," *The New York Times* (June 24, 1988). https://www.nytimes.com/1988/06/24/us/global-warming-has-begun-expert-tells-senate.html.

145 Ibid.

146 Ibid.

Although the IPCC didn't declare man-made climate change real until 2001, all the developed countries and many of the developing countries had signed the Kyoto Protocol in 1997 to commit to GHG emission reductions.

Early in the twentieth century a pattern was emerging. Through the UN and other bodies, the governments of the world had agreed to begin to solve the problem. But there was a clear lack of follow-up and coordinated action. This pattern continues today. Surrounded by peers and TV cameras, no delegates to any of the many UN climate change gatherings ever disputed why they were there, what they were trying to do, or disagreed with a final communiqué promising immediate and effective action and remedies.

Except that when their delegates got home, and the participating governments confronted the political and economic reality of what their electorates thought of using less fossil fuel or paying more for the privilege of damaging the environment, they seemed to contract amnesia. When comparing the collective commitments of every international climate change conference against what actually happened in the countries that signed the final agreements, it is impossible to conclude that the process has been successful.

The concerned figured this out and realized that the only way to change the behaviour of the politicians was to change the views and demands of the voters. The best way to do this was to ring the carbon-catastrophe alarm bells even louder. This issue was too important for diplomats and scientists. The public had to become engaged.

The single most pivotal event to move climate change from a high-level global political discussion item and into the craniums of voters and motivating them to action, was the 2006 release of a movie by Al Gore, *An Inconvenient Truth*.

11

CARBON CATASTROPHE – FOSSIL FUELS BECOME A MAINSTREAM MENACE

"Since the time of Malthus over 200 years ago, the idea has been growing that mankind is incapable of protecting the world we live in and our relentless pursuit of life, health, wealth, comfort, happiness, and procreation will ultimately lead to our own destruction. That we are genetically programmed to do so is a human flaw that must be overcome."

Al Gore was already famous when his movie, *An Inconvenient Truth*, was released. Gore served as the 45th Vice-President of the United States from 1993 to 2001. Elected as a Senator in 1984, Gore ran for the Democratic nomination for US president in 1988. Bill Clinton chose him Vice-President in 1992. Gore became leader of the Democratic Party in 2000 when Clinton stepped down after serving two terms as president.

Al Gore lost the 2000 election to Republican George W. Bush. Having sought the top political job in the US twice, Gore moved away from politics. He resumed life in the private sector as partner in an investment firm and founded a cable news channel in 2005, which was subsequently sold to *Al-Jazeera*, the Arab news network based in Qatar. Gore is generally regarded as quite wealthy.

An Inconvenient Truth was released in 2006. Its purpose was to raise awareness of the issue of climate change. A low-budget production, the movie contained several charts, graphs and images that visually depicted the looming disastrous changes caused by climate change. For example, if oceans rose by 20 feet because the ice caps on Greenland and Antarctica melted, Gore depicted what parts of Florida—like Miami—would be submerged. He didn't say exactly when this would occur, but a picture is worth a thousand words. To reinforce his messages, he incorporated clever visual images such as using a forklift to elevate him while he talked about rising temperatures and sea levels.

In 2005 Hurricane Katrina flooded New Orleans. With this fresh in people's minds, Gore used footage from that disaster to reinforce the idea that climate change was causing more frequent and intense hurricanes. The movie predicted temperatures would rise at such a pace that "within a decade there would be no more snows of Kilimanjaro." This was loosely based on research in 2002, which indicated it was snowing less and the famous glaciers on that African mountains were, like every other glacier in the world, receding. Climate change would cause more heat waves in the summer and more severe cold snaps in the winter.

One can only speculate what reaction Gore anticipated when he produced the movie, but it is unlikely he was disappointed. *An Inconvenient Truth* became one the most successful documentaries in history. It was awarded the Nobel Peace Prize in 2007 and an Oscar for best documentary the same year. One prominent climate scientist commented that Gore's movie "had a much greater impact on public opinion and public awareness of global climate change than any scientific paper or report."

On the 10th anniversary of its release there were dozens of articles examining *An Inconvenient Truth*. It was generally agreed Gore was a bit loose with the facts. No climate change model saw the ocean rising that much, or the snows of Kilimanjaro disappearing. There was no measurable change in hurricane intensity and frequency, although terrible events like Hurricane Sandy in 2012 caused enormous damage. But so had every other major hurricane to reach the coast of the US in history.

In the court of public opinion, if your intentions are noble (as alerting all mankind to the perils of future climate disaster most certainly is), that the facts supporting the argument are not 100% correct becomes secondary.

On May 2, 2016 *npr,* the website of National Public Radio, published an article summing up the impact of the movie as well as any. Besides raising the level of public awareness on the risks of climate change, "… it also politicized global warming to an unprecedented level. It brought the spotlight to an issue that, as the title says, many investors and politicians find inconvenient. If nothing is done to curb the emissions of carbon dioxide and other greenhouse gases, temperature will rise, ice caps will melt, ocean levels will rise and weather patterns across the globe will be disrupted. This truth remains unchanged."[147]

---

147 Marcelo Gleiser, "After 10 Years, 'An Inconvenient Truth' Is Still Inconvenient," *NPR. org* (May 4, 2016). https://www.npr.org/sections/13.7/2016/05/04/476717308/ after-10-years-an-inconvenient-truth-is-still-inconvenient.

Winning a Nobel Price and an Academy Award made Gore's documentary about the weather not only acceptable but, in many quarters, essential. Teachers usually enjoy a break from delivering a prepared classroom lesson plan by showing a movie, particularly a documentary. *An Inconvenient Truth* immediately found its way into schools and was often the first exposure students have ever had to the climate change issue.

Climate change was no longer confined to scientists droning on about fluctuations in the atmospheric content of carbon dioxide, or diplomats and politicians at yet another UN climate change conference politely agreeing about everything, then doing nothing. Gore's work was an action movie with hurricanes, floods, devastation, and destruction. It was closer to a video game than a textbook.

A generation of parents wonder what happened to their children when the kids spout the mantra of fossil fuel-induced climate change without question. Thank Al Gore.

The move also energized the climate change-alarmist community which, if you'll forgive the pun, turned up the heat on what could be said and claimed about the impending carbon catastrophe. The ENGOs gauged the public interest and realized where the money was. Those who had not yet seized on climate change as the mother of all environmental challenges soon got with the program.

In 2006 British economist Sir Nicholas Stern released an extensive report analyzing the economic cost tomorrow of not doing something about climate change today. Stern had been a senior economist for Britain's Treasury and the World Bank. Like Gore's movie, Stern's report was a very political document presented in a manner that looked objective and scientific. They key assumptions were repeated from the climate change models of the day. A 3°C to 4°C increase in the world's average temperature could cause major flooding in the twenty-first century, displacing 200 million people. Somewhere between 15% and 40% of the world's species would face extinction. Stern called climate change, "the greatest market failure the world had ever seen," ignoring the fact that it hadn't happened yet.

Stern then monetized the future disasters using what critics regarded as a very low discount rate to increase the current cost of future events. He wrote that unchecked climate change could cost the world 5% of GDP each year but doing something now, like increasing the cost of carbon immediately to reduce consumption, would cost only 1%. The cost could rise to 20% of GDP if the impact turned out to be more serious than the worst-case scenarios at the time. The poorer countries would suffer more because of their limited capacity to adapt. Stern even went as far to talk about the unborn, lamenting the fate of the people of the future who didn't exist yet but nevertheless "lacked representation" (Foster 2014).

One of the impacts of Stern's work in monetizing the future impacts of climate change was changing investment behaviour and disclosure of future risks for public companies. This is discussed in detail in the next section.

About the same time as *The Inconvenient Truth* was receiving widespread attention and global acclaim, Alberta's oil sands were getting more attention. In 2006 the Alberta government parked a giant bitumen-ore-hauling truck in public view in Washington, DC, to promote Alberta's enormous oil-production potential. While the oil sands were already on the radar of the world's oil companies, they were not that known publicly. That would all change in the next few years.

By 2008 a full-blown oil sands-development boom was underway, new export pipelines were under construction, and the general view was there was much more to follow. For the ENGOs, the objective was not just to warn the public about the impending damage of climate change but to galvanize the efforts towards stopping at least one growing source of fossil fuel production and hopefully emerge victorious in stopping something, somewhere, somehow. In the US, opposing economic development in another country that had no collateral impact on America was ideal. The target became the oil sands.

It started in October 2008 with a discussion document authored by Corporate Ethics International titled "The Tar Sands Campaign 2.1." It opens with, "We stand at a crossroads. The path we choose may well determine the fate of the earth. To the right, the path leads to a dirtier fuel future of coal-fired generating plants, liquid coal, tar sand and oil shales, with certain catastrophic global warming. To the left, the path leads to internalized carbon costs through cap and trade, clean technologies, and energy conservation with the hope of eventually returning our atmosphere to normalcy."[148]

American ENGOs were enjoying reasonable success opposing coal but, "(w)hat we lack is a comparable transportation fuels campaign to address the challenges posed by this industry. Transportation contributes 25% of the annual global emissions ... (w)hile non-conventional fuels, like tar sands oil from Canada, are a small percentage of the US annual fuel consumption today, the percentage is

---

148 Michael J. Marx, "Tar Sands Campaign 2.1," *Corporate Ethics International* (October 2008). http://www.offsettingresistance.ca/TarSandsCoalition-StrategyPaper2008.pdf.

projected to rise. *Stopping the flow of tar sands oil now ... is critical if we are to force government and industry to pursue a clean and sustainable energy future.*[149] (Italics in original.)

The document continued. "Why a campaign on tar sands oil? ... Extraction of these 'bottom of the barrel fuels' is energy intensive, emits much higher greenhouse gases, and poses grave risks to air, water, land and wildlife while placing severe pressure on the health and infrastructure of communities."

Under the heading "Theory of Change," the strategy is discussed in detail. "The question is, 'How in the face of nearly unlimited gas and oil industry money can we ever hope to stop tar sands production?' We are confident that this can be accomplished through the strategy outlined below. Even in the nascent stages of this campaign, we have seen the power of raising the visibility of tar sands negatives, educating the public, and legal challenges. What gives us additional optimism is the following:

- The potential to break the chain for delivery of vital inputs to tar sands operations

- The legal potential to block vital links in the tar sands oil delivery infrastructure

- The potential costs to the industry associated with mitigation and legal fights

- The fact that Barack Obama, the potential next president, has criticized tar sands oil

- The growing potential for national carbon legislation that pre-empts tar sands oil"

That's how the courts overturned the Trans Mountain pipeline decision in August 2018 and the Northern Gateway decision in 2016. Significant resources were made available to opponents such as Canadian ENGOs and Indigenous groups.

West coast researcher and writer Vivian Krause has done a remarkable job of following the money from the ENGOs back to its source. She tracks the formation and funding from the first anti-oil sands strategy in 2008 right through to the

---

149 Ibid.

various Canadian pipeline and oil sands opponents, including Canadian ENGOs and First Nations.

Krause started her investigations years ago. Only recently has her worked begun to receive the broad recognition it deserves.

The strength of American ENGOs grew when the oil sands battles started taking place on US soil as the Keystone XL (KXL) pipeline approval became a public issue in 2011. Former pipeline executive Dennis McConaghy wrote in *Dysfunction* about how anti-oil sands momentum increased when ENGO hopes of a Barack Obama presidency came true later that year. McConaghy explained how Obama wanted to be remembered if he won. In the acceptance speech for his nomination as Democratic candidate for president in June 2008 '... Obama intoned that future generations would look upon his presidency as "the moment when the rise of the oceans began to slow and our planet began to heal." With Obama in the White House and Keystone XL's future on American soil, the ENGOs began to mobilize" (McConaghy 2017).

A key player in the opposition to KXL was Bill McKibben who was already head of an anti-carbon ENGO 350.org. McConaghy wrote about the state of the carbon-catastrophe industry after the US had failed to introduce a carbon cap and trade system in 2009. The commitments of the Copenhagen Accord were faltering, just as Kyoto's had 10 years earlier. McKibben "... appreciated perhaps more than most the need for a tangible cause that could reinvigorate the entire environmental movement. A cause that could not only be depicted as significant in the context of dealing with the climate change risk but one that had some reasonable prospects of success. He discovered KXL [Keystone XL] for that purpose" (McConaghy 2017).

McKibben had been in contact with James Hansen who was still active in the anti-carbon movement. In 2009 Hanson had given a speech in England where he called coal, "the single greatest threat to civilization and all life on our planet."[150] He reiterated the issues of rising temperatures and ocean levels and how the oceans could only absorb a certain amount of carbon dioxide. Being in England, home of the coal-fuelled Industrial Revolution, Hansen attacked coal, stating, "Coal is not only the largest fossil fuel reservoir of carbon dioxide, it is the dirtiest fuel.

---

150 James Hansen, "Coal-fired power stations are death factories. Close them," *The Guardian* (February 15, 2009). https://www.theguardian.com/commentisfree/2009/feb/15/james-hansen-power-plants-coal.

Coal is polluting the world's oceans and stream with mercury, arsenic and other dangerous chemicals. ... The trains carrying to power plants are death trains. Coal-fired power plants are factories of death."[151]

Back in the US, by 2011 Hansen was on the anti-oil sands crusade when he commented publicly on Washington Draft Supplemental Environmental Impact Assessment on KXL. In a blog titled "Silence is Deadly," Hansen wrote how the "tar sands monster," if fully developed, would be "game over" for the climate.[152] McConaghy wrote how he believed Hansen was trying to invigorate the ENGOs, explaining how big the oil sands were and how much carbon would be released if all of it were produced. It would be impossible for the world to meet any GHG ceiling if the oil sands were not stopped. McConaghy argued it would take more than 2,000 years for Hansen's projections to be realized, but by 2011 the ENGO movement no longer depended on facts or science (McConaghy 2017).

Pictures of ugly oil sands tailings ponds had appeared in *National Geographic* in 2009 helping to strengthen public opinion against the resource. McConaghy figured that even though the oil sands had momentum and developers were cognizant of the need to do a better job, "... Hanson pointed out that oil sands-related transportation infrastructure could be just the Achilles heel the environmental movement needed to fundamentally alter its prospects" (McConaghy 2017)."

Hansen's opposition to the oil sands continued. Possibly his most famous article on this subject appeared in *The New York Times* under the title, "Game Over for the Climate." Discouraged by US President Obama's acknowledgement that Canada was a sovereign country and thus there were limits on what the White House could do to curtail oil sands development, Hansen wrote, "If Canada proceeds, and we do nothing, it will be game over for the climate. Canada's tar sands, deposits of sand saturated with bitumen, contain twice the amount of carbon dioxide emitted by global oil use in our entire history." If all the oil sands were produced along with coal, conventional oil, and natural gas, CO2 levels "... would eventually reach higher levels than the Pliocene era, more than 2.5 million years ago, when the sea level was at least 50 feet higher than it is now."[153]

---

151 Ibid

152 Ibid

153 James Hansen, "Game Over for the Climate," *The New York Times* (May 9, 2017). https://www. nytimes.com/2012/05/10/opinion/game-over-for-the-climate.html.

Hansen's message was clear. Burn your oil, gas, and coal New York. Then watch The Big Apple disappear under rising oceans. He urged US President Obama to do something immediately to save the planet from certain destruction.

Over the next few years the anti-KXL demonstrations would rise. Famous people chained themselves to the White House fence. Eventually it worked, but it took a while. In late 2015, President Obama formally killed KXL by refusing to issue the cross-border permit to import Canadian oil into the United States.

Corporate Ethics has rebranded itself as CorpEthics, "Strategic Advisors to Environmental Campaigns." One of the signature success stories it advertises on its website is "The Tar Sands Campaign." Today it reads, "In 2008 two major US foundations asked CorpEthics to recruit the groups, develop the strategy, create a coordinated campaign, and act as a re-granting agency for the North American Tar Sands Campaign. … From the very beginning, the campaign strategy was to land-lock the tar sands so their crude could not reach the international market where it could fetch a high price per barrel. This meant national and grassroots organizing to block all proposed pipelines. This strategy is successful to this day. All the proposed pipelines in Canada have effectively been blocked, as have those proposed in the US."[154]

Bragging about its power to influence massive economic investments and even Canadian elections, CorpEthics states, "The Tar Sands Campaign jump started the climate movement in the US as major political figures, celebrities, and a diverse group of NGOs came together to pressure the Administration to reject this pipeline (Keystone XL). It also played a role in helping to unseat the Conservative Party in Alberta and nationally."[155]

(On January 25, 2019 Vivian Krause alerted the world via Twitter that CorpEthics had modified the foregoing, perhaps in response to growing Canadian media coverage including a CBC TV documentary released January 20, 2019 narrated by Wendy Mesley. The references to land-locking oil sands and influencing Canadian elections no longer appears.)

In Canada a similar war of words was underway and the reaction of the public and politicians to the oil sands was similar. In April 2008 1,606 migrating ducks died when they landed in a Syncrude oil sands tailings pond. The outrage was

---

154 "Tar Sands Campaign," *CorpEthics* (2018). https://corpethics.org/the-tar-sands-campaign/.
155 Ibid.

immediate as environmentalists and government officials demanded to know why. The incident got international attention. Greenpeace's Canadian oil sands vilifier-in-residence, Mike Hudema, released a statement saying the incident was ".... yet another example of the deadly implications of Alberta's rampant tarsands development. The fact that this devastating incident wasn't reported by Syncrude begs the question of how many other incidents have been quietly covered up to safeguard their image." Syncrude was eventually levied and paid a $3-million fine.

In 2013 the Avian Conservation and Ecology, a Canadian online journal about birds, reported an average of 269 million birds are killed each year in Canada by a variety of means. The main culprits are wild and domestic cats which kill about 200 million birds annually. Power lines, buildings, homes, vehicles, hunting, forestry, electricity, and communications towers are the next largest group: over 66 million. Wind turbines accounted for 16,700. There are no bird deaths directly related to oil sands production but deforestation and construction activity results in the loss of 2,939 nests each year. While 1,606 dead ducks in the oil sands were mainstream-media front page news, the other 269 million dead birds were not.

Canadian writers were by no means slackers when describing how awful oil sands were to everyone and everything. In 2009, Alberta author Andrew Nikiforuk published a book titled *Tar Sands: Dirty Oil and the Future of Continent*. On his own promotional website about the book, the advertisement expands the reach of destruction and despair writing, "The region has become a global Deadwood, complete with rapturous engineers, cut-throat cocaine dealers, Muslim extremist, and a huge population of homeless individuals. In this award-winning book. ... Nikiforuk exposes the disastrous environmental, social and political costs of the tar sands, arguing forcefully for change."[156]

The oil sands weren't just dirty oil. The resource was also part of dirty government. *Tar Sands* opens with Declaration Of A Political Emergency, 22 statements highlighting the urgency of doing something about the oil sands disaster. Number 16 states, "Oil hinders democracy and corrupts the political process through the absence of transparent reporting and clear fiscal accounting. Alberta, a classic petrostate has one of the least accountable governments in Canada as well as

---

156 "Books," *Andrew Nikiforuk* (2015). http://andrewnikiforuk.com/Books/books.html.

the lowest voter turnout" (Nikiforuk, Tar Sands: Dirty Oil and the Future of Continent 2009).[157]

Petrostate is defined as, "a small oil-rich country in which institutions are weak and wealth and power are concentrated in the hands of a few."[158]

To ensure the reader does not miss the point of the awfulness of oil sands, read this. "Bitumen is what a desperate civilization mines after it's depleted its cheap oil. It's a bottom-of-the-barrel resource, a signal that business as usual in the oilpatch has ended. ... Calling the world's dirtiest hydrocarbon 'oil' grossly diminishes the resource's huge environmental footprint." Looking ahead in 2009 at the forecast growth in oil sands production, Nikiforuk wrote, "By most estimates, the tar sands will pump out 140 megatonnes of global heaters by 2020. By that time, the megaproject will be making more greenhouse gases than current output from the world's volcanoes" (Nikiforuk, Tar Sands: Dirty Oil and the Future of Continent 2009).[159]

Oil sands carbon emissions in 2017 were about 70 million megatonnes. According to a research report from IHS Markit, reported in the *Calgary Herald* September 13, 2018, emissions intensity from oil sands production measured per barrel produced declined 21% from 2009 to 2017 and will fall another 16 to 23% by 2030. However, the total number of barrels produced has risen.[160]

In a review of Nikiforuk's book in *The Globe and Mail* by environmental writer Alanna Mitchell, the author heaped more scorn on Alberta's most valuable resource and the people who support or even permit its development. "The book is, in essence, a revolting, blush-making case for Canada to develop integrated energy and environmental regulation suitable for the post-carbon age. And then swiftly enforce it ... the massive and growing project gulps fresh water, destroys valuable

---

157 Excerpt from *Tar Sands* by Andrew Nikiforuk reprinted with permission from Greystone Books Ltd.

158 "Petrostate," *Collins Dictionary* (2018). https://www.collinsdictionary.com/dictionary/english/petrostate.

159 Excerpt from *Tar Sands* by Andrew Nikiforuk reprinted with permission from Greystone Books Ltd.

160 The Canadian Press, "Report Forecasts 16 to 23 per cent Emissions Intensity Drop for Oilsands," *The Calgary Herald* (September 13, 2018). https://calgaryherald.com/business/energy/ihs-report-forecasts-16-to-23-per-cent-emissions-intensity-drop-for-oilsands.

boreal forest, poisons air, water, and soil, and uses up a substantial portion of the energy it produces. ... That they exist at all is a scandal."[161]

By now the reader understands that the concerned can write anything they want about the oil sands with impunity and without fear of repercussion.

Decades of climate alarmism have proven facts don't matter if the intentions appear noble. In December 2017, *National Geographic*, a respected magazine, released video footage of an emaciated polar bear in the Canadian Arctic accompanied by, "This is what climate change looks like ... scientists warn that as temperatures rise, and sea ice melts, polar bears lose access to the main staple of their diet – seals." This was one of many reports that global warming was wiping out the cute, but deadly, polar bear population.[162]

In August 2018, *National Geographic* issued an apology confessing, "*National Geographic* went too far in drawing a definitive connection between climate change and a particular starving polar bear. ...There is no way to know for certain why this bear was on the verge of death." Kudos to at least one source of climate change disaster disseminators for admitting a mistake.[163]

Climate change alarmists have long claimed polar bears are threatened but University of Victoria adjunct professor Susan Crockford claims otherwise. So instead of saving bears, they attack Crockford. In a column in the *Financial Post* on February 2, 2018, columnist Terrance Corcoran wrote about one published attack in which Crockford was accused of "unsubstantiated statements and personal attacks on scientists, using names like eco-terrorists, fraudsters, green terrorists and scammers."[164]

---

161 Alanna Mitchell, "Review: Tar Sands: Dirty Oil and the Future of a Continent, by Andrew Nikiforuk," *The Globe and Mail* (January 16, 2009). https://www.theglobeandmail.com/arts/books-and-media/review-tar-sands-dirty-oil-and-the-future-of-a-continent-by-andrew-nikiforuk/article4216210/.

162 Sarah Gibbens, "Heart-Wrenching Video Shows Starving Polar Bear on Iceless Land," *National Geographic* (December 2017). https://news.nationalgeographic.com/2017/12/polar-bear-starving-arctic-sea-ice-melt-climate-change-spd/.

163 Cristina G Mittermeier, "Starving-Polar-Bear Photographer Recalls What Went Wrong," *National Geographic* (August 2018). https://www.nationalgeographic.com/magazine/2018/08/explore-through-the-lens-starving-polar-bear-photo/.

164 Terence Corcoran, "Terence Corcoran: Canadian finds polar bears are doing fine — and gets climate-mauled," *Financial Post* (December 7, 2017). https://business.financialpost.com/opinion/terence-corcoran-canadian-finds-polar-bears-are-doing-fine-and-gets-climate-mauled.

Meanwhile, there is other evidence the polar bear population is in fact increasing. On the website *The Global Warming Policy Foundation,* in February 2017, Crockford wrote, "In 2005, the official global polar bear estimate was about 22,500. Since 2005, however, the estimated global polar bear population has risen by more than 30% to about 30,000 bears, far and away the highest estimate in more than 50 years. A growing number of observational studies have documented that polar bears are thriving, despite shrinking summer sea ice."[165]

Worried about reports of declining polar bear numbers, the US declared the animals endangered in 2008. Crockford asked for this ban to be reviewed in light of more recent data, concluding, "A thorough external review of the polar bear status issue is now required – not only because it's the right thing to do but because it may help restore public support for science and conservation."[166]

The zenith of carbon-catastrophe predictions in the future of mankind is so hopeless the ultimate and only surefire solution lies back with Ehrlich and *The Population Bomb*: depopulation. Maurice Strong, a Canadian and former CEO of Petro-Canada, organized the 1972 UN conference about the environment in Stockholm on behalf of Canada. Strong was the secretary general of the Rio Earth Summit in 1992.

In his book *Why We Bite the Invisible Hand,* author Peter Foster writes extensively about Maurice Strong after reading Strong's autobiography published in 2000, *Where On Earth Are We Going?*

> In Strong's imagined future "the human tragedy is on a scale hitherto unimagined." There have been unprecedented extremes of weather. The North American prairies are experiencing their 10th year of drought. "Water vendors with armed guards roam the streets of Los Angeles." Hundreds of thousands of people have just died in Washington from a heat wave. There are plagues of insects and rodents. Malaria has turned New Orleans into a "shrinking fortress held with only poisonous amounts of lethal pesticides." (Foster 2014)

165 "As Polar Bear Numbers Increase, GWPF Calls For Re-assessment Of Endangered Species Status," *Global Warming Policy Foundation* (February 27, 2017). https://www.thegwpf.org/as-polar-bear-numbers-increase-gwpf-calls-for-re-assessment-of-endangered-species-status/.
166 Ibid.

In Foster's view, Strong's deepest desire was to impose a dictatorship to save a (rightfully) dwindling global population. This was reflected in Strong's autobiography. "Not all is bleak," wrote Strong. "There is a good guy running Germany, a 'benevolent dictator' seeking 'to ensure all Germans work together for the common good and share equitably in 'scattered islands of sanity and order' ... The brightest prospect—according to Strong—lay in forecasts that two-thirds of the world's already diminished population might be wiped out, 'a glimmer of hope for the future of our species and its potential regeneration.' Surely somebody who regards two-thirds of the human race being wiped out as a 'glimmer of hope' betrays a very ambivalent attitude towards humanity" (Foster 2014).

Foster also writes about how Strong influenced the outcome of the 1972 and 1992 UN environment conferences by supplementing the usual political attendees with ENGOs who were more likely to articulate stronger positions than professional diplomats. Foster found an article written in the 1990s, titled "The Protest Ethic" written by British journalist John Lloyd, who credited Strong with, in Foster's words, "unleashing the NGO."

> "The UN began a series of major conferences on global issues; at the first of these, on the environment in 1992, the event's secretary general, the Canadian Maurice Strong, allowed all interested NGOs to participate." Quoting the head of an Italian news agency Lloyd continued, "At a conference where the majority of delegates were diplomats operating according to the instructions of their governments, and frequently without any emotional involvement in their actions, the arrival of a sea of people with passions and deeply held commitments upset the traditional mechanism of diplomatic consensus" (Foster 2014).

The modern ENGO—which says whatever it wants about climate change so long as the objectives appear to be altruistic—was born 27 years ago. Perhaps even 47 years ago if you include Stockholm in 1972.

The idea of saving the world by decreasing its population is not new. Faced with massive population growth and poverty, China introduced its one-child policy in 1979. The country has faced enormous international criticism for this barbaric approach to human rights, but it is consistent with the writings of Malthus and Ehrlich.

The notion that mankind is destroying the world and must hold itself account-able is quite common when you know what to look for. In 2010 there was a proposal put forth for an international UN climate conference. It was called, The Universal Declaration Of The Rights Of Mother Earth. It stated a recognition "that the capitalist system and all forms of depredation, exploitation, abuse and contamination have caused great destruction, degradation and disruption of Mother Earth, putting life as we know it today at risk through phenomena such as climate change."[167]

The message is that the personified Earth has more rights than the people who live on it. Put another way, the land, air, water, oceans, plants, and animals rank above the needs, activities, and aspirations of human beings in the greater order of things.

By now you must be starting to understand that being alive is much more complicated than you ever imagined. Not feeling guilty enough yet? Keep reading.

The clearest example of what must be done to save the world came from the city council in Berkeley, California in June 2018. Carbon taxes be damned. Depopulate. A June 13, 2018, article published by The Daily Caller reported the council unani-mously supported a resolution that read, in part, "Reversing global warming and restoring a safe and stable climate requires an emergency mobilization to reach zero greenhouse gas emissions across all sectors at wartime speed, to rapidly and safely drawdown or remove all the excess carbon from the atmosphere, and to implement safe measures to protect all people and species from the consequences of abrupt warming in the near-term."[168]

But not everything will be protected. To do this, "… requires an effort to pre-serve and restore the earth's biodiversity in interconnected wildlife corridors and to humanely stabilize the population."[169]

Wow. Humane inhumanity. But after all, there is no other choice.

This could go on for pages. The simple fact is that since the time of Malthus over 200 years ago, the idea has been growing that mankind is incapable of protecting

---

167 "Universal Declaration of Rights of Mother Earth," Global Alliance for the Rights of Nature (April 22, 2010). http://therightsofnature.org/universal-declaration/.

168 Michael Bastach, "Berkeley Declares a 'Climate Emergency,' Calls for Population Control," The Daily Caller (June 13, 2018). https://dailycaller.com/2018/06/13/berkeley-climate-emergency-population-control/.

169 Ibid.

the world we live in and our relentless pursuit of life, health, wealth, comfort, happiness, and procreation will ultimately lead to our own destruction. That we are genetically and uncontrollably programmed to do so is a human flaw that must be overcome.

And in the past 30 years, climate change caused by fossil fuels has replaced nuclear war as the greatest single cause of "mutually assured destruction" of life on Earth. It would be reckless for Albertans to not take note of what an ever-growing number of our fellow human beings believe is the problem and the solution.

12
POLITICIANS BOARD THE CLIMATE CHANGE BANDWAGON

"This created unlimited opportunities for the government to protect the environment by reducing CO2, or at least be seen by voters as reducing CO2. ... As public concerns grew, so did the potential solutions. While it isn't measured this way, going green was surely the greatest expansion of the role of government in history, with the possible exception of the implementation of public healthcare."

In 1986 American economist James Buchanan won a Nobel Prize in economics for articulating the Public Choice Theory, a combination of political science and economics. Human behaviour in the form of self-interest is a key element of economics. Writing about Buchanan and the subject for the website *Library of Economics and Liberty,* economist Jane S. Shaw wrote, "Although most people base some of their actions on their concern for others, the dominant motive in people's actions in the marketplace – whether they are employers, employees, or consumers – is concern for themselves. Public choice economists make the same assumption – that although people acting in the political marketplace have some concern for others, their main motive, whether they are voters, politicians, lobbyists or bureaucrats, is self-interest."[170]

Buchanan wrote about how the Public Choice Theory replaces the "romantic and illusory" concept that governments and politicians act for the greater good. In fact, Buchanan asserted that most people in the political process behave just like buyers and sellers in the rest of the world.

The reality today is that a growing number of voters are not interested in politics once they figure out how little impact one vote has on what happens next. Witness declining voter turnouts. Similarly, politicians are not that interested in any single voter because their objective is to get elected and stay in office and one vote doesn't count for much. Organized groups of voters, or interest groups, are another matter.

---

170 Jane S. Shaw, "Public Choice Theory," *Library of Economics and Liberty* (2002), http://www.econlib.org/library/Enc1/PublicChoiceTheory.html.

Politicians may intend to spend taxpayer money wisely. Efficient decisions, however, will neither save their own money nor give them any proportion of the wealth they save for citizens. There is no direct reward for fighting powerful interest groups in order to confer benefits on a public that is not even aware of the benefits or who conferred them. Thus, the incentives for good management in the public interest are weak. In contrast, interest groups are organized by people with very strong gains to be made from governmental action. They provide politicians with campaign funds and campaign workers. In return they receive at least the 'ear' of the politician and often gain support for their goals.[171]

Everybody knows this. What you didn't know was that explaining it in the lingo of economics was worthy of a Nobel Prize. It is better known as "influence peddling" or "vote buying." The Public Choice Theory explains why politicians are inclined to migrate to where the votes are. Spending and deficits rise because the players are not using their own money. This is how relatively small groups of people as a percentage of the population can punch well above their weight to create favourable policy and programs and generate grants.

Interest groups like ENGOs vote as a block and help politicians win elections. Politicians reciprocate with preferential treatment. The fact ENGOS claim to represent the interests of all humanity helps considerably. In this way ENGOs publicly align themselves with politicians who carry the same message.

Once again we see a key element of the politics of carbon is the demand that governments "do something" when problems emerge. When climate change became real in the minds of voters, it moved to the top of the pile to become the mother of all environmental calamities.

The environment movement had grown significantly in the latter decades of the twentieth century. But the problems to be solved were primarily regional; polluted air in urban centres (smog), dirty rivers and lakes, and industrial contamination near plants and facilities.

By comparison, global warming (later renamed climate change) was caused not by local instances of pollution, but by emissions of carbon dioxide from fossil

---

171 Ibid.

fuels. And just about everything was made with or from materials that contained or emitted CO2.

This created unlimited opportunities for the government to protect the environment by reducing CO2, or at least be seen by voters as reducing CO2. As public concerns grew, so did the potential solutions. While it isn't measured this way, going green is surely the greatest expansion of the role of government in history, with the possible exception of the implementation of public healthcare. Who in their right mind could possibly support damaging the environment or oppose protecting it?

Global warming changed more than the climate. It changed how politicians and public servants stood up for something everybody cared about. They received significant publicity for their dedication to improving the human condition, aligning their political futures with growing blocks of supportive lobbyists and interest groups, expanding their importance and departments, writing laws, introducing bills, hiring people, spending money, and saving the world—all at the same time. Perfect.

Whether or not these actions actually succeeded was secondary to the sheer political attractiveness of jumping on the green bandwagon. Just trying to do something about climate change was regarded as laudable. Even talking about it was praised.

Before you think this view is excessively cynical, be assured the intention is not to diminish the important role governments can and must play in protecting the public interest. Great things have been accomplished by government intervention to reduce pollution and protect the environment. The world is a safer and cleaner place today thanks to governments and concerned citizens.

A good example of government action to reduce pollution is the remarkable improvements in air quality from vehicle fuel efficiency. Today's problem with vehicles emitting GHGs is not the emissions per vehicle or per kilometre, but the improvement in the quantity of vehicles in service. Cleaner-burning gasoline, exhaust scrubbers (catalytic converters), elimination of polluting additives like lead, and more efficient engines have all contributed.

In 1970 the US passed the Clean Air Act. The website of the Environmental Protection Agency (EPA) reads, "Dense, visible smog in many of the nation's cities and industrial centres helped to prompt passage of the act in 1970 at the height of the national environmental movement." The act was expanded and strengthened in 1977 and 1990. "The subsequent revisions were designed to improve its effectiveness and to target newly recognized air pollution problems such as acid rain and damage to the stratospheric ozone layer."[172]

The timing was excellent. When the first oil crisis emerged in 1973, legislating that vehicles must consume less fuel became good for the air *and* the wallet. America's national speed limit was reduced to 55 MPH. Fuel-economy standards were raised. While automobile manufacturers raised concerns about how much this would cost, Japanese vehicles that virtually sipped fuel had entered the US market almost unnoticed as Detroit focused on muscle cars. Rising fuel prices and new fuel-economy standards put pressure on domestic vehicle sales, particularly when buyers had options. In the end, progress was made not only because of government. Market forces from offshore manufacturers exerted a powerful influence on the vehicles Detroit would build in the future.

Government-initiated legislation and efforts to preserve or reclaim the environment and protect it for future generations have been significant and measurable. The environment of developed countries is so vastly improved that many of initiatives already undertaken are often overlooked or taken for granted.

The air and water are cleaner. The infant mortality rate is falling and people are enjoying longer and healthier lives. Protected regions are expanding. Industrial development is being restricted. Recycling paper and carboard is saving trees. Recycling glass, plastic, and discarded metals is reducing resource harvesting and associated development impacts. Hazardous wastes are being disposed of properly and safety.

Therefore, when climate change moved to the front of the line, voters assumed that, based on past success, governments could also solve this problem. But reducing pollution from known sources in a finite geographical area is much different than replacing proven fossil fuels with new forms of energy. Pollution control is effective because it is regional and within the jurisdiction and control of the government making the decision. Climate change is global and requires the cooperation of multiple governments.

A different challenge requires different solutions completely out of the reach of local or even national governments. But people demanded action and the

---

172 "Clean Air Act Requirements and History," *United States Environmental Protection Agency* (2017). https://www.epa.gov/clean-air-act-overview/clean-air-act-requirements-and-history.

politicians would never publicly admit they couldn't solve the problem. That would cost votes.

Eliminating climate change required replacement fuels and technologies that were new, expensive, restricted, and unproven on a broad scale. For many applications, non-carbon fuels and energy sources didn't even exist, and neither did substitutes for plastics and petrochemicals. The issue proved far more complicated than anticipated, resulting in significant unintended consequences.

In Canada provincial governments were quick to jump on the file. Nobody went further, sooner, with more determination than the Liberal government elected in Ontario in 2003. By 2005 Ontario had announced its intention to eliminate coal-fired electricity generation by 2009. The official news release read, "The McGuinty government's aggressive plan to replace coal-fired generation with cleaner sources of energy and conservation will clean up our air, improve the health of our citizens, and contribute to the sustainability of our environment while ensuring a reliable supply of electricity to improve air quality."[173] Coal use ended in 2014, five years later than projected.

The government information continued, "A cost/benefit analysis released in April uncovered massive health and environmental costs from coal-fired generation. The study found emissions from all coal-fired stations were responsible for up to 668 premature deaths, 928 hospital admissions and 1,100 emergency room visits in Ontario per year. It also found that with an annual cost of $4.4 billion, coal-fired generation is significantly more expensive than other sources of electricity. The plan is receiving praise from environment and healthcare experts and regulatory agencies in charge of the reliability of Ontario's electricity system."[174]

The life-saving and -extending properties of cheap and reliable electricity are never actually enumerated in this type of cost/benefit analysis.

So dirty, polluting, dangerous (but economical and plentiful) coal would be replaced by renewables. A Liberal policy document from 2005 promised "… to bring clean, renewable energy to Ontario. … In 2003 there were only 10 windmills in the province; today we have 700 planned or in place … Contracts awarded will

---

173 Ministry of Energy, "McGuinty Government Unveils Bold Plan To Clean Up Ontario's Air," *Ontario Newsroom* (June 15, 2005). https://news.ontario.ca/archive/en/2005/06/15/McGuinty-Government-Unveils-Bold-Plan-To-Clean-Up-Ontario039s-Air.html ©Queen's Printer For Ontario, 2005. Reproduced with permission.

174 Ibid.

potentially lead to over $4.5 billion in renewable energy investment in Ontario. We have contracted to have five per cent of new generating capacity to come from renewable sources by the end of 2007, and 20 per cent by 2010."[175]

The key to coal phase out was subsidies for renewables. These were called "feed-in tariffs," whereby small generators could sell wind and solar power into the grid at higher prices that reflected the cost. "We delivered a standard offer for renewables. It will allow small local renewable energy producers to get into the energy market. We're harnessing renewable energy from the farm, small business, and consumers that can now receive credit for the excess power they produce with the net metering program we have created."[176]

In the US in 2005 the George W. Bush government introduced the Energy Policy Act. This included a wide range of initiatives ranging from loan guarantees to innovative technology to reduce GHGs. It provided support for more natural gas development by exempting hydraulic fracturing fluids from existing regulations. Additionally, it paid cash for clean coal, offered tax credits for wind power, funded geothermal and tidal power, and extended daylight savings to conserve energy.

But possibly the most significant component of the act was legislation mandating major increases in the amount of biofuel—primarily ethanol—that was required to be mixed with gasoline. Biofuel, which is gasoline made from plants or plant material, was regarded as a lower-carbon replacement for oil because the plants removed carbon dioxide from the atmosphere during growth (so did fossil fuels but that was some time ago). This had an immediate and massive impact on growing and processing corn into ethanol. Crops were replaced, trees were cut down to create farmland, and new ethanol plants opened across the US Midwest. The climate change issue was so serious that it became law that food was to be used for fuel.

In 2007 12 states, led by Massachusetts, had asked the Supreme Court to rule on whether or not carbon dioxide was a "pollutant" and if the EPA had the legal right to regulate CO_2 emissions from new cars under the Clean Air Act.

Washington's EPA, in a position supported by 10 states and several corporate trade associations representing automobile and engine manufacturers, insisted it did not have this power because CO_2 was not a declared pollutant. The regulator

---

175 *Ontario Liberals: Change That's Working* (2005),http://www.ontla.on.ca/library/repository/ monoth/10000/277162.pdf.

176 Ibid.

also believed that the linkage between global warming and GHGs was still a theory, not a scientific fact.

But the US Supreme Court determined CO_2 was indeed a pollutant and the EPA had the legal power to enforce legislated reductions. It was a remarkable decision because American people were CO_2 emitters with every breath. Taken to the extreme, the case could be made for the EPA to order people to breathe less, or to stop breathing entirely.

Relevant to Canada today is which states wanted federal authority to regulate CO_2 emissions and which did not. Those that forced a Supreme Court opinion included California, Connecticut, New Jersey, New York, Vermont, Rhode Island, Oregon, and Washington: states without major carbon resources and devoid of vehicle manufacturing. Those opposed were Michigan, Ohio, Alabama, Nebraska, North Dakota, Texas, and Utah: all carbon-resource producers or manufacturers of carbon-fuelled transportation machines.

In Canada, support for oil sands production is similarly divided provincially between producers and consumers.

In 2007 Alberta joined the climate-policy-action initiative through the New Royalty Framework, in part to deflect growing criticism of the oil sands. Using the new revenue from higher royalties, Alberta committed $2 billion to carbon-capture-and-storage projects and a CO_2 pipeline to carry, then inject, emissions captured from hydrocarbon processing into old oilfields. The province also introduced a $15-per-tonne large-emitter carbon tax. The carbon taxes were earmarked for GHG emission reductions strategies and technologies.

A year later BC introduced its Climate Action Plan with a $10-per-tonne carbon tax starting July 1, 2008. This would rise by $5 per tonne per year. This was accompanied by offsetting reductions in personal and corporate taxes to be revenue neutral. The tax was originally designed to impact consumption behaviour but not cost taxpayers or businesses any extra money.

In 2009, when Barack Obama was sworn in as US president, climate change was a growing issue in American politics. He had campaigned on doing better. In May 2009 the American Clean Energy and Security Act, also called the Waxman-Markley bill after its sponsors, was passed in the US House of Representatives. This legislated GHG emission caps and targets for 2050 and included a system for emissions-trading allowances and credit, or "cap and trade". It was rejected by the senate.

Later that year the Obama administration exercised what the press called Plan B, by formally declaring carbon dioxide a public danger. In this way the administration could enact policies to reduce GHG emissions without approval of the senate. This was timed with the latest COP climate change convention in Copenhagen at which the Obama administration could display its "green creds" on the global stage.

On December 7, 2009, news website *The Daily Signal* reported "EPA Formally Declares CO2 a Dangerous Pollutant." Noting that "85 percent of the US economy runs on fossil fuels that emit carbon dioxide," the article predicted significant costs for the US economy. The article quoted noted American columnist George Will, who wrote that the emission targets were unattainable. "Barack Obama, understanding the histrionics required in climate change debate, promises that US emissions in 2050 will be 83 percent below 2005 levels. If so, 2050 emissions will equal those of 1910, when there were 92 million Americans. But there will be 420 million Americans in 2050, so Obama's promise means that per capita emissions then will be about what they were in 1975. That. Will. Not. Happen."[177]

By 2009 the EPA had also changed its messaging from two years earlier. Lisa Jackson, head of the EPA, declared, "Climate change has now become a household issue. This administration will not ignore the responsibility we owe to our children and our grandchildren."[178] New EPA regulations being considered at that time would apply primarily to large emitters releasing more than 25,000 tonnes of CO2 per year.

Conservative news website *The New American* ran the headline, "EPA Declares Human Breath (CO2) a Pollutant" reporting, "In fact, about 8 percent of man-made worldwide carbon dioxide emissions are due to simple human breathing. The EPA says they do not want to regulate this activity ... for now. But there's

177 Nicholas Loras, "EPA Formally Declares CO2 a Dangerous Pollutant," *The Daily Signal* (December 7, 2009). https://www.dailysignal.com/2009/12/07/epa-formally-declares-co2-a-dangerous-pollutant/.

178 Suzanne Goldenberg, "US Climate Agency Declares CO2 Public Danger," *The Guardian* (7 December 2009). https://www.theguardian.com/environment/2009/dec/07/us-climate-carbon-emissions-danger.

no chemical difference between CO2 emitted from a gasoline engine and that emitted from a human lung."[179]

In 2010 Canada embraced the biofuel solution for climate change by introducing the Ethanol Expansion Program "to increase domestic production and use of ethanol, a renewable transportation fuel, and reduce transportation-related greenhouse gas emissions." The program including financial support for ethanol production facilities. As of December 15, 2010, federal law required gasoline contain 5% ethanol.

By 2010 it was also becoming apparent that the growth of ethanol fuel production was directly linked to world food prices. A 2010 article in *The Guardian* reported in 2009 US Department of Agriculture data indicated one-quarter of US grain crops were now being used as automotive fuel, not food. Lester Brown of Washington's Earth Policy Institute estimated "The grain grown to produce fuel in the US (in 2009) was enough to feed 330 million people for one year at average world consumptions levels." America at one point was the source of 25 percent of the global maize (corn) exports.[180]

Brown stated, "Continuing to divert more food to fuel, as is now mandated by the US federal government in its renewable food standards, will likely only enforce the disturbing rise in world hunger. ... The worst economic crisis since the great depression has recently brought food prices down from their peak, but they will remain well above their long-term averages." The article continued, "The US is by far the world's leading grain exporter, exporting more than Argentina, Australia, Canada, and Russia combined. In 2008, the UN called for a comprehensive review of biofuel production from food crops."[181]

This problem had been identified two years earlier. A *Reuters* article July 28, 2008 was titled, "Biofuels Major Drive of Food Price Rise: World Bank." It opened, "Large increases in biofuel production in the United States and Europe are the main reason behind the steep rise in global food prices, a top World Bank economist said in research published Monday." Obviously controversial, the allegation

---

179 Thomas R. Eddlem, "EPA Declares Human Breath (CO_2) a Pollutant," *The New American* (April 20, 2009). https://www.thenewamerican.com/tech/environment/item/6685-epa-declares-human-breath-co2-a-pollutant.

180 John Vidal, "One quarter of US grain crops fed to cars - not people, new figures show," *The Guardian* (January 22, 2010). https://www.theguardian.com/environment/2010/jan/22/quarter-us-grain-biofuels-food.

181 Ibid.

was supported by an earlier International Monetary Fund (IMF) study "... which estimated in May that biofuels accounted for 70 percent of the increase in maize prices and 40 percent in soybean prices ... the use of maize for ethanol in the United States has global implications because the US produces about one-third of the world's maize and two-thirds of global exports, and used 25 percent of its production for ethanol in 2007/08."[182]

According to an annual global food price index published by the Food and Agricultural Organization of the United Nations (FAO), food prices began to rise significantly in 2006 and by 2008 the index was nearly double the average for the 2002 to 2005 period. Prices moderated in 2009 thanks to the global economic collapse, but by 2011 were 120% higher than early in the previous decade. There were a lot of factors at play. There always are. For example, oil prices, important for transportation, peaked in 2008 at US$147 per barrel, double the levels of a few years earlier.

In early 2011, the Middle East experienced what would be known as the "Arab Spring," open rebellion that either overthrew or destabilized the governments of Tunisia, Egypt, Yemen, Syria, and Libya. Most of these countries have been in turmoil ever since. By fall, American news network *PBS* ran a story titled, "Did Food Prices Spur the Arab Spring?" after rising food prices were identified as one of several trigger points. An article by *Scientific American* linking food prices to the Arab Spring in 2013 concluded higher food prices were exacerbated by "... droughts in Russia, Ukraine, China and Argentina and torrential storms in Canada, Australia and Brazil – all major wheat and grain producers."[183]

Scientific American also blamed climate change but made no direct linkage to biofuels or the reduction in US maize exports. But since the biofuel policy was in place to mitigate climate change and others had warned this was a problem three years earlier, perhaps the author didn't have to. In 2011 *The Guardian*, under the headline "Biofuels Are Driving Food Prices Higher" wrote, "Rising food prices contributed to the uprising in Tunisia and Egypt and sparked riots in several other countries." Citing World Bank sources, the article states, "... between June and

---

182 Lesley Wroughton, "Biofuels major driver of food price rise: World Bank," *Reuters* (July 28, 2008). https://www.reuters.com/article/us-worldbank-biofuels-idUSN2849730720080728.

183 Inez Perez, "Climate Change and Rising Food Prices Heightened Arab Spring," *Scientific American* (March 4, 2013). https://www.scientificamerican.com/article/climate-change-and-rising-food-prices-heightened-arab-spring/.

December 2010 an additional 44 million people fell below the poverty line due to rises in food prices." World Bank President Robert Zoellick urged the world to "put food first."[184]

Americans spend about 6% of their income on food. In Egypt it is nearly 43%. Combatting global warming with renewables was clearly not as simple as reducing tailpipe emissions from vehicles in Los Angeles.

Ontario's bold and aggressive conversion from coal to renewables would have the same impact on electricity prices. Fortunately, its citizens were much better equipped to absorb the price shock. While Ontario did not experience open revolution, it did make major changes at the ballot box in the June 7, 2018 election. The Liberal government, which introduced all the climate policies, was replaced by the Progressive Conservative Party, which had campaigned in part on a platform to undo many of the Liberal policies. The Liberals suffered their worst electoral thumping in history and lost official party status after winning only 7 of 124 seats.

That the 2005 Liberal energy policy overhaul was in trouble was apparent years earlier. Issues and complaints escalated as electricity prices rose more than anticipated. Renewable-project costs were soaring and the negative impact on business and consumers was measurable. Plus the promised job benefits were not materializing. By 2013 one of the agreements with Samsung for wind and solar power was off the rails as the province backed off on purchase commitments and Samsung advised it would invest less than previously planned. The Progressive Conservative opposition declared the Liberal green energy strategy was a "colossal failure" and the cause of skyrocketing electricity prices.

In November 2016, Ontario's auditor general issued a report revealing that since 2006 the program had cost $37 billion more than planned ($9.2 billion of the increase was for green energy projects) and would cost an additional $133 billion from 2015 to 2032. A *Global News* article backed this up saying electricity prices for homes and small businesses in Ontario had jumped 70% between 2006 and 2014 as coal was being phased out.[185]

---

184 Tim Rice, "Biofuels are driving food prices higher," *The Guardian* (June 1, 2011). https://www.theguardian.com/global-development/poverty-matters/2011/jun/01/biofuels-driving-food-prices-higher.

185 Keith Leslie, "Ontario Advises Other Provinces to Keep an Eye on Hydro Bills as Coal Phased Out," *Global News* (November 22, 2016). https://globalnews.ca/news/3081125/ontario-advises-other-provinces-to-keep-an-eye-on-hydro-bills-as-coal-phased-out/.

After viewing the report, the Liberal government warned other provinces to be careful as they switched electricity sources on a broad scale. Coal at one time supplied 20% of Ontario's power. Noting Alberta, Saskatchewan, New Brunswick, and Nova Scotia burned coal for electricity, "Ontario Energy Minister Glenn Thibeault urged those provinces ... to make sure they have a diverse supply mix – wind, solar, biomass and nuclear power – to replace coal-fired generation."[186] Of the four provinces cited, only New Brunswick has a nuclear reactor.

In the spring of 2016, Ontario Premier Kathleen Wynne had announced a new $7-billion climate change plan that, according to a *Globe and Mail* news report, "... will affect every aspect of life – from what people drive to how they heat their homes and workplaces – in a bid to slash the province's carbon footprint."[187] It also targeted that by 2025 12% of the vehicles sold would be electric. Financing the program would come from the pending cap and trade system, a form of carbon tax. Electric vehicles would receive cash rebates of up to $14,000 and $1,000 to install a home charging system.

A new Tesla would qualify for the $14,000 rebate. Depending on the model, a Tesla could cost well over $100,000 in Ontario. In 2016 the average annual family income in Ontario was about $80,000, meaning only the very wealthy would be putting a new Tesla in the driveway. Using everybody's tax dollars to make a Tesla more affordable to the province's wealthiest citizens seems like a peculiar way to address global climate change.

Stung by the auditor general's report and with an election looming in the next year, the government announced that in March 2017 it would slash electricity rates by up to 25% to make electrical power more affordable. When reviewed by the auditor general in the fall, it was revealed the government had monetized $26 billion of future power-purchase commitments by converting them into 30 years of long-term debt through Ontario Power Generation, a crown corporation and electricity generator. A *CBC News* report on October 17, 2017 stated, "The auditor

---

186 Ibid.

187 Adrian Morrow and Greg Keenan, "Ontario to spend $7-billion on sweeping climate change plan," *The Globe and Mail* (May 16, 2016). https://www.theglobeandmail.com/news/national/ontario-to-spend-7-billion-in-sweeping-climate-change-plan/article30029081/.

has already expressed concerns that the government is not showing that $26 billion on its books, making the province's bottom line look healthier than it really is."[188]

Poof. Gone. Twelve years of poorly conceived and unsuccessful renewable energy policy off the electricity bills and, at least optically, off the government's books.

As the reality of Ontario's failure was unfolding, Alberta's new NDP government introduced its Climate Leadership Plan in November 2015, just before Premier Notley and a large delegation attended COP 21, the Paris Climate Change conference. The background materials for the conference began by reminding delegates of the enormity of the problem the world faces with, "increasing temperatures, rising ocean levels, more frequent droughts, floods and forest fires."

Alberta's plan was multi-faceted. Coal-power generation would be phased out by 2030 and by that date 30% of the province's electricity would come from solar, wind, and hydro. Like Ontario, Alberta framed burning coal as harmful to Albertans today and to the world in the future. In a document titled "Phasing out Coal Pollution," the government wrote, "Poor air quality has been linked to a number of health condition including chronic and acute respiratory diseases, ... heart disease, ... and stroke and diabetes."[189] Apparently ridding Alberta of coal would prevent 600 premature deaths, 500 emergency room visits and nearly $3 billion in negative healthcare outcomes.

The plan called for a broad-based carbon tax would to be introduced January 1, 2018. Further, existing large-emitter carbon taxes would increase, and oil sands GHG emissions would be capped at 100 million megatonnes per year, about 50% higher than current output. For reasons that have not yet been fully grasped by Albertans who watched the televised announcement, the CEOs of four major oil sands operators—Suncor, Cenovus, Shell and Canadian Natural Resources— appeared on TV with Premier Notley in support of the plan. Sixteen months later Shell sold its oil sands operations and exited the bitumen-extraction business.

The Climate Leadership Plan was introduced one year after the world price of oil had collapsed. Shortly after being elected in May 2015, the government followed through with its election promises and raised corporate taxes and personal taxes

---

188 Mike Crawly, "Ontario auditor examines Liberal plan to cut hydro bills," *CBC News* (October 17, 2017). https://www.cbc.ca/news/canada/toronto/ontario-hydro-bills -auditor-general-report-1.4356744.

189 "Phasing Out Coal Pollution," *Alberta.ca* (2019). https://www.alberta.ca/climate-coal- electricity.aspx.

on higher income earners. Oil prices continued to fall through early 2016 with the bottom being 75% below mid-2014 levels. In the 2015/16 fiscal year, the NDP's first year in government, the deficit was $10.8 billion, the highest in history. In 2016/17 it declined to $8.0 billion, the second highest in history.[190]

Alberta's Climate Change Leadership Plan included free energy-efficient light bulbs for Albertans to demonstrate how to conserve electricity, and low-volume shower heads to use less water. These elements of the NDP plan were widely ridiculed.

In the fall of 2015 a federal Liberal government under Justin Trudeau was elected. That government also raised corporate taxes and income taxes for high income earners. After attending COP 21 in Paris with a large delegation, Trudeau's administration pledged to meet its Paris 2030 GHG-reduction commitments. A national carbon tax would be introduced effective January 1, 2019. Any province that did not its own in place would be stuck with the federal program.

This was only a sample of the various policies and initiatives introduced in the US and Canada this century. Europe was early to embrace renewables but has since cut back subsidies. Countries like France are announcing long-term plans to phase out fossil fuel-powered vehicles. But the reality is all Western countries have made aggressive commitments to "do something" for decades and virtually all have fallen short or changed plans due to rising costs and consumer resistance.

All these policies and programs have been introduced by politicians who were undoubtedly well intentioned. All have been well received by those pushing for climate change-mitigation action.

Starting in 2016, several foreign oil companies sold all or most of their Canadian oil sands assets and exited the business. This included Statoil from Norway, and Arkansas-based Murphy Oil, which sold its stake in Syncrude, Shell, Marathon, and Conoco-Phillips. The total value was estimated at over $30 billion.

In November 2016 Republican Donald Trump was elected US president. He immediately criticized most of the climate change commitments of the previous administration. Trump revived the Keystone XL pipeline, reversed policies at the EPA, stimulated the coal industry, and publicly announced his intention for the US to withdraw support for the Paris GHG-reduction commitments of the prior year. He also cut corporate taxes and introduced accelerated write-offs for investment.

---

190 Government of Alberta, Ministry of Treasury and Finance Annual Reports, https://www.alberta.ca/government-and-ministry-annual-reports.aspx#toc-0.

The combination of continued low oil prices, increased corporate tax rates, carbon taxes, and added regulations motivated CAPP to make a submission to the federal government prior to the 2018 budget to highlight Canada's sagging competitiveness on the world stage as an attractive destination for investment capital. Industry trade associations like CAPP usually prefer to keep a low profile and deal directly with governments behind the scenes. Noteworthy is that in this case it was also released through the media.

But as the old Bob Dylan song goes, "When you ain't got nothing, you got nothing to lose." CAPP highlighted the growing costs of operating in Canada compared to other jurisdictions, the more attractive tax regime in the US, and the inability to gain low-cost market access to growing oil production through new pipelines.

By May 2018 the gloves were off. Respected University of Calgary economist Jack Mintz wrote a column in the *Financial Post* that challenged the federal government, and Finance Minister Bill Morneau, to admit Canada has a major competitiveness issue regarding new investment. "Our current level of business investment ... is lower than business investment for every quarter going back to mid-2012. Intellectual property investment ... is the lowest since 2005, except during the recessionary crisis of 2008-09, when it was comparable."[191]

Part of the problem, according to Mintz, is the common political tendency to replace hard cold numbers with political dogma. "It is often claimed by the left – and unfortunately publicly repeated by the Minister of Finance – that Canada's reforms to our business taxes has no effect on investment. Not only does this undermine the most credible international scholarly studies (including an excellent paper published by Finance Canada) it also goes against experience."[192]

Recovery in Alberta's oil industry should go hand in hand with higher oil prices but money generated from production isn't being reinvested here. ARC Energy Research Institute reported in January 2019 that after-tax cash flow from production—the amount of money available for reinvestment after taxes, royalties and

191 Jack Mintz, "Jack Mintz: End the denial and admit it: Canada has a competitiveness problem," *Financial Post* (May 9, 2018). https://business.financialpost.com/opinion/jack-mintz-end-the-denial-and-admit-it-canada-has-a-competitiveness-problem.
192 Ibid.

operating costs—in 2016 was only $23 billion,[193] the lowest in 10 years and $13 billion less than during the world financial collapse of 2009.

For 2018, ARC had estimated the cash flow figure figures had more than doubled to over $47 billion.[194] This is by any measure a massive recovery. But even with another $24 billion to reinvest, producers aren't spending it in Canada. The "reinvestment ratio," the percentage of available cash flow being reinvested in Canada, is only 0.86 or 86% of cash flow.[195] This is the lowest in a decade and well below the 2009 to 2017 average of 1.28 or 128%.

But what is remarkable about these figures is how much they declined in only four months. From July 16, 2018, to the end of 2018, estimated revenue from production had declined from $125 billion[196] to $110 billion and after-tax cash flow from $59 billion[197] to $47 billion. According to ARC, this had been caused by the collapse of Canadian oil prices, commencing in October 2018 due to overproduction, restricted transportation capacity, and the temporary maintenance shutdown of US refineries engineered to process Canadian oil sands production.

However commendable the intentions of politicians were in responding to the climate change challenge, their policies have resulted in unintended consequences, higher-than-anticipated economic costs, and significant taxpayer resistance. This is evidenced by elections in the US and Ontario, court challenges by provinces against federal carbon taxes, and the vow of the opposition in Alberta to undo much of the Climate Leadership Plan if elected.

The purpose of this section of the book is to review the history of carbon politics and policy and illustrate that local governments cannot solve global problems.

This is not to say reducing global GHG emissions cannot be achieved.

But when the next politician campaigning at your door promises to change the weather if elected, you too should become a climate change skeptic ... of a different variety.

---

193 ARC Energy Research Institute, *ARC Energy Charts*, November 19, 2018.
194 Ibid.
195 ARC Energy Research Institute, *ARC Energy Charts*, January 7, 2019.
196 ARC Energy Research Institute, *ARC Energy Charts*, July 16, 2018 and January 7, 2019.
197 Ibid.

Section 3

CARBON FUTURE

Introduction
THE POLARIZATION OF DIVERGENT VIEWS
FOR FUTURE FOSSIL FUELS

"... for the billions of people in the world still waiting to join modern civilization with access to clean water, proper sanitation, enough food, medicine, education, and perhaps even electricity, the last thing to worry about is what the temperature may or may not be in 2050 or 2100. A solution that does not include everyone is no solution at all."

In its history mankind has faced great challenges and has managed to overcome most of them. As it is framed today, climate change could be the most significant. In the absence of the great crises of the recent past—such as the First and Second World Wars and the Great Depression—perhaps climate change looms largest by default.

Once thing is certain: the discussion about the future of the planet is decidedly one-sided. Virtually nobody says, "Things are great and sure to get better" even if the weight of a significant amount of evidence supports this view. So much of modern society's focus is on what is going wrong, that people don't spend enough if any time to reflect upon, or even notice, the many things that are going right.

By every quantitative measure the collective well-being of the human species is steadily improving and has never been better. There are fewer wars and less death due to armed conflict. People live longer and healthier lives. Diseases that once killed many have been controlled or eradicated. The percentage of the population that is literate and educated is rising while the number who live in poverty or go hungry is declining. The nuclear Armageddon that threatened the planet following World War II never materialized.

However, doubts remain. What mankind has accomplished had once been declared impossible by prognosticators such as Malthus. No matter how many problems we solve or challenges we overcome, our worst enemy remains ourselves. Our confidence in our ability to sustain ourselves remains in question despite all the evidence things usually work out better than that we thought they would at the time.

This is not due to a shortage of advice. It come from all directions all the time.

Which brings us to the challenge of addressing climate change, the latest human-induced disaster that we're assured will lead to massive dislocation, starvation, and the end of life as we know it. What makes climate change so daunting is the most recently proposed cures appear to be worse than the disease. The only current prescription is the immediate and significant reduction—and ultimate elimination—of fossil fuels as the world's primary source of energy. We cannot add any more GHGs to the atmosphere or the temperature will continue to rise. Many say it is already too late.

But at present there is no substitute for carbon, as the following chapters will evidence. No alternative fuel for airplanes, ships, trains, or big trucks. No cheap and reliable way to generate electricity anywhere without a water supply on a small or large scale, 24/7/365. No other source of plastics, petrochemicals, and a myriad of ubiquitous and invaluable items from carbon resources.

The good news, we're told, is the end of fossil fuels may halt climate change before the world gets any warmer. We're assured this will solve many problems.

The generally unacknowledged bad news is that in the absence of affordable and reliable alternatives, billions will suffer. Ground every airplane, dock every ship, park every truck, and stop every train tomorrow and the economy will collapse. Civil disobedience and mass starvation are assured, a societal apocalypse of a scale once confined to science fiction movies. This is an outcome the climate change alarmists never mention as they continue to make disingenuous claims the world can run on renewable energy sources like wind, solar and biomass. Tomorrow. All it takes is enough politicians with the spine to make a few tough decisions.

The battle over climate change has taken some interesting directions. It has overshadowed the progress man has made in other areas. Since most people appear reluctant to voluntarily reduce carbon consumption, there are initiatives to squeeze fossil fuel producers through the capital markets. But following the money in these efforts has revealed some remarkable twists.

Climate change has become intertwined with other issues not necessarily focused on solving the problem. This includes the inequality of the economic progress of developed and underdeveloped countries, and as a proxy for the long-overdue overthrow of a free market-based financial system. There is big money to be made in replacing fossil fuels, particularly if governments can be persuaded to make carbon more expensive through legislation while simultaneously subsidizing low-carbon alternatives.

Taxing carbon has been identified, at least in theory, as the most effective way of changing consumer consumption. But support for financial penalties to change the climate shrinks considerably when people must pay them. Carbon taxes are shaping up to be one of Canada's number one political issues in 2019.

Whatever Alberta claims to be, or whatever people think it should be, modern Alberta was built on carbon and remains that way today. An upcoming chapter revisits the impact on coal, oil, and natural gas on provincial finances and what they mean today and tomorrow. If the whole world must look at the carbon economy differently, how can Alberta not do the same?

There is sufficient evidence that the world is warming and GHGs are a factor. Therefore, something should be done. At a minimum, mankind should be clever enough to not make a bad situation worse.

There are many advantages to not having the world warm much more than it has. A lot of people depend upon winter snow and melting glaciers for their water supply. It would be very useful if the glaciers could continue to be reloaded in the winter so they could supply water in the summer, as has been the case for centuries. Millions live near oceans. Not having the polar ice caps melt and raise the levels of the seas to the point that people are dislocated is surely a worthwhile endeavour.

That mankind can change the climate of the future remains to be seen. In a world that still cannot accurately predict tomorrow's weather, a lot of what is stated about this phenomenon seems a trifle far-fetched.

But reduction of all emissions into the atmosphere is pure common sense.

Since the UN's first environment conference in Stockholm in 1972, there have been 30 international gatherings at which global warming has been raised as a serious issue; seven United Nations summits and 23 COPs. The IPCC has issued five reports since 1990, with each successive report more convinced that climate change is real and GHGs are the primary cause.

Not one of these gatherings has ended without near-unanimous commitment among hundreds of delegates representing nearly 200 countries that action must be taken. But, as will be explained in Chapter 1 of this section, nothing material has changed. Current forecasts indicate fossil fuel production and consumption and GHG emissions will be significantly higher in 2040 than today.

The vocal segment of an increasingly anxious public demands their governments deal with the problem of climate change. Governments across the developed world have for most of this century spent billions if not trillions of dollars "doing something." Virtually every politician running for public office, from municipal council to the leader of a country, pays homage to climate change action, replete with a vast assortment of programs, policies, subsidies, grants, taxes and incentives to help solve the problem.

But the collective impact of all the conferences, all the commitments, all the speeches, all the promises, all the money, and all the collective anxiety has been zero, *zero* defined as no measurable reduction in year-over-year global GHG emissions. One can argue that these efforts have slowed the rate of increase, which is good. But in the absence of a major energy technological breakthrough that at the time of writing does not exist, fossil fuel consumption and GHG emissions are forecast to rise for the next 20 years.

Millions of people in the Western world are "virtue signalling" by saying the right things. The more committed buy carbon offsets, solar panels or, if they don't have to drive very far, an electric vehicle. Some ride bikes or take public transit.

But for the billions of people in the world still waiting to join modern civilization with access to clean water, proper sanitation, enough food, medicine, education, and perhaps even electricity, the last thing to worry about is what the temperature may or may not be in 2050 or 2100. A solution that does not include everyone is no solution at all.

This section concludes with the non-debate on climate change. It is polarized to the point of paralysis. If climate change could be eliminated by fear, rhetoric, and posturing, it would be resolved by now. It ends with a real plan to reduce carbon emissions over a realistic period of time. It won't be attractive to many of the current vested interests, but it will solve the problem. If, of course, that's what people truly desire.

At some point, those who really care about climate change must take an objective look at what has been promised versus what has been accomplished and start looking for Plan B, something that might actually work.

1

THE FUTURE OF CARBON

"An increasing proportion of the population regards climate change forecasts as absolutely true. This may be the only forecast in existence where the certainty of the outcome is unquestioned by so many. The only exception is death and taxes, and climate change promises more of both. Even those who think the climate predictions are likely wrong are nervous in case they are right"

When people comment, read, think, or talk about the future they should be careful. Because when you're dealing with something that hasn't happened yet and is subject to a myriad of unknown circumstances and variables, the probability of being wrong is very high. As the old saying goes, the only things certain are death and taxes.

Regardless, the world demands and depends upon forecasts and predictions. It is an integral part of life in the twenty-first century. Those who don't plan for the future—be it individuals, families, businesses, or governments—are regarded as reckless. Budgets. Risk analysis. Probabilities. Even a wrong forecast is considered better than no forecast.

An underlying characteristic of human behaviour is the desire to know about things that haven't happened yet, the "unknown" in the well-understood "fear of the unknown." Worrying about what might go wrong is a life-preserving element of the human condition. What lurks around the corner? Behind that tree? Under that rock? People who "think ahead" are generally regarded as wise. The *c'est la vie* or "that's life" crowd—those who roll with the punches life throws at them—are regarded as taking greater and often avoidable risks.

Anyone who can accurately forecast anything is sought out, idolized, and often becomes very wealthy. Being a fortune teller using a crystal ball, tarot cards, or tea leaves has been a paying profession for years. "Seers" are consulted by more people than will admit to their friends and relatives. Religions have relied upon prophets for centuries. This path leads to heaven, the opposite to hell. The future is certain; the path is up to you.

Forecasting has become a science of sorts. The veracity of predictions of future events is awarded more credibility when developed from a foundation of historical data. The forecasting process doesn't permit input variables on future outcomes not supported by something that hasn't happened yet. You can't include something that hasn't existed before or doesn't yet exist, no matter what it is.

When a forecast is released, those affected pay attention, wondering how it will impact them. But when a forecast is wrong, most simply shrug their shoulders. After all, nobody can predict the future.

If we accept forecasts as an essential part of human existence, that they are entirely about things that haven't happened yet, and the rest of this book is about the future, here are a few nuggets from the past to put your brain in the right frame of mind.

Nobel laureate Nils Bohr said, "Prediction is very difficult, especially if it's about the future."[198] Centuries ago William Shakespeare wrote, "If you can look into the seeds of time, and say which grain will grow and which will not, speak then unto me."[199] Mark Twain added, "Prophesy is a good line of business, but it is full of risks."[200] Writing in *Newsweek* magazine in 1966, famous economist Paul A. Samuelson commented, "Wall Street indices predicted nine of the last five recessions!"[201] Warren Buffet's contribution was, "Forecasts may tell you a great deal about the forecaster; they tell you nothing about the future."[202]

An increasing proportion of the population regards climate change forecasts as absolutely true. This may be the only forecast in existence where the certainty of the outcome is unquestioned by so many. The only exception is death and taxes, and climate change promises more of both. Even those who think the climate predictions are likely wrong are nervous in case they are right.

198 Niels Bohr, "Niels Bohr Quotes," *BrainyQuote* (2019). https://www.brainyquote.com/quotes/ niels_bohr_130288

199 William Shakespeare, *Macbeth*, 1.3.59–61. https://www.sparknotes.com/nofear/shakespeare/ macbeth/page_14/

200 Mark Twain, "Prophesy Quotes," *BrainyQuote* (2019). https://www.brainyquote.com/topics/ prophesy.

201 "Entry from December 13, 2008: 'Wall Street indexes predicted nine out of the last five recessions' (Paul Samuelson)," *The Big Apple* (December 13, 2008). https://www.barrypopik.com/index.php/ new_york_city/entry/wall_street_indexes_predicted_nine_out_of_the_last_five_recessions.

202 Viktor Reklatis, "5 Quotes that Tell You Everything You Need to Know About Forecasting," *MarketWatch* (March 9, 2017). https://www.marketwatch.com/ story/5-quotes-that-tell-you-everything-you-need-to-know-about-forecasting-2017-01-11.

The climate change models most commonly cited are the multi-decade work of the IPCC, created in 1998 in part by the UN. We're repeatedly told this is the work of thousands of climate scientists from around the world who have analysed the future of the climate relative to past and current conditions and are in near-unanimous agreement the foundation of their models is sound and their view of the future is accurate. The mantra is, "the science is settled." Those who question the IPCC climate models and forecasts are called "deniers." This was discussed in the previous section.

If you think this has religious overtones, you're absolutely right.

For proof of the veracity of and confidence in the IPCC forecasts, the Government of Alberta provides an excellent example. On its official website "Climate Change in Alberta." the province mixes some history and the future stating:

> The climate is changing and globally we are experiencing impacts, such as:
>
> - Increasing temperatures
> - Rising ocean levels
> - More frequent droughts, floods and forest fires
>
> More extreme weather is creating greater challenges ... 97% of climate scientists now agree that human activity is responsible for most temperature increases over the past 250 years. ... Climate change will likely result in long-term changes in temperature and precipitation, as well as increased frequency and severity of weather events such as droughts, floods, forest fires, and severe storms.

The Government of Alberta wouldn't lie, would it? Mislead its citizens? This message is serious business. Every problem the world has with the climate and weather is certain to get worse. Except the government snuck in the word "likely" because, after all, it is about something that hasn't happened yet. Careful examination of all that is written about climate change and future weather events from responsible sources (defined as in some way accountable such as governments, mainstream media, academia, research entities, and corporate economic forecasters) will reveal most contain words such as "likely," "believe," "theorize," or "speculate."

However, the myriad of unaccountable commentators such as ENGOs, environmentalists, climate activists, or oil-haters are not bound by convention to qualify their commentaries. They talk about the future like it has already happened; not *if*, but *when*. This has even spilled over into the financial and legal arena, as will be discussed in an upcoming chapter. The IPCC forecasts may be the only predictions in history used as a basis for litigation in which plaintiffs are seeking damages from fossil fuel producers for climate change damages that haven't happened yet.

IPCC climate change models are widely accepted as credible because predictions about the future are based on data from the past. While the invention of the thermometer dates back several centuries, accurate temperature readings to make comparisons to today only date back to the late 1800s. Accurately measuring CO_2 concentrations directly in the atmosphere is much more recent. The so-called "Keeling Curve," which tracks atmospheric CO_2 from the top of a volcano in Hawaii only dates back to 1950.

Much of the historic temperature and CO_2 data used to model future climate change from rising CO_2 levels is imputed from other sources, such as ice cores, tree rings, and lake-bottom sediments. If you were trying to raise money in the private sector using this type of historical data to build a future business case, you'd be laughed out of the room and possibly arrested.

But the purpose of this book is not to dispute climate change forecasts from accountable sources. Again, "the science is settled." But to better to understand the enormity of the challenge we must also review forecasts of carbon-resource consumption and GHG emissions. These are also heavily modelled on history. As with climate change data, the statistics upon which energy-consumption forecasts are based have been improving continuously in the last few decades. And it is not necessary or possible to go back several million years to build a case for what is likely to happen in the next 25.

There are multiple professional forecasters prognosticating about the future of energy and fossil fuels. They include multinational oil giants Royal Dutch Shell, BP, and ExxonMobil. Their work is widely regarded as credible. They are publicly traded, broadly owned, and heavily regulated. It is illegal for these companies to knowingly release information that is false.

Recent litigation in the US by governments suing ExxonMobil has been based upon allegations the company withheld information that it knew its products caused climate change and this was concealed for commercial gain. The basis of the lawsuits is that ExxonMobil did not disclose to shareholders material information that could negatively affect the future value of their investments. Public disclosure by broadly held, publicly traded companies is serious business.

Two non-commercial agencies that study this subject are the International Energy Agency (IEA) and the Energy Information Administration (EIA). Based in Paris, the IEA was formed in the mid-1970s during the global oil crisis. It is supported by 30 member countries with a mandate to research energy in an objective manner and disseminate information as it is created. Based in Washington, DC, the EIA was formed in the US in 1974 for the same reason; ensure the American government, industry, and citizens had the best information possible about the essential and, at that time, volatile world of energy.

IEA and EIA data are regularly used by the global investment community, governments, and policy-makers to shape decisions, write policies, and respond to events in domestic and global energy markets. Each week or month when these organizations release information, they affect global commodity prices and the value of companies in the energy business. Serious stuff undertaken by serious people.

All five of these entities rub their computerized, data-driven crystal balls and peer into the future, generally out 20 to 30 years. They all have released public information to 2040 and beyond. They don't always agree. Following is a summary of the general findings, with the exceptions noted. Be assured this information is collected, sorted, and analysed with a staggering degree of granularity and accuracy. No extrapolated data from tree rings, fossils, ice cores, or lake-bottom sediments are used or permitted.

The fundamental drivers of future energy demand are population and economic growth. ExxonMobil, for example, sees world GDP in 2040 double that of 2010. Its economic growth forecast sees OECD countries (Organization for Economic Cooperation and Development, the 34 countries widely considered to be "developed" in Europe, North America, and the wealthiest countries in South America and Asia) growing at an average of 1.9% per year through to 2040 while the rest of the world's annual growth rate is more than twice that at 4.1%. All five forecasters see total energy consumption in 2040 being as much as 50% higher than in 2010 and 30% higher than 2015 due to these key demand drivers.

The growth in energy demand will not come from OECD member countries. They already have most of their energy and transportation infrastructure in place;

electricity grids to supply virtually every building and dwelling, roads, railways, public transit, and airports. Only a few, like Austria, Luxemburg, and Switzerland, don't have ocean access via a port. These countries have electricity on demand, heat in the winter, air conditioning in the summer, a gas station within a few kilometres, and airline service to the rest of the world. Few people freeze, fewer starve.

In 2016 OECD members had a combined population of 1.28 billion while the world's population was 7.4 billion. OECD countries account for only 17% of the population.

Energy demand growth will come from the rest of the world where the other 83% live, about 6.2 billion people. Big countries that are not OECD members include Argentina, Brazil, China, Columbia, Costa Rica, India, Indonesia, Russia, Saudi Arabia, and South Africa. They have a total population of 3.48 billion or nearly 47% of the world's people. The other non-OECD countries are the rest of Africa and the Middle East, most of South America, and the other countries of Asia, except South Korea and Japan, which are OECD members.

These are the "have-not countries" that in many ways are playing energy catch-up with Western nations. In the US there is a car or light truck for almost every citizen, almost 300 million vehicles. China, with more than four times the population, has only half that many vehicles. The World Economic Forum estimates about 1.1 billion people, almost 15% of the world's population, still don't have electricity. That's almost as many people as those who live in all 34 OECD member countries. There are only 210 million cars registered in India for a population of 1.3 billion people. India is widely regarded as being one of the world's fastest growing markets for all types of energy as it modernizes and industrializes. Canada, with a population of 36 million, has about 22 million vehicles on the road that weigh less than 4,500 kilograms.

The EIA demand-growth forecast is geographically granular. It sees the Americas, Europe, and Eurasia flat from 2015 to 2040. Asia will see its energy needs grow by 50% from 2015 in the 25 years to 2040, accounting for half of the total demand increase of 29%. The other areas requiring more energy are Africa and the Middle East.

The composition of the anticipated global-energy-demand growth varies somewhat, but overall the role of carbon energy remains very strong. BP, for example, predicts less than 30% supplied by non-fossil fuels in the form of renewables or hydro-electricity or carbon-free nuclear power. Coal as a percentage of the total energy and fossil fuel mix will decline as it replaced by natural gas, or more electricity is generated by renewables such as wind and solar. All models anticipate a growth in natural gas as a replacement for coal and, in some countries, for nuclear power.

The elephant in the fossil fuel room is crude oil and its pivotal role in petrochemicals and transportation.

One-quarter of crude oil production is used as feedstock for petrochemicals, which are used in products such as plastics and chemicals essential to everyday life. There is no substitute at this time. Demand for oil as petrochemical feedstock will grow with the economy and transportation.

About 70% of oil is used as transportation fuel in three main areas.

Airplanes and ships consume about 25% of all petroleum. A growing population and economy will place further demands on intercontinental and intra-continental transportation served by air and water. Current movements to use lower-sulphur fuel for ships is a good idea, but at present nobody has any idea how to power these machines by another means on a large scale. This percentage of oil consumption looks very strong. Progress, defined as reduced demand and emissions per unit of freight or distance travelled, may come from fuel efficiency, meaning more planes and boats can operate using the same amount of fuel as today.

The next big block of transportation fuel powers heavy trucks and trains, another 25%. Growth in population and GDP will result in more things moving everywhere. In the absence of a replacement energy source, the only gain may be in fuel economy by 2040—moving way more goods with less energy per tonne of material transported.

The most intriguing opportunity in materially reducing oil consumption in the next 20 years comes from light cars and trucks, currently consuming about 20% of the world's petroleum. This is where the greatest deviation in future oil consumption forecasts exists and probably the riskiest element in future oil demand forecasts. Depending upon the rate of adoption of electric vehicles (EVs), peak oil demand for light-vehicle transportation could be reached by 2030 if not sooner.

An outfit called Carbon Tracker Initiative from the UK advertises its mandate on its website as "Aligning capita market actions with climate reality." It says oil consumption for vehicles will peak in 2023. The September 18, 2018 edition of *Forbes* magazine carried an article titled, "Forecasts of Peak Oil Demand Overstated" which cites the Carbon Tracker study and that of a Norwegian risk analysis firm DNV GL. *Forbes* reported, "These two groups have an axe to grind with the oil industry, with agendas predicated on the demise of the fossil fuel business. Anything they can do the hasten the process, all the better."[203]

BP is of the view EVs will have a big impact by 2040. There will 100 million EVs in service, a combination of battery electric and plug-in hybrids. There were only about two million EVs on the road in 2016. This growth in 22 years can only be described as meteoric. The auto manufacturers all see this number and are modifying their product offerings accordingly. The sobering part of BP's EV forecast is that the total number of cars on the road will reach 2 billion, about twice that of 2015. However, continued progress in fuel economy will help shrink this portion of oil demand. BP figures a world passenger-car fleet that averages seven litres of fuel for every 100 kilometres in 2015 will go twice as a far on a litre refined petroleum by 2040.

Shell thinks light-vehicle carbon fuel demand will peak as early 2025. But this is qualified. In the same *Forbes* article cited above, the author says Shell's projection will only occur if there is significant compliance with the 2015 Paris climate commitments, which is not taking place. "The energy transition will happen over the coming decades, but an array of political, economic and social considerations will determine how long it takes to 'de-carbonize' the entire global energy system, if ever."[204]

Despite the unknown element of the variables around EVs, the overall forecasts of the three oil companies plus the EIA and IEA are similar. Shell is the most aggressive on EVs but realistic about heavy transport. The November 2017 *World Energy Outlook* from the IEA was titled, "A World in Transformation." The opening line in the website link read, "The resurgence in oil and gas production from the United States, deep declines in the cost of renewables and growing electrification are changing the face of the global energy system and upending traditional ways of meeting energy demand. ... A cleaner and more diversified energy mix in China is another major driver of this transformation."[205]

203 Dan Eberhart, "Forecasts of Peak Oil Demand Overstated," *Forbes* (September 18, 2018). https:// www.forbes.com/sites/daneberhart/2018/09/18/forecasts-of-peak-oil-demand-overstated/

204 Dan Eberhart, "Forecasts of Peak Oil Demand Overstated," *Forbes* (September 18, 2018). https:// www.forbes.com/sites/daneberhart/2018/09/18/forecasts-of-peak-oil-demand-overstated/

205 "A World in Transformation: World Energy Outlook 2017," *International Energy Agency* (November 14, 2017). https://www.iea.org/newsroom/news/2017/november/a-world-in-transformation-world-energy-outlook-2017.html.

The IEA sees the next quarter century (from 2015 to 2040) unfolding as follows. "Over the next 25 years, the world's growing energy needs are met first by renewables and natural gas, as fast-declining costs turn solar power into the cheapest source of new electricity generation. Global energy demand is 30% higher by 2040 – but still half as much as it would have been without energy efficiency improvements. The boom years for coal are over... and rising oil demand slows down but is not reversed before 2040, even as electric-car sales rise steeply."[206]

The IEA summarized oil this way. "... it is too early to write the obituary of oil. Global oil demand continues to grow to 2040, although at a steadily decreasing pace ... other sectors – namely petrochemicals, trucks, aviation and shipping – drive up oil demand to 105 million barrels a day by 2040."[207]

This summarizes the predictions shared by all five forecasts. Despite pleas from climate change alarmists to abandon fossil fuels as soon as possible and predictions of vested interest like Carbon Tracker that this source of energy will be heading for obsolescence sooner than most think, oil demand which will exceed 100 million b/d in 2018 or early 2019 is predicted to be 105 to 110 million b/d by 2040.

Natural gas, the most benign of the fossil fuels based on carbon molecules, has a very promising future leading to 2040. The IEA wrote, "... a strong emphasis on cleaner energy technologies, in large part to address poor air quality, is catapulting China to a position as a world leader in wind, solar, nuclear and electric vehicles and the source of more than a quarter of projected growth in natural gas consumption."[208]

Shell is particularly bullish on natural gas, having made significant investments in global gas supplies and LNG while abandoning Alberta's oil sands. On its website, Shell reports only 100 million tonnes of LNG were produced in 2000 versus 300 million tonnes in 2017. That's just the beginning. Shell sees 40% of energy demand growth in the next 20 years coming from natural gas with an average demand growth of 4% per year for LNG. This would take the global LNG market to over 600 million tonnes per year by 2037. In its 2018 natural gas outlook disclosure Shell wrote, "Final investment decisions on new LNG supply projects are required

206 "A World in Transformation: World Energy Outlook 2017," *International Energy Agency* (November 14, 2017). https://www.iea.org/newsroom/news/2017/november/a-world-in-transformation-world-energy-outlook-2017.html.

207 Ibid.

208 Ibid.

soon to avoid a supply shortage in the 2020s." This explains why Shell and its partners sanctioned the Canada LNG project at Kitimat on October 2, 2018.[209]

The good news is that the economy of the developing world is forecast to grow at over 4% per year through to 2040 and the energy mix will be making the largest transition in history, away from high-carbon fuels such as coal and oil which, at least percentage-wise, will shrink. To put things in historical perspective, BP figures coal's percentage of total world energy supply will fall from 95% in 1900 to only about 20% in 2040. Oil, which peaked at 50% in about 1970s, will be down to 20% in 2040.

The bad news is that despite all the improvements that have been made and will continue to be made, GHG emissions will rise for the next 20 years. The IEA wrote, "While carbon emissions have flattened in recent years, the report finds that global energy-related CO2 emissions increase slightly by 2040, but at a slower pace than in last year's projections. Still, this is far from enough to avoid severe impacts of climate change." ExxonMobil sees CO2 emissions from energy demands continuing to rise with transportation, electricity generation and industrial demand (petrochemicals) being the primary drivers.[210]

Are the forecasts of the IPCC and these five energy players/analysts correct? Undoubtedly, no. Because they involve things that haven't happened yet, they will never be right. Success in the forecasting business is outcomes that are the least wrong.

But so long as the methodology is reasonable, one must either believe them all or reject them all. It is disingenuous to support the forecasts that agree with your view of the world and reject the ones that don't.

The Alberta government publicly justifies the massive changes to the provincial economy and its carbon energy industries in its 2015 Climate Leadership Plan based on the certainty of damages and dislocation predicted by climate change forecasts. The federal government regularly uses the same approach when it justifies a national carbon tax and commitments to reduce emissions. All governments introducing carbon taxes, emission caps, subsidizing renewables, phasing out coal

---

209 "Shell Launches LNG Outlook 2018," *Shell* (February 26, 2018). https://www.shell.com/energy-and-innovation/natural-gas/liquefied-natural-gas-lng/lng-outlook.html.

210 "A World in Transformation: World Energy Outlook 2017," *International Energy Agency* (November 14, 2017). https://www.iea.org/newsroom/news/2017/november/a-world-in-transformation-world-energy-outlook-2017.html.

power, or providing cash incentives for energy-conserving home improvem... are using public funds to reduce or mitigate the future damage, calamity, and dislocation predicted in climate change forecasts.

The consensus conclusions of the foregoing forecasts on world energy demand are based on the certainty that the 6.2 billion people who don't live in OECD countries want access to electricity, food, shelter, clothing, and transportation and they need cheap energy to do it. These forecasts are also based on the certainty that renewable energy in the form of hydro, solar, or wind or carbon-free nuclear power will not replace carbon resources for petrochemicals or transportation fuel for airplanes, ships, trains, and big trucks.

At the Paris COP 21 conference in 2015, 195 countries publicly and collectively agreed to ensure the future was not going to unfold the way BP, Shell, ExxonMobil, IEA, and EIA have forecast because they were going to change it. Canada committed to a 28% reduction in GHG emissions by 2030, 517 megatonnes compared to 722 megatonnes in 2015. In March 2018, Minister of Environment and Climate Change Catherine McKenna discussed in her department's annual report to the UN its progress and targets, stating that it wasn't going to make them. Canada's revised 2030 target was now 66 megatonnes higher than estimated a year earlier.

While the COP 21 agreement didn't actually state what the 2030 annual emission target was among its 195 participating countries, there has never been a year-over-year reduction in total GHG emissions. In early December 2018, The Global Carbon Project, an international research organization, reported GHG emissions in 2018 were likely to rise by 2.7% in 2018 following a 1.7% increase in the previous year.[211]

On a website titled "Energy Matters," energy consultant and blogger Roger Andrews wrote "At the Paris climate conference ... 195 countries adopted the first-ever universal, legally binding global climate deal. The agreement sets out a global action plan to put the world on track to avoid dangerous climate change by limiting global warming to well below 2°C." This is the target for global warming by 2100.[212]

---

211 Mark Tutton, "Carbon Emissions to Hit All-Time High, Says Report," *CNN* (December 5, 2018). https://www.cnn.com/2018/12/05/world/emissions-global-carbon-budget/index.html.

212 Roger Andrews, "Global CO2 emissions forecast to 2100," *Energy Matters* (March 7, 2018). http://euanmearns.com/global-co2-emissions-forecast-to-2100/.

This was based on the most recent IPPC report from 2014, which Andrews summed up, "...the goal is likely to be met if cumulative emissions (including the 535 GtC emitted by the end of 2013) do not exceed 1 trillion tonnes of carbon (PgC)."[213]

This means that even if Paris achieves full cooperation and the 2°C temperature-increase cap by 2100 in the IPCC models is achieved—and the energy growth forecasts are true—it can only be done by removing CO_2 from the atmosphere in the latter half of the current century. Carbon emissions would have to be zero by 2100.

Removing CO_2 from the atmosphere is currently only a theory on a large scale.

Since Paris the IPCC has doubled down and released another forecast, asking for accelerated CO_2 reductions by 2030. To keep the temperature rise to 1.5°C and avoid a long list of weather disasters, CO_2 emissions should be reduced by 45% by 2030 and be "net zero" by 2050. Countries grappling with their 2015 commitments are surely wondering how to meet the latest guidelines.

Meanwhile, at the January 2019 World Economic Forum in Davos, Switzerland, where the rich and powerful gather to discuss and pontificate about the future of mankind, carbon emissions and climate change were again on the agenda. Two comments are trenchant.

IEA boss Fatih Birol was quoted as commenting that electric vehicles alone were not going to move the needle on global GHG emissions. "This year we expect global oil demand to increase by 1.3 million b/d. The effect of 5 million cars is 50,000 b/d. 50,000 versus 1.3 million."[214]

Saudi Aramco's CEO Amin Nassar said at Davos, "I don't see peak [oil] demand happening in 10 years or even by 2040. There will continue to be growth in oil demand. ... We are the lowest cost producer and the last barrel will come from the region."[215]

What does this all mean? These are, after all, just forecasts.

---

213 Ibid.

214 Tsvetana Paraskova, "IEA Chief: EVs Are Not The End Of The Oil Era," *OilPrice.com* (January 22, 2019). https://oilprice.com/Energy/Energy-General/IEA-Chief-EVs-Are-Not-The-End-Of-The-Oil-Era.html.

215 Zahraa Alkhalisi, "Saudi Arabia: We'll pump the 'last barrel' of oil," *CNN Business: What to Watch on the Final Day at DAVOS* (January 25, 2019).https://edition.cnn.com/business/live-news/davos-2019-live-updates/index.html.

Regarding IPCC forecasts, the actual temperature rise in the past 100 years is about 1°C. The reports of changes in ocean levels vary on the source; the range is 10 to 20 centimetres over the same period.

Every bad weather event is deemed by the doomsayers to the worst ever, and only a portend of the terrible things to come. But most rational people with an understanding of history know it is highly unlikely the world's worst drought or hurricane took place in the last 20 years.

The energy demand forecasts are also just forecasts. Everybody accepts they are wrong. Just a question of how much and in which direction.

But if you believe the climate change models and the dire predictions that accompany them, then you must, by extension, also believe the predictions that energy demand, fossil fuel consumption, and emissions will continue to rise unless something major happens on the technology or policy front that is currently unknown.

The desires and aspirations of billions of our fellow citizens are in their own way a force much more powerful than any storm or weather event the world's climate may unleash on us. Climate alarmists say if we don't stop global warming, hundreds of thousands will suffer, starve, and die. These same people fail to understand that if the world is denied cheap energy or lose what energy they have because non-carbon alternatives are unaffordable, hundreds of thousands will suffer, starve, and die.

At some point people must look at all the facts and make some tough decisions. They must not base their actions on the facts they agree with while ignoring the facts they don't agree with.

Canada is going down two opposing and irreconcilable paths simultaneously.

On the one hand we are a safe, honest, reliable, ethical and transparent supplier of the carbon energy that, according to all credible forecasts, the world needs. Today. This not a forecast, it is a fact. The removal of Canada's 7.5 million barrels of oil equivalent oil, natural gas liquids, and natural gas from the market would have an immediate impact. Prices would spike overnight, resulting in massive damage to the world's economy and measurable suffering to the billions of people who depend on carbon energy to survive in their daily lives. Excessively high energy prices have been a problem in the past. It can happen again.

At the same time, too many Canadians citizens and politicians believe one small country can make a difference on global GHG emissions and the climate of the future by implementing a patchwork of poorly conceived policy initiatives that serve no purpose but to affecting our global economic competitiveness. Canada is hobbling its domestic carbon-resource industries in a purely symbolic, enormously expensive, and ultimately meaningless attempt to save the world.

Everyone agrees Canadians should do their part. But Canadians cannot do it alone. Canada is responsible for 1.6% of the global GHG emissions, but among major carbon-energy suppliers has chosen to bear 100% of the responsibility for the problem. Ninety-six percent of the world's oil is produced elsewhere, but no other major producer is being subjected to the same combination of taxation, regulation, and market access restrictions.

The foreseeable future of carbon energy in the world is clear. Climate change or not, the world runs on coal, oil, and gas today, and for the most likely for some time to come.

The future of carbon energy in Canada is confused, punitive, and regionally divisive.

Unless of course the forecasts are incorrect. Depending on which forecast is wrong and in what direction, it could result in Canada's self-inflicted and contradictory situation being much worse.

2
THE BEST NEWS IN THE WORLD YOU'VE NEVER HEARD

"What climate change alarmism has done is marry mankind's genetic preoccupation with the weather with the media's insatiable appetite for bad news. Bad weather is good news, at least from a commercial standpoint. Every day. All the time. Extreme weather. Hot weather. Cold weather. … Meanwhile, is there anything else going on in the world?"

The old saying goes when you're worried about something is, "No news is good news." In the media this phrase is reversed: "Good news is no news." Disasters sell newspapers or capture eyeballs. Ordinary life does not.

Fortunately, there's enough going wrong every day to keep the myriad of information-distribution services in business and support a multi-billion-dollar industry worldwide.

Think about the headlines you have not, nor will ever, encounter. "7.4 billion people did not die yesterday." "Ninety-four percent of Canadians employed." "Most people very happy."

People are fixated on bad news. The closer it is, the more impact it has. You're driving down the street. There is a car crash on one side the road and a happy family in the playground on the other. What playground? Drivers almost instinctively drive slowly by a collision to see how awful it is, or if anybody is hurt. Could have been me. Terrible.

The other thing people are preoccupied with is the weather. It is the most popular conversation icebreaker in human history, particularly in confined spaces like elevators. The weather changes continuously. Everyone in the same place has the same experience. "Nice day if it doesn't rain." "Hot/cold/warm (pick one) enough for you?" Utterly banal, totally engaging.

The weather has become a news industry unto itself. Smart phones offer access to hundreds of free weather apps with current and forecast conditions from your backyard to the ends of the earth. Every device today has a weather-network app installed at purchase. Do a Google search on "TV weather channel" and you'll get

432 million possible results. Will that suffice? Or will your own curiosity require a more extensive investigation?

One of the best quips about the weather is attributed to Mark Twain, but was actually from one of his friends, Charles Dudley Warner. It went, "Everybody talks about the weather, but nobody does anything about it." That's ancient news nowadays. Changing the future weather is what the entire climate change movement is all about.

What climate change alarmism has done is marry mankind's genetic preoccupation with the weather with the media's insatiable appetite for bad news. Bad weather is good news, at least from a commercial standpoint. Every day. All the time. Extreme weather. Hot weather. Cold weather. Be assured the weather in your neighbourhood is going to get worse. Not if, but when.

Meanwhile, is there anything else going on in the world? The lot of mankind is quietly and steadily improving in multiple ways at astonishing speed. It is all good news, so it never makes headlines. Nobody brings it up in conversation. "Wow, isn't life great?" "Isn't the progress of mankind incredible?" "Are we fortunate or what?"

Here is the state of the human condition when you're not fretting about the weather.

The first popular book about how good things are was *The Ultimate Resource* by Julian L. Simon in 1981. This was 37 years ago. Almost half of all Canadians weren't born yet. The world was in the middle of a severe recession while trying to adapt to sky-high oil prices. Interest and inflation rates were at record levels. Things were awful. This was the most difficult period economically since World War II, which had ended 36 years earlier.

The topics Simon covered were all the hot buttons of day, including food, land, natural resources, pollution, pathological effects of population density, the standard of living, immigration, human fertility, future population growth, world population policy, domestic population activities, and involuntary sterilization. The fixation on population had followed the 1968 release of the book *The Population Bomb* and the 1972 book *Limits to Growth*. Both were best-sellers and were based entirely on scaring the heck out of anybody who could read about how life on Earth was unsustainable because of too much of one thing (people) and not enough of everything else (food, energy, resources).

Simon would have no part of it, devoting the next 340 pages to explaining why so many things were much better than people believed, and how the human condition would continue to improve. Particularly disturbing to Simon at that time was the public discussion about the birth rate, depopulation, and government-mandated sterilization to arrest population growth. In the introduction he wrote, "The longer I read the literature about population, the more baffled and distressed I have become that one idea is omitted. Enabling a potential human being to come into life and enjoy life is a good thing, just as enabling a living person's life not to be ended is a good thing. Yet I find no logic implicit in the thinking of those who ... are positively gleeful with the thought that 1 million or ten times that many lives will never be lived..." (Simon, The Ultimate Resource 1981).

In 1996 Simon updated his 1981 book with *The Ultimate Resource 2*. Nothing had changed his overall view, but he was able to review how events had unfolded since his original thesis 15 years earlier. The data was better, and Simon was able to quantify what had previously been qualitative observations.

Ever the humanist, Simon was determined to force individuals to acknowledge facts in the face of public opinion and address people's predilection to believe the worst about the human condition. He wrote, "The common assertions that resources are growing more scarce, that environmental conditions are worsening, and that the poor suffering from economic freedom are the gravest danger to the poor and environment. And many of the government interventions that are predicated on these false assertions constitute fraud because they benefit only the well-off and the vested interests that many constricting regulations protect from competition for a good life of cheap, high-quality housing and consumer goods, easy access to wilderness, and upward mobility in society" (Simon 1996).

Simon's message to the youth of the world—22 years ago—was inspirational. And, of course, it did not make headlines anywhere.

> I part company with the doomsayers in that they expect us to come to a bad end despite the efforts we make, whereas I expect a continuation of humanity's successful efforts. And I believe that their message is self-fulfilling, because if you expect your efforts to fail because of inexorable natural limits, then you are likely to feel resigned, and therefore to literally resign. But if you recognize the possibility – in fact the probability – of success, you can tap large reservoirs of energy and enthusiasm (Simon 1996).

In 2001 Danish environmentalist Bjorn Lomborg published the English version of *The Skeptical Environmentalist—Measuring the Real State of the World*. In the preface Lomborg confessed to being influenced by the work of Julian Simon. Lomborg wrote, "I was provoked. I'm an old left-wing Greenpeace member and had for a long time been concerned about environmental questions. At the same time I teach statistics, and it should therefore be easy for me to check Simon's sources. Moreover, I always tell my students how statistics is one of science's best ways to check whether our venerable beliefs stand up to scrutiny or turn out to be myths. Yet, I had never really questioned my own belief in an ever-deteriorating environment – and here was Simon, telling me to put my own beliefs under the microscope" (Lomborg, The Skeptical Environmentalist—Measuring the Real State of the World 2001).

Once he looked for himself, Lomborg came to the same conclusion as Simon— all the data indicated mankind was doing much better than most thought and certainly better than was reported in the media. The first chapter of Lomborg's book was titled "Things Are Getting Better" followed by a section called "The Litany." Lomborg recited the prevailing belief that the world was going to hell in a handcart and then began to refute much of this collective wisdom point by point. Under "Exaggeration and good management" Lomborg wrote, "The constant repetition of the Litany and the often heard environmental exaggerations has serious consequences. It makes us scared and it makes us more likely to spend our resources and attention solving phantom problems while ignoring real and pressing (possibly non-environmental) issues. This is why it is important to know the real state of the world."

Which was why "Measuring the Real State of the World" was the subtitle of the book (Lomborg 2001).

Lomborg devoted 66 pages, about one-fifth of the book, to global warming, opening with, "Climate change and especially global warming has become the overriding environmental concern since the 1990s. Most discussions about the environment end up pointing out that, despite all other indicators that may show us to be doing better and better, we still have to change our current lifestyle dramatically because our way of life is now changing the climate and causing global warming (Lomborg 2001)."

Lomborg took the climate change issue very seriously but he also compared it to other issues impacting the human condition. At that time 1997's Kyoto Protocol was the GHG reduction commitment of the day and Lomborg analysed the cost of reduced emissions versus the estimated cost of the damages caused by global warming. Since developing countries were exempt at that time, OECD nations were going to have to pay the full cost. Lomborg wrote, "This means that the cost to the OECD countries of complying with Kyoto will – each year – by 2050 cost about as much as global warming will cost in 2100 (that is about 2 percent of GDP)." Further, he calculated these GHG reductions would only delay warming by six years (Lomborg 2001).

He reviewed several scenarios and concluded, "Despite our intuition that we naturally need to do something drastic about such a costly global warming, economic analysis clearly show that it will be far more expensive to cut CO_2 emissions than to pay the costs of adaptation to the increased temperatures." Lomborg was of the view that other problems, like clean drinking water, sewage treatment, medicines, healthcare, nutrition, and more economic opportunities were a more compelling investment to improve the world than making drastic and expensive reductions in carbon emissions, primarily by making essential energy more expensive (Lomborg, The Skeptical Environmentalist—Measuring the Real State of the World 2001).

The reaction to *The Skeptical Environmentalist* was swift and harsh. After the Danish version (*Verdens sande tilstand*) was released in 1998, there were attempts by the environmental movement to prevent its translation into English and subsequent second publication by Cambridge University Press. Lomborg was accused of "scientific dishonesty" in his home country. The format of the book made an objective assessment difficult for purists because it is a mixture of Lomborg's opinions, science, and economics. Unrepentant, Lomborg went on a global speaking tour, which sold a lot of books and markedly improved his English.

Two years later another book was released by author Gregg Easterbrook, *The Progress Paradox—How Life Gets Better While People Feel Worse*, which took a very non-scientific approach to much of the same subject matter. The opening line was, "Suppose your great-great-great grandparents, who lived four generations ago, materialized in the United States of the present day" (Easterbrook 2003).

Assuming a generation is about 25 years (the average age of people when they replace themselves), that would be the period from about 1900 to when the book was published.

You know the rest. The increase in lifespan and standard of living, as well as improvement in health and eradication of many diseases would be impossible for our ancestors of 100 years ago to comprehend. Statistics Canada data comparing 1921 to 2005 shows men and women are both living about 20 years longer, an increase of 130% over previous life expectancy. Most of what we take for granted—running water, central heating, reliable automobiles, computers, fresh produce in the winter— wasn't commonly available anywhere but in temperate climates at the beginning of the last century.

Having defined progress, Easterbrook then pondered the paradox, writing, "Yet how many of us feel positive about our moment, or even believe that life is getting better? Today Americans tells pollsters that the country is going downhill; that their parents had it better; that they feel unbelievably stressed out; that their children face a declining future – and Americans were telling pollsters this even during the unprecedented boom that preceded the tragedy of September 11, 2001" (Easterbrook 2003).

As did Simon and Lomborg before him, Easterbrook rolled out the data about extended lifespans, falling child mortality rates, diseases cured or controlled, pollution declines, and reductions in illiteracy, poverty, and hunger. But the author was more psychologist than scientist and environmentalist and dove deeply into why people think the way they do. In the introduction, he promised his book explained:

- Why huge numbers of people do not appreciate the fact that Western life grows steadily better, or even deny this is happening.

- Why the prosperous, free, and basically decent societies of the United States and western Europe produce so many citizens who are unhappy.

- Why rapid progress against "unsolvable" problems, such as pollution and crime, should give us hope that "unsolvable" problems of the present, such as global warming and developing world suffering, can be overcome.

- Why even overcoming every problem that exists might not make many us any happier.

With so many of the problems of 100 years ago solved, Easterbrook concluded, "society is undergoing a fundamental shift from 'material want' to 'meaning want,' with ever larger numbers of people feeling they lack significance in their lives." He ends with, "That ultimately we should be glad society is creating the leisure and prosperity that allows people by the millions to feel depressed, for it's better to be prosperous, free and unhappy than other possibilities" (Easterbrook 2003).

This is a peculiar definition of progress indeed, but it is regrettably very accurate.

The book's title is brilliant. The progress is glaringly obvious, but the paradox is one of the "unintended consequences" of all the fabulous improvements that leave more people with the wealth, health, food, shelter, and time to worry more about the future. "Arriving at this moment, your great-great-great grandparents would be surrounded by people who take for granted circumstances our ancestors would view as astonishing progress in the ancient quest to banish privation and establish a Golden Age. ... We live in a favored age yet do not feel favored. What does this paradox tell us about ourselves and our future?" (Easterbrook 2003)

By 2007, the controversy generated by his first book had died down to the point Bjorn Lomborg was willing stick his head out of his shell-shocked environmentalist foxhole one more time with a book titled *Cool It—The Skeptical Environmentalists' Guide to Global Warming*. In 2006 Al Gore's movie *An Inconvenient Truth* had put the perils of climate change in everyone's living room and tens of thousands of classrooms. The first words in Lomborg's preface read, "Global warming has been portrayed recently as the greatest crisis in the history of civilization. ... In the face of this level of unmitigated despair, it is perhaps surprising – and will be seen by many as inappropriate – to write a book that is basically optimistic about humanity's prospects" (Lomborg 2007).

After three years of dodging green bullets, Lomborg decided to take an entirely different approach. Starting in 2004 he assembled what he described as "the smartest economists in the world" (including four Nobel Laureates) and asked them to identify mankind's biggest challenges and attempt to rank them by priority. This would ultimately be called the Copenhagen Consensus. The highest ranking went to the greatest of amount of improvement for the greatest number of people at the lowest cost (Lomborg 2007).

Lomborg's point was that while global warming was indeed an issue, it was by no means the only issue. He cited World Health Organization statistics at the time, which indicated that climate change was killing about 150,000 a year people in the developing world. But four million were dying annually from malnutrition, three million from HIV/AIDS, 2.5 million from pollution, two million from a shortage of micronutrients (iron, zinc, vitamin A), and nearly two million from a shortage of clean drinking water (Lomborg 2007).

Lomborg's group came up with the following list of challenges and ranked them in descending order.

VERY GOOD OPPORTUNITIES

- Diseases (HIV/AIDS)
- Malnutrition (providing micronutrients such as vitamins to growing children)
- Subsidies & Trade (trade liberalization to improve the world's economy)
- Diseases (malaria)

GOOD OPPORTUNITIES

- Malnutrition (improved agricultural techniques)
- Sanitation & Water (small-scale water technology for livelihoods)
- Sanitation & Water (community-managed water supply and sanitation)
- Government (lowering the cost of starting a new business)

FAIR OPPORTUNITIES

- Migration (lowering barriers for skilled workers)
- Malnutrition (improving infant and child nutrition)
- Malnutrition (reducing the prevalence of low birth-weights)
- Diseases (improved basic health services)

BAD OPPORTUNITIES

- Migration (guest workers programs for the unskilled)
- Climate (optimal carbon tax, $25 to $300 per tonne)

- Climate (Kyoto Protocol)

- Climate (value-at-risk carbon tax, $100 to $450 per tonne) (Lomborg 2007)

Cool It was not attacked as vociferously as *The Skeptical Environmentalist*. Perhaps that is because *An Inconvenient Truth* received a Nobel Prize and an Oscar the same year *Cool It* was released.

The Copenhagen Consensus Centre lives on and continues to research ways to solve all the other problems the world has besides climate change. The challenges ranking had not changed much by 2012 because the process hinged on a cost/benefit analysis. Lomborg still believed more impact can be achieved for more people with less money by first dealing with disease, water, nutrition, medication, healthcare, and sanitation.

Today the Copenhagen Consensus calls itself a "think tank." Lomborg is the president and founder. Under "Education", the website states the following; "The book *The Skeptical Environmentalist* is used as a standard text in environmental policy studies at universities across the world. It was called 'probably the most important book on the environment ever written' by the Daily Telegraph, UK."[216]

Two books were released in 2018 about the progress of mankind, both of which carry plugs from Microsoft founder, billionaire and philanthropist Bill Gates on their cover.

Gates called *Factfulness—Ten Reasons Why We're Wrong About the World, and Why Things Are Better Than You Think* by Hans Rosling, "One of the most important books I've ever read – an indispensable guide to thinking clearly about the world."

Gates describes *Enlightenment Now—The Case For Reason, Science, Humanism and Progress* by Steven Pinker as, "My new favorite book of all time."

Both Rosling and Pinker acknowledge climate change and global warming as very serious issues. But has Lomborg and Easterbrook had said a decade earlier, attention also had to be paid to numerous other issues facing mankind.

From Sweden, Rosling has been a doctor, academic, and public educator. His book cover bio reads, "He was an adviser to the World Health Organization and UNICEF, and he co-founded Médecins Sans Frontières (Doctors Without

---

216 "Our Impact," *Copenhagen Consensus Center* (n.d.). https://www.copenhagenconsensus.com/our-impact.

Borders) in Sweden … he as listed as one of *Time* magazine's one hundred most people in the world." He died after finishing this book.

Having spent years saving lives in the Third World and more years in the class-room, Rosling wrote about the world from first-hand experience. In 2005 he founded the Gapfinder Foundation to explore and explain the "gap" between what people thought about the world and what was actually occurring. Rosling would give his classes or audiences a 10-question quiz about the state of the world. Each question had three multiple-choice answers. To make it entertaining, he reminded people a chimpanzee making a random choice would be correct one-third of the time. He then compared their scores to the chimp.

Rosling broke human behaviour into 10 main categories: the same 10 questions he'd put to his live audiences and which were referenced on the book cover. He maintained there was a gap between perception and reality, and this was an interesting, if unscientific, way to illustrate it.

He observed there was tremendous negativity surrounding the human condition compared to the facts. People took trends and extrapolated them in straight lines despite this being the exception, not the norm. Fear was and remains a big driver on what people think so it clouds judgement and analysis. When numbers are provided and the size looks large, this should be put into context. *Generalization* and *destiny* were listed as ways people take historical or insufficient information and reach the wrong conclusions about the future. Using only one a single informa-tion source for anything makes it easy to make mistakes. The tendency to blame others or events for a certain outcome or problem leads people away from the best conclusion or solution. And treating complex issues with excessive urgency increases the likelihood of doing the wrong thing.

The book has many examples like, "In low income countries across the world today, how many girls finish primary school?" The correct answer is 60%. The chimps beat the respondents by a wide margin. Only 7% got the correct answer, immediately concluding it could not be that high. When he asked if the majority of people lived in high income, middle income, or low-income countries, only the US and Korea scored higher than the chimps with the right answer—middle income (Rosling 2018).

A similar question involved extreme poverty. "In the past 20 years, the propor-tion of people living in extreme poverty has A: almost doubled B: remained more or less the same C: almost halved." The answer was C but in online polls conducted by Gapfinder only an average of 10% of respondents got the answer right. Again, this was a combination of generalization and perceptions of destiny, that if you're poor you'll always be poor. Rosling includes a graph showing 80% of the world in extreme poverty (income of $2 a day or equivalent) in 1800, 50% in 1966, and only 9% in 2017. (Rosling 2018).

It is not likely you read this in the newspaper yesterday.

In the chapter "The Negativity Instinct," Rosling listed 16 ways life in which the human condition had vastly improved when compared to the past. The improvements were in decreased instances of legal slavery, oil spills, expensive solar panels, HIV infections, child mortality, battle deaths, death penalty, leaded gasoline, plane-crash deaths, child labour, deaths from disaster, nuclear arms, smallpox, smoke particles, ozone depletion, and hunger (Rosling 2018).

Humans instinctively fear harmful things to a point that anxiety can overwhelm reality. Deaths and injuries from disaster decline significantly with wealth, assisted greatly by notification, preparation, and adaptation. One chart shows annual disaster deaths per million people at 453 in the 1930s, a figure that had declined to 10 from 2000 to 2016. Death from wars has been flat since 1950. Rosling pointed out that 85% of the population has electricity and 90% of all girls in the world attend primary school.

Size, or a big number, mean nothing without perspective. The Spanish flu of 1918 killed somewhere between 20 million and 50 million people (records weren't that good 100 years ago, particularly after the upheaval of World War I). This was serious. So was the swine flu in 2009, but only in the media. Rosling wrote, "… I got tired of the hysteria and calculated the rate of news stories versus fatalities. Over a period of two weeks, 31 people had died from swine flu, and a news search on Google brought up 253,442 articles about it. That was 8,178 articles per death" (Rosling 2018).

Do not hold any preconceived notions about the destiny of races, regions, or countries. Rosling notes the birth rate is lower in Iran than the US and Africa is modernizing at a rapid pace. Even in the US, things are changing fast. In 1996 only 27% of Americans supported same-sex marriage. By 2018 the number was 72% and rising. Get a second opinion. Or a third. He wrote, "Forming your worldview by relying on the media would be like forming your view about me by looking only at a picture of my foot" (Rosling 2018).

Under urgency, Rosling expanded on climate change. "'We need to create fear!' That's what Al Gore said to me at the start of our first conversation about how to teach climate change." While the author admired Gore for his efforts and agreed climate change was a huge problem, Rosling didn't believe instilling fear was the right approached. He wrote, "… the volume on climate change keeps getting turned up. Many activists, convinced it is the only important global issue, have made it a practice to blame everything on the climate, to make it the single cause of all other global problems, … the war in Syria, ISIS, Ebola, HIV, shark attacks, almost anything you can imagine – to increase the feeling of urgency about the long-term problem" (Rosling 2018).

In *Enlightenment Now*, Steven Pinker takes a different approach. Also an academic and a psychologist, Pinker starts his book by reviewing the state of mankind. He states, "More than ever, the ideals of reason, science, humanism, and progress need a wholehearted defense. We take its gifts for granted; newborns who will live more than eight decades, markets overflowing with food, clean water that appears at a flick of a finger and waste that disappears with another, pills that erase a painful infection, sons who are not sent off to war, daughters who can walk the streets in safety, critics of the powerful who are not jailed or shot, the world's knowledge and culture available in a shirt pocket" (Pinker 2018).

Pinker devotes a lot of space to climate change and has real concerns. "When the Industrial Revolution released a gusher of usable energy from coal, oil and falling water, it launched a Great Escape from poverty, disease, hunger, illiteracy and premature death, first in the West and increasingly in the rest of the world. … And the next big leap in human welfare – the end of extreme poverty and spread of abundance, with all its moral benefits – will depend on technological advances that provide energy at an acceptable economic environmental cost to the entire world" (Pinker 2018).

Pinker takes an uncommon view of the climate change challenge. He attacks both extremes from the right and left. For all those who question the global warming models and science, Pinker writes, "Anthropogenic climate change is the most vigorously challenged scientific hypothesis in history. By now all major challenges … have been refuted, and even many skeptics have been convinced".

But Pinker figures one of the challenges is cognitive. "People have trouble thinking in scale; they don't differentiate among actions that would reduce CO_2 emissions by thousands of tons, millions of tons, and billions of tons." Another is moral. They gravitate towards virtue as a substitute for actual progress. Working with other psychologists, Pinker concluded, "... people esteem others according to how much time or money they forfeit in their altruistic acts rather than how much good they accomplish. ... Much of the public chatter about mitigating climate change involves voluntary sacrifices, recycling, reducing food miles, unplugging chargers, and so on. But however virtuous these displays may feel, they are a distraction from the gargantuan challenge facing us" (Pinker 2018).

Pinker wants to actually change things, therefore, he supports market-based carbon taxes as a means of achieving meaningful reductions. "Having billions of people decide how to best conserve, given their values and the information conveyed by prices, is bound to be more efficient and humane than having government analysts try to divine the optimal mixture from their desks (Pinker 2018)".

In a world in which climate change is framed in outcomes that are only bad, awful, and worse, Pinker is downright inspirational.

> Despite a half-century of panic, humanity is not on an irrevocable path to ecological suicide. The fear of resource shortages is misconceived. So is the misanthropic environmentalism that sees modern humans as vile despoilers of a pristine planet. An enlightened environmentalism recognizes that humans need to use energy to lift themselves out of poverty to which entropy and evolution consign them. It seeks the means to do so with the least harm to the planet and the living world. History suggests that this modern, pragmatic, and humanistic environmentalism can work. As the world gets richer and more tech-savvy, it dematerializes, decarbonizes and densifies, sparing land and species. As people get richer and better educated, they care more about the environment, figure out ways to protect it, and are better able to pay the costs. (Pinker 2018)

The last tranche of good news people rarely hear or read about it is in Alex Epstein's 2014 book *The Moral Case For Fossil Fuels*. He calls himself an "alternative environmentalist" and founded the Centre for Industrial Progress, a California think tank devoted to exploring "the new industrial revolution." "Eighty-seven percent of the energy mankind uses every second, including most of the energy

I am using as I write this, comes from burning one of the fossil fuels; coal, oil or natural gas" (Epstein 2014).

He then spends much of the rest of the book explaining why carbon became the number one energy resource and how it built the world in which we live today. Epstein also highlights the challenges that alternatives like wind and solar must overcome. It is not that they aren't much cleaner, but their immediate application as a viable alternative has been oversold and overstated. He calls crude oil "(t)he greatest energy technology of all time" and revisits the obvious; how the Industrial Revolution, powered by fossil fuels, brought the greatest advancements in life, health, and happiness to the greatest number of people in history in, relatively speaking, a very short period of time" (Epstein 2014).

When you're dealing with 87% of the energy supply for 7.4 billion people, asking a few questions about what everyone is supposed to do in the post-carbon era is hardly a radical idea. He writes, "Where did the thinkers go wrong? One thing I have noticed in reading most predictions of doom is the 'experts' almost always focus on the *risks* of a technology but never the benefits – and on top of that, those who predict the most risk get the most attention from the media and from politicians who want to 'do something.' But there is little to no focus on the *benefits* of cheap, reliable energy from fossil fuels" (Epstein 2014).

When thinking of the world, think of the instincts Rosling stressed in *Factfulness*. Knowledge gap. Negativity. Fear. Size. Generalization. Destiny. Multiple sources. Blame. Urgency.

Climate change is indeed a problem. But it is a lot more complicated than the last story you saw or heard in the media.

3

CARBON CAPITAL CRUNCH – FINANCIAL PRESSURE ON FOSSIL FUELS

"One of the strategies cooked up by the ever-imaginative climate change alarmist industry years ago was persuading as many people as possible that providing capital—equity or debt—to fossil fuel companies was morally reprehensible and should be discontinued. This was called 'divestment.' Pools of money … were encouraged to not hold or buy shares in fossil fuel companies."

The climate change debate has moved beyond science and the environment to include money. Big money. It is difficult to extrapolate the future impact of climate-related events by monetizing the impact today. But we do know that as the global financial system is increasingly intertwined with the future of the climate, the world of world of finance will never be the same.

The 1997 Kyoto Protocol conceived an international carbon-emissions trading system to monetize carbon. Once in place, the system would punish countries producing more emissions reward those that reduced emissions. This will be discussed further in the next chapter.

In the 2001 updated report from the IPCC, it was declared certain that climate change would impact the future weather with severe consequences. This would, of course, cause extensive and expensive damages resulting in higher insurance payouts. The 2007 IPCC report was awarded a Nobel Peace Prize, which to many made its veracity beyond question. The report increased the acceptance that future personal, property, and asset damages were inevitable.

Not surprisingly, the financial sector most likely to worry about money and the weather are the insurers that have been doing this for centuries. The original and venerable Lloyd's of London—today Lloyd's—began providing insurance protection from bad weather 330 years ago. Its website reads, "In the 17th century, London's importance as a trade centre led to an increasing demand for ship and cargo insurance. Edward Lloyd's coffee house became recognized as the place for obtaining marine insurance and this is where the Lloyd's that we know today began."[217] The major threat to ships, sailors, and cargo was severe storms, although pirates and navigational blunders also created damage and losses.

Munich RE, the global reinsurer based in Munch, Germany, has been in the global insurance business since 1880. In the 2017 fiscal year its 42,000 world-wide employees generated 49.1 billion Euros ($73.2 billion) in premiums. On its website tab "Climate Change" Munich RE states, "For more than 40 years, Munich RE has been dealing with climate change and related risks and opportunities for the insurance industry."[218]

Under "Mission," Munich RE states climate change has long been part of its risk-analysis process. Because Munich RE has always insured people, companies, assets and communities against weather damage, it has been studying the weather to better assess and mitigate risk for decades. "Geoscientists at Munich RE have been analysing the effects of climate change on the insurance industry since the 1970s. Because of natural and anthropogenic changes to probability distributions for meteorological and hydrological parameters, a risk of change has resulted in the portfolios of Munich RE and its clients. Based on our analyses over a period of more than 40 years, we have developed a comprehensive strategy whereby we identify and assess the risks and reflect them in our business processes."[219]

It was insurance companies that first started issuing warnings about the financial risks of climate change once the IPCC made it clear that there was a new factor involved in insuring against future damages from the weather. While the financial success of the insurance industry has long been dependent on the weather—and risk analysis dependent upon trying to figure out will happen next—now insurers were confronted with elaborate scientific models that predicted storms would become stronger and more frequent. This meant more risk.

The next consideration was rising ocean levels, which were predicted to result in more severe damage on land from ocean storms such as hurricanes and typhoons. If it rose enough, higher oceans would, over time, submerge coastal communities to the point that people couldn't live there any longer.

---

217 "History," *Lloyd's of London* (2018). https://www.lloyds.com/about-lloyds/history.

218 "Climate Change & Climate Protection," *Munich RE* (n.d.). https://www.munichre.com/en/group/focus/climate-change/index.html.

219 "Mission & Vision," *Munich RE* (n.d.) https://www.munichre.com/en/group/focus/climate-change/mission-and-vision/index.html.

Remember how insurance works. Through insurance companies, individuals or corporations contribute funds to acquire a share in a syndicated pool of capital and associated syndicated risk. Those who want to protect themselves financially from physical or personal damage from bad weather buy insurance. The pooled money is invested to grow in value; the fund waiting to pay claims that everyone—insurer and client alike—hope never occurs. Insurance providers are only profitable if the total premiums they collect, plus investment returns, exceed the amount they will eventually pay out.

The business succeeds because the likelihood of every insurance policy holder getting clobbered at the same time is low. This is the risk-analysis element. So the cash cost of insurance is a fraction of what you may collect if damages occur. The price of insurance is based entirely on the likelihood of collecting, or conversely, the likelihood an insurer's claims exceed premiums plus investment returns.

When man-made climate change entered the picture, the increased risk of future payouts made the insurance industry plenty nervous. A CBC radio broadcast in 2017 about the impact of climate change on the global insurance industry was titled, "There Are No Climate Change Deniers in the Insurance Industry."

In 2009 a paper was published by The Geneva Association titled "A Global Review of Insurance Industry Responses to Climate Change." The Geneva Association has for 44 years been "the leading international think tank of the insurance industry." The document opened with, "A vanguard of insurers is adapting its business model to the realities of climate change. In many ways, insurers are still catching up to both mainstream science and their customers …" The paper was based upon reviewing 300 documents and the responses of 244 "insurance entities" from 29 countries.[220]

Customers, primarily corporate, were looking at insuring different types of assets and investments, from buildings and ships to the types of energy they produced. They were also looking to insurers for different products than before, hopefully at an affordable price. Insurers were seeking to stay in the business by properly analysing the risks, which they believed were changing more and faster than ever before.

---

220 Evan Mills, "A Global Review of Insurance Industry Responses to Climate Change," *The Geneva Papers on Risk and Insurance – Issues and Practice* (July 2009). DOI: https://doi.org/10.1057/gpp.2009.14.

The paper is peppered with quotes from leading insurers and underwriters, all recognizable brands today.

Marsh & McLennan's quote was all motherhood: "A healthy and sustainable environment is a precursor to the long-term well-being of society, the strength of the economy and the continuing success of our business. We recognize that climate change is one of the most significant risks facing the world today ..."[221]

Munich RE was more business-like: "Climate change is a fact. Countering it is a must. We are convinced that climate protection makes economic sense, as it would be more expensive in the long term to pay for the damage it causes. It offers companies and national economies that react quickly great opportunities ..."[222]

Allstate was the most direct, declaring that at some point the risk would render some insurance unaffordable. The reported stated, "Allstate – insurer of one in nine vehicles and one in eight homes in the United States – recognizes the onset of climate change the presence of human fingerprints: 'Allstate recognizes the emerging scientific consensus that the world is getting warmer and that this trend is influenced to some extent by emissions of greenhouse gases. Climate change, to the extent that it produces rising temperatures and changes in weather patterns, could impact the frequency and severity of extreme weather events and wildfires. Such changes could also impact the affordability and availability of homeowner insurance... (Response to a 2008 Carbon Disclosure Project Survey).'"[223]

In 2009—a decade ago—the global insurance industry came to the following conclusion:

> Progress in scientific understanding is no doubt driving the growing engagement of insurers. The scientific debate is over, with the IPCC ... now using the considered term 'unequivocal' in describing the certainty of climate change is here. IPCC has also pinpointed human activity as the main driver of observed and projected warming.
>
> The economic analysis has shifted as well. Reports (such as the U.K.'s government's 'Stern Review') turn on its head the conventional wisdom that taking action on climate change will

---

221 Ibid.
222 Ibid.
223 Ibid.

harm the economy. Economy and investors now increasingly realize that, in fact, it is the lack of action to combat climate change that is the true threat to the economy, while engaging with the problem and mounting solutions, represents not only a duty to shareholders but also a boon to economic growth.[224]

The paper continued, "Many in the insurance world share the concern. In the words of an associate editor at *National Underwriter* (the *Sports Illustrated* of the insurance world); 'Given the stakes for insurers covering catastrophic losses, waiting for proof instead of taking action now would amount to just plain foolish behavior.'"[225]

In the Geneva Association's report there was some debate among those surveyed as to whether the insurance industry should be proactive or reactive, quietly mitigating risk and adjust their business model, or jumping on the climate-catastrophe bandwagon. "Most agree that reducing vulnerability to weather extremes should be a higher priority, but some dispute the need for insurers to engage in addressing the core drivers of climate change or the need to discern the relative roles of human influence and natural factors."[226]

Insurance companies can make more money if they can persuade governments to spend more public funds reducing their risk. In a *Financial Post* article on October 17, 2018, columnist Terence Corcoran questions their motives. "The latest insurance industry initiative wanders even deeper into the quagmire of green policy advocacy. 'Combating Canada's Rising Flood Costs,' a new report from the Insurance Bureau of Canada (IBC), urged governments across the country to adopt 'natural infrastructure' to limit escalating climate change risks."[227]

Flooding could be reduced by protecting existing streams, ponds, trees, and other natural infrastructure to the greatest degree possible because, unlike parking lots and roads, they hold or absorb water. When nature is disturbed but the land use goes unused, it should be reclaimed. And governments should pay for it.

---

224 Ibid.

225 Ibid.

226 Ibid.

227 Terence Corcoran, "Terence Corcoran: Why Insurers Keep Hyping 'Climate Risks' that don't Materialize," *Financial Post* (October 17, 2018). https://business.financialpost.com/opinion/terence-corcoran-why-insurers-keep-hyping-climate-risks-that-dont-materialize.

Fortunately for the insurance industry, it was in no way alone in warning about future monetary risks. Monetization of future financial climate risks would take several other avenues. The ENGO movement, legitimized as global actors at the 1972 UN environment conference in Stockholm, was a driver behind all four.

One of the strategies cooked up by the ever-imaginative climate change alarmist industry years ago was persuading as many people as possible that providing capital—equity or debt—to fossil fuel companies was morally reprehensible and should be discontinued. This was called "divestment." Pools of money like trust funds, mutual funds, sovereign wealth funds, or university endowment funds were encouraged to not hold or buy shares in fossil fuel companies. If they sold or refused to own them, this was ideally to be accompanied by the appropriate virtue signalling of telling the world what was done and why.

Exactly when it started is hard to pinpoint, but it picked up speed after *An Inconvenient Truth* won international acclaim in 2007. The easiest place to start was university endowment funds, particularly in the US Large American universities collectively hold hundreds of billions of dollars in investments, primarily donations from wealthy graduates. Plus university students are historically environmentally engaged and will actively push for immediate change once presented with an appropriate pressing cause. Attacking fossil fuel providers for their role in creating the global climate change catastrophe was an ideal platform for action and vocal dissent.

Other pools of capital were encouraged to participate included organizations such as churches philanthropic foundations, governments, pension funds, and non-government organizations.

Needless to say, there is a website devoted to this subject called *gofossilfree.org*, a project of well-known climate change activist and oil sands critic *350.org*. Without passing judgement on the accuracy of the information, the website reports "institutions divesting" had a total of $US17.7 trillion under administration. A total of 988 institutions had agreed to divest all or part of their holdings in oil or coal companies. They were joined by over 58,000 individuals with a cumulative worth of about $US5.2 billion. The website breaks out the organizations by number, not the amount they hold or have divested. For example, faith-based organizations account for 29% of the participants. Education institutions, pension funds, and governments account for 45% at 15% each. For-profit corporations are only 3%.[228]

There is a long list of participating organizations. Canada has about 40 names, but the majority are single churches or municipal groups of churches such as the Anglican Diocese of Ottawa, the Bathurst Street United Church, The First Unitarian Church of Victoria, and the Unitarian Church of Calgary. Others include Laval University, the Council of Canadians and Citizens for Public Justice. The size of funds they administer is not disclosed, the percentage of funds they once had invested in fossil fuel companies (if any) is unknown, and the monetary impact of their divestment is not reported.[229]

Operating on a larger scale is *Oil Change International—Exposing the True Costs of Fossil Fuels*. It can be found at *priceofoil.org*. It is supported by six main ENGOs including the Rainforest Action Network, Indigenous Environmental Network, Sierra Club plus about 60 others. It pursues lenders, primarily banks, because they are often publicly traded and broadly held. Unlike endowment funds of all types (which rarely issue refunds once the cheque has cleared), deposits of capital at banks to support lending activities can easily be moved or withdrawn. Banks will respond to client pressure.

Oil Change International publishes a yearly report titled "Banking on Climate Change." The 2018 fossil fuel finance report card grades banks on "extreme fossil fuel" financing and reports loans in this area from 2015 to 2017. "The report assesses 36 private banks from Australia, Canada, China, Europe, Japan, and the United States, with policies from additional banks in these countries and Singapore included for comparison."[230]

> The report states: …extreme fossil fuels refer to extreme oil (tar sands, Arctic, and ultra-deepwater oil), liquefied natural gas (LNG) export, coal mining, and coal power. … It is environ-mentally, reputationally, and often financially risky for banks to back these fossil fuel projects and companies. More and more,

228 *Fossil Free: We Can Build a Fossil Free World* (n.d.). https://gofossilfree.org/.

229 Ibid.

230 Lorne Stockman, "Banking on Climate Change: Fossil Fuel Finance Report Card 2018," *Oil Change International* (March 28, 2018). http://priceofoil.org/2018/03/28/banking-on-climate-change-2018/ Authoring Organizations: Rainforest Action Network, BankTrack, Indigenous Environmental Network, Sierra Club, Oil Change International an Honor the Earth.

the public is tying the impacts of fossil fuels to the financial institutions backing the sector.

The report goes on to list by name the offending banks. Under "Tar Sands" the report reads, "The biggest single biggest driver of the overall increase in extreme fossil fuel financing came from the tar sands sector, where financing grew by 111 percent from 2016 to 2017."[231]

This is most likely due to the acquisitions of foreign-owned oil sands assets by Suncor, Cenovus and Canadian Natural Resources from companies that either no longer wanted to be in the oil sands business or operate in Canada.

The strategy of pressuring banks seems to work. In late 2017 The World Bank announced it would no longer provide financial support for companies in oil and gas production by the end of 2019. According to an article in *The Guardian* December 12, 2017, this was due to climate change. It ceased lending to coal-fired generating plants in 2010. It reported at the time only 1% to 2% of the bank's US$28 billion portfolio was loaned to oil and gas exploration and production companies.[232]

On its website The World Bank describes itself as follows: "With 189 member countries, staff from more than 170 countries, and offices in over 130 locations, the World Bank Group is a unique global partnership: five institutions working for sustainable solutions that reduce poverty and build shared prosperity in developing countries."[233]

In April of 2018 HSBC Holdings plc, Europe's largest bank, announced its intention to exit the fossil fuel finance business. In a *Financial Post* article April 20, 2018, the newspaper reported, "HSBC … will no longer finance new oil sands projects or pipelines, a move that could lead to increased borrowing costs for domestic players as European banks scale back exposure to the sector." The article reported that HSBC is joining "Paris-based BNP Paribras and Amsterdam-based ING Group." The head of HSBC in Canada told the newspaper, "… we have been

231 Ibid.

232 Larry Elliott, "World Bank to End Financial Support for Oil and Gas Extraction," *The Guardian* (December 12, 2017). https://www.theguardian.com/business/2017/dec/12/uk-banks-join-multinationals-pledge-come-clean-climate-change-risks-mark-carney.

233 "enGENDER IMPACT: Increasing Economic Opportunities," *World Bank* (2018). http://www.worldbank.org/en/topic/gender/publication/engender-impact-increasing-economic-opportunities.

here for close to 40 years and we've invested $200 million here in the last two years alone. At the same time, we are committed to doing our part in supporting the transition to a low-carbon economy."[234]

A more interesting divestment announcement was when Norway's sovereign wealth fund advised in August of 2018 that it was not selling its equity investments in oil and gas companies. On August 24, 2018 *The Globe and Mail* reported, "Norway's 1-trillion kronor (US$147.2-billion) sovereign wealth fund is meeting resistance for its plan to dump more than 40 billion kronor (US$5.7 billion) in oil and gas stocks." It originally announced its intention to no longer hold shares in fossil fuel producers in late 2017. What made this remarkable was that the fund is comprised entirely of Norway's profits from decades of oil and gas production in the North Sea.[235]

When the 2017 decision was independently reviewed, the conclusion was that the fund's primary purpose was to make money for Norwegians to cushion the economy after the oil ran out or no longer had commercial value. The report stated, "This investment strategy is simple, well-founded and has served the fund well." Selling its shares in oil companies would, "… have little effect on protecting Norway against falling crude prices and alter what has been a successful investment philosophy."[236]

In March of 2019 Norway reversed its prior position and decided to sell a portion of its oil and gas industry equities portfolio, primary exploration and production operators without refining or downstream operations. The decision was announced as having more to do with protecting future investment returns than climate change and was carried in the world business media.

As this book goes to press the Norwegian offshore oil and gas industry has 16 active drilling rigs searching for more oil, about the same average level as the past five years. Although output will decline again in 2019, it is expected to rise in the next four years thanks in part to new discoveries coming on production.

---

234 Geoffrey Morgan, "HSBC Joins Other European Banks in Scaling Back Oilsands Financing," *Financial Post* (April 20, 2018). https://business.financialpost.com/news/hsbc-joins-other-european-banks-in-scaling-back-oilsands-investments.

235 Sveinung Sleire and Mikael Holter, "Norway's $1-trillion Wealth Fund Meets Resistance as it Plans to Dump Oil, Gas Stocks," *The Globe and Mail* (August 24, 2018). https://www.theglobeandmail.com/business/article-norways-1-trillion-wealth-fund-meets-resistance-as-it-plans-to-dump/.

236 Ibid.

McGill University in Montreal endured a very public struggle with its endowment fund. Led by activist staff and students, there was pressure in 2015 for it to not hold any fossil fuel company shares in its fund, estimated at $1.4 billion. However, in 2016 the decision was made publicly not to do so. The university's administration was, according to a CBC News report, subjected to intense criticism including being called "not only myopic, but shameful." A student said, "They say they agree with the science, but they don't seem to know what it means to act accordingly." A professor added, "This is the sort of thing that leads people to lose faith in the integrity of the institution and its governance." About 6% of the funds were invested in fossil fuels.[237]

Those in favour of divestment wanted more money put to work in financing renewable energy. The CBC reported that other Canadian universities dealing with divestment pressure were the University of Ottawa, University of Toronto, University of British Columbia and Dalhousie University in Halifax.

Perhaps concerned about their future pensions, there has been no known public pressure by Canadian ENGOs for the Canada Pension Plan (CPP) to join the divestment movement. On March 31, 2018, CPP held shares in all the major Canadian oil sands producers and pipeline shippers: Suncor, Cenovus, Canadian Natural Resources, TransCanada and Enbridge. Internationally, CPP held a diverse group of oil and gas producers including Repsol of Spain, Royal Dutch Shell, Malaysia's Petronas and Russia's Rosneft.

Other public-sector pension funds pay more attention to maximizing investment returns than to moral guidance dispensed by climate change activists. In August 2018 the Ontario Municipal Employees Retirement System (OMERS) announced it was paying US$1.4 billion for a 50% interest in a pipeline that carries oil to market from wells in the Permian Basin in Texas. *The Globe and Mail* reported on August 23, 2018—this investment by OMERS was, "... its third big US energy acquisition this year."[238]

With the future damages caused by climate change now quantifiable, it should come as no surprise that in the US, parties experiencing current damages or

---

237 Benjamin Shingle, "McGill University to Stick with Fossil Fuel Investments," *CBC News* (March 24, 2016). https://www.cbc.ca/news/canada/montreal/divest-mcgill-rejected-1.3505347.

238 Jeffery Jones, "OMERS Buys Pipeline Stake for $1.4-billion," *The Globe and Mail* (August 21, 2018). https://www.theglobeandmail.com/business/article-omers-buys-pipeline-stake-for-14-billion/.

anticipating future damages are using the courts to seek financial restitution. On January 10, 2018 the municipal government of New York City sued five big oil companies—ExxonMobil, Royal Dutch Shell, ConocoPhillips, BP, and Chevron—blaming them for climate change.

This joined lawsuits already filed in California against the same five companies by cities and counties including San Francisco, San Mateo County, and Imperial Beach. They claim rising sea levels caused by climate change are causing or will cause future physical and economic damage. San Mateo, for example, claims to be "particularly vulnerable to sea level rise" and that by 2050 there is a 93% likelihood this will cause a devastating flood.

The NYC statement of claim read, "In this litigation, the City seeks to shift the cost of protecting the City from climate change impacts back onto the companies that have done nearly all they could to create this existential threat." NYC claims it will be "spending billions of dollars" to protect itself from rising oceans. "To deal with what the future will inevitably bring, the City must build sea walls, levees, dunes and other coastal armament, and elevate and harden ... structures, properties, and parks along its coastline."[239]

Announcing the legal action, NYC mayor Bill de Blasio called Hurricane Sandy in 2012, "a tragedy wrought by the actions of the fossil fuel companies." Expanding on the damages he added, "That is the face of climate change. The city of New York is taking on these five giants because they are the central actors, they are the first ones responsible for this crisis and they should not get away with it anymore. We're going after those who have profited. And what a horrible, disgusting way to profit – the way it puts so many peoples' lives in danger."[240]

The legitimacy of the litigation was allegedly enhanced by news articles revealing that ExxonMobil knew in the 1980s its products affected the climate but suppressed the information, choosing instead to question climate science. ExxonMobil rejects this allegation. Some commentators have linked current litigation against oil companies to the historical tobacco lawsuits, where the courts agreed cigarette producers knew their product was harmful but conspired to conceal the information.

---

239 David Yager, "The 'Big Apple' Sues Five Big Oil Companies For Causing 2012's Hurricane Sandy – Has it Come to This?" The Washington Post (January 10, 2018).
240 Ibid.

The idea that five oil companies are solely responsible for wrecking the whole world is, of course, preposterous. The only driver of oil production growth has been growing demand from billions of customers. One must wonder whether this litigation is driven by political ambition, virtue signalling, or both.

Fortunately, US courts are not permitted to do either. On June 20, 2018, *Reuters* news agency reported, "A federal court in California dismissed climate change lawsuits by the cities of San Francisco and Oakland against five oil companies, saying the complaints required foreign and domestic policy decisions that were outside its purview."[241]

While the carbon-resource producers never denied fossil fuels and climate change were a problem, whether or not Big Oil intentionally destroyed the world was well beyond the scope of one court and these companies. The judge said, "... the problem deserves a solution on a more vast scale than can be supplied by a District Judge or jury in a public nuisance case." During the trial the judge also insisted that while the dangers of fossil fuels were becoming increasingly evident, they obviously had some benefit, something he had to take into consideration even if the plaintiffs did not.[242]

Seeking compensation today for damages that may or may not occur tomorrow has moved to Canada. Last year in British Columbia an ENGO called West Coast Environmental Law (WCEL) persuaded multiple BC municipalities to write letters to several multinational oil companies demanding funding to help pay for community protection for upcoming climate and weather damages. WCEL's website lists Saanich, Victoria, Powell River, Rossland, and West Vancouver as some of the more recognizable municipalities that have done so and publishes links to their letters.[243]

This campaign remained fairly low key until the mayor of mountain resort community Whistler, BC sent a request for funds to Calgary's Canadian Natural Resources Limited in November of 2018. This letter in turn found its way into the press. Whistler's mayor later publicly apologized but not before oil and gas

---

241 "USU.S. court dismisses climate change lawsuits against oil companies," *Reuters* (June 20, 2018). https://www.reuters.com/article/us-oil-climatechange-lawsuits/u-s-court-dismisses-climate-change-lawsuits-against-top-oil-companies-idUSKBN1JM0EP.

242 Ibid.

243 "Campaign Update," *West Coast Environmental Law* (2018). https://www.wcel.org/campaign-update.

industry pressure forced the Canadian Imperial Bank of Commerce (CIBC) to move the location of that particular segment of its annual investment conference out of Whistler to a different location after several oil company executives refused to attend. It is nearly impossible to travel to Whistler, or do anything once arriving there, without consuming fossil fuels. The outrage over Whistler's hypocrisy was long and loud.

In Quebec a group called ENvironnement JEUnesse (the latter word which translates to English as, "a time when you are young") announced it was suing the federal government for inadequate action in protecting Canada's youth from climate change. A *Globe and Mail* report on December 3, 2018 states the group was suing Ottawa "for violating the rights of young people by failing to tackle climate change." The article says the lawsuit was on behalf of all 3.5 million people in Quebec aged 35 or younger. The damages are set at $100 per person, or $350 million.[244]

Some of the positions taken by elected politicians appear to have been developed in a place where carbon-resource substitutes for transportation fuels and plastics already exist. Except they don't. In January 2019, the municipal council in the City of Victoria endorsed the concept of a class-action lawsuit to force oil- and gas-producing companies to help pay adaptation costs that current and future climate change may cause to be incurred in places like Victoria.

Access to Vancouver Island depends entirely on diesel-powered ferries for car and truck traffic or airplanes. Yet residents have yet to even imagine abandoning their leisure boats not powered by wind or paddles.

Victoria is a major port of call for cruise ships burning bunker oil, and the community is actively pursuing more cruise-line business. Other visitors arrive by ferry or plane. Tourism is somehow considered a benign form of economic activity but when the destination is an island, carbon energy remains the only means of access.

Across the Strait of Georgia on mainland BC is the huge Roberts Bank coal-export facility that ensures the power plants and coking mills of Asia never run out of high-carbon fuel.

---

244 Ingrid Peritez, "Quebec Group Sues Federal Government Over Climate Change," *The Globe and Mail* (November 25, 2018). https://www.theglobeandmail.com/canada/article-quebec-group-sues-federal-government-over-climate-change/.

And, of course, there is the legendary story of how Victoria still dumps raw sewage into the Pacific Ocean. Not exactly indicative of an environmentally sensitive city.

One can only speculate what Victoria's municipal politicians are thinking when they undertake this type of legal initiative. Except it is not unreasonable to conclude there are many more local votes to be gained from suing faceless, evil oil companies than from introducing policies and taxes that cost voters money, drive away valuable tourism revenue, or force serious lifestyle changes.

The scorecard is always the same. Local politicians 1, atmosphere 0.

On to subsidies. Somewhere along the way tax deductions by fossil fuel developers for investments and capital expenditures incurred in the normal course of business were redefined as government subsidies. The politicians bought into this argument and the crusade to end "fossil fuel subsidies" as reckless and dangerous taxpayer support for destroying the world entered the lexicon of the climate change movement. At most recent international conferences on climate change there have been commitments by countries to end "fossil fuel subsidies." Put it in the pile.

The issue of government subsidies resurfaced in early June of 2018 when an article appeared from "the Reuters Thomson Foundation – the charitable arm of Thomson Reuters that covers humanitarian issues, conflicts, land and property rights, modern slavery and human trafficking, gender equality, climate change and resilience." The title reads, "Rich Nations Spend $100 Billion A Year On Fossil Fuels Despite Climate Pledges."

This story was quickly picked up by dozens of online media redistribution services. This is how news travels in the twenty-first century, and how easy it is to disseminate anything when it involves climate and comes from a reputable source like *Reuters*.

The article claimed, "The world's major industrial democracies spend at least $100 billion each year to prop up oil, gas and coal consumption, despite vows to end fossil fuel subsidies by 2025, a report said on Monday (May 28) ahead of the G7 summit in Canada."[245]

The source was Britain's Overseas Development Institute (ODI). Its website states, "ODI promotes global progress and prosperity by focusing on improving

---

245 Lin Taylor, "Rich nations spend $100 billion a year on fossil fuels despite climate pledges," *Reuters* (June 3, 2018). https://www.reuters.com/article/us-global-climatechange-subsidies/rich-nations-spend-100-billion-a-year-on-fossil -fuels-despite-climate-pledges-idUSKCN1J000A.

the lives of the world's poorest people. Our research and policy work covers a wide range of development and humanitarian issues." This includes, "Climate, environment and natural resources."[246]

The article continued, "Britain scored the lowest on transparency for denying that its government provided fossil fuel subsidies, even though it supported tax breaks for North Sea oil and gas exploration."[247]

The idea fossil fuels are subsidized was easy to introduce into the climate change controversy because in some countries they are. It is the definition of the word *subsidy* that is in question, not the activity. The IEA follows the issue but does not employ the ODI definition. The IEA, "… estimates subsidies to fossil fuels that are consumed directly by end-users or consumed as inputs to electricity generation. The price-gap approach, the most commonly applied methodology for quantifying consumption subsidies, is used for this analysis. It compares average end-user prices paid by consumers with reference prices that correspond to the full cost of supply."[248]

Governments ensure their citizens are protected financially by absorbing the cash cost of the difference. Oil producing countries like Saudi Arabia, Iran, and Venezuela have long provided citizens with cheap gasoline. Poor and developing countries, including China and India, do this to encourage economic development and industrialization to lift their citizens out of poverty.

Using this method, the IEA subsidy estimate for the entire world is US$260 billion for 2016, down from US$320 billion in 2015. However, there were no G7 countries on the IEA's list. What G7 countries do is support investment in fossil fuel development and extraction through tax deductions, the same deductions available to all businesses investing money on expanded or future income. This is not considered a subsidy by the IEA definition.

Another article on subsidies was published by *The Guardian* August 7, 2017, titled, "Fossil Fuel Subsidies Area Staggering $5 Trillion per Year." This study was published by the Journal of World Development from work done by an arm of

---

246 "Our Work," *ODI: Evidence. Ideas. Change.* (n.d.). https://www.odi.org/our-work-0.

247 Lin Taylor, "Rich nations spend $100 billion a year on fossil fuels despite climate pledges," *Reuters* (June 3, 2018). https://www.reuters.com/article/us-global-climatechange-subsidies/rich-nations-spend-100-billion-a-year-on-fossil -fuels-despite-climate-pledges-idUSKCN1J000A.

248 "Fossil Fuel Subsidies," *International Energy Agency* (2018). https://www.iea.org/weo/energysubsidies/.

the IMF. To arrive at this enormous figure, the IMF attributed 22% of the cost to "undercharging for global warming," 46% to "air pollution," and 13% to "broader vehicle externalities."[249]

Undercharging for global warming is the future cost of climate change. Air pollution, which is caused by activities such as burning poor-quality coal to generate electricity, is a public health problem. Vehicle externalities include traffic accidents, which cause injuries and property damage, and traffic jams, which impair productivity. These account for 81% of the total, over $4 trillion. The opening line read, "A new study finds 6.5% of global GDP goes to subsidizing dirty fossil fuels."[250]

Another example of alleged subsidies is in a document titled, "Unpacking Canada's Fossil Fuel Subsidies" from *iisd.org*, the International Institute for Sustainable Development (IISD). The IISD mission is "to promote human development and environmental sustainability."[251]

It states, "A subsidy is a financial benefit the government gives, usually to a specific business or industry." It confesses to stretching the definition of subsidy by stating, "Economists can debate the difference between a subsidy and "support" for hours, *but that's a pretty good plain-English definition* [italics added]."[252]

The more one studies this subject, the harder it is to conclude which has suffered more damage: the climate or the English language.

IISD totals Canadian subsidies to $1.9 billion, $1.6 from Ottawa and $289 million from the provinces. The alleged subsidies include the traditional deductions for Canadian Development Expense (CDE), Canadian Exploration Expense (CEE), Crown royalty deductions, deep drilling credit, and Atlantic Investment Tax Credit.

For those who don't understand accounting, these are the bread-and-butter deductions that make capital investment in oil and gas economic. Global accounting firm KPMG says CEE expenses include tax deductions for geophysics, geology, coring, drilling, completion, roads, leases, etc. Eligible CDE deductions are drilling,

249 John Abraham, "Fossil fuel subsidies are a staggering $5 tn per year," *The Guardian* (August 7, 2017). https://www.theguardian.com/environment/climate-consensus-97-per-cent/2017/aug/07/fossil -fuel-subsidies-are-a-staggering-5-tn-per-year.

250 Ibid.

251 "Unpacking Canada's Fossil Fuel Subsidies," *International Institute for Sustainable Development* (n.d.). https://www.iisd.org/faq/unpacking-canadas-fossil -fuel-subsidies/.

252 "Ibid.

completion, roads, leases, waste disposal, secondary recovery, or converting a well from a producer to an EOR injector or observation well.

Every business in the Western world is entitled to deduct capital investment expenses for business expansion from taxable income. Only in the carbon-resource business are these routinely and disparagingly labelled subsidies.

CAPP tries to balance the debate, but it is a voice in the wilderness nowadays. CAPP's statement on fossil fuel subsidies reads, "One of the tenets of the G-20 commitment was to eliminate inefficient fossil fuel subsidies that 'encourage wasteful consumption, and undermine efforts combat the threat of climate change.' It is critical to distinguish between subsidies targeted to the production of fossil fuels, versus subsidies targeted to the consumption of fossil fuels … In Canada fossil fuel consumption subsidies are not prevalent, and in fact, the consumption of fossil fuels are heavily taxed which discourages consumption, the opposite of a subsidy."[253]

CAPP further states, "… it is erroneous and misleading to deem measures of the federal tax framework that seek to enable economic activity and maintain the neutrality of the tax system, a subsidy. In Canada, all businesses can deduct certain expenses and the oil and gas industry is no different."[254]

The October 2, 2018 announcement by LNG Canada that it was proceeding with a $40-billion project in Kitimat, BC to export liquefied natural gas to Asia was made economically viable by tax relief in the form of taxes not paid. This included provincial sales tax breaks by the province of British Columbia and a reduction in tariffs on steel imports by the federal government. These were levies not charged, not cash fossil fuel subsidies as defined by the IEA.

The last area where pressure is being applied through financial markets is forcing publicly traded companies to disclose the future risk climate change will have on their core business. On June 27, 2018 *The Globe and Mail ROB Magazine* published a major article by European Bureau Chief Eric Reguly titled, "Adapt or Die; How Big Name Investors are Pushing-Canadian Companies on Climate Change." The article focused on how shareholders were demanding greater accountability on how climate change and the phasing out of fossil fuels would impact future financial performance.

253 CAPP, *Fossil Fuel Subsidies* (November 2017). https://www.iisd.org/sites/default/files/publications/fossil_fuel_subsidies_WTO.pdf
254 Ibid.

Reguly related the story about the 2018 Annual General Meeting (AGM) of shareholders of Kinder Morgan, the pipeline company that sold the Trans Mountain pipeline to the federal government. At the meeting two shareholder resolutions passed. The first was better reporting on environmental, social and governance (ESG) issues. The other requested a report on how the international goal of capping global warming at 2°C would affect the company as it adapted its business model to a low-carbon future.[255]

The resolution was driven by a Boston-based fund's "manager of socially responsible investing," who called the move "historic." Something similar happened at the 2018 AGM of Anadarko Petroleum, a leading independent oil and gas producer. Other funds pushing for the changes at Kinder Morgan included the California Public Employees' Retirement Fund and the New York State Common Retirement Fund. The article also named another major investment fund that demanded more companies comply: the Caisse de dépôt et placement du Québec, "one of the world's top 20 pension funds, with about $300 billion assets under management."

The issue is gaining traction internationally. *The Globe and Mail* wrote, "One crucial ESG event this decade saw Canada's own Mark Carney, governor of the Bank of England and chairman of the Financial Stability Board, and Michael Bloomberg launch the Task Force on Climate-Related Financial Disclosures (TCFD). It urged companies to report their climate change risks to help them make the transition to the low carbon economy and allow investors to put a price on their investments' potential liabilities, from flood risk to stranded assets, as once-in-a-century climate disasters turn into once-in-decade catastrophes."[256]

That was three years and many meetings ago. Today TCFD has a global oversight board that includes representatives from major financial organizations and institutional investors from all over the world. Its website reads, "The TCFD seeks to develop recommendations for voluntary climate-related financial disclosures that are consistent, comparable, reliable, clear, and efficient, and provide decision-useful information to lenders, insurers, and investors. ... Better access to data will enhance how climate-related risks are assessed, priced, and managed. Companies

---

255 Eric Reguly, "Adapt or die; How big-name investors are pushing Canadian companies on climate change." *The Globe and Mail ROB Magazine* (June 27, 2018). https://www.theglobeandmail.com/business/rob-magazine/article-adapt-or-die-are-canadian-companies-ready-for-climate-change/.

256 Ibid.

can more effectively measure and evaluate their own risks and those of their suppliers and competitors. Investors will make better informed decisions on where and how they want to allocate their capital. Lenders, insurers and underwriters will be better able to evaluate their risks and exposures over the short, medium, and long-term."[257]

TCFD pays homage to the ENGOs that started this crusade by motivating students on university campuses and church groups to express their concern with their wallets. "Though great work has been done by NGOs in this space, as an industry-led initiative, the TCFD represents an opportunity to bring climate-related financial reporting to a mainstream audience." The chairman remains financial media tycoon Michael Bloomberg, a very public believer in climate change caused by fossil fuels.[258]

Financial Post reported September 20, 2018 that, "More Canadian banks measuring carbon risk as clients demand more analysis." This was in response to a report from a shareholder-proxy services company that "'ESG' had arrived in Canada in light of the growing number of such proposals brought forward by investors. 'This year, we saw a record number of majority-supported ESG proposals in the United States and a strong reception to proposals in Canada,' the proxy advisor said in a report ..."[259]

The move of some European banks to exit the sector was already public. Now Canadian banks like the CIBC and RBC were paying attention by developing tools and policies to better report on and manage risk in the sector. The article concluded that while Canadian banks are not yet passing judgement on the economic impacts of carbon and climate change on their lending practices or their customers, like the rest of the world, Canadian banks are certainly watching carefully the rapid changes in patterns of global capital flows and climate change.

A company trying to do the right thing while staying in the oil sands business is Suncor Energy Inc. In its public disclosure it acknowledges the growing pressure

---

257 Michael R. Bloomberg, "Final Report: Recommendations of the Task Force on Climate-Related Financial Disclosures," *Task Force on Climate-Related Financial Disclosures* (June 2017). https://www.fsb-tcfd.org/wp-content/uploads/2017/06/FINAL-TCFD-Report-062817.pdf.

258 Ibid.

259 Geoffrey Morgan, "More Canadian Banks Measuring Carbon Risk as Clients Demand More Analysis," *Financial Post* (September 20, 2018). https://business.financialpost.com/commodities/more-canadian-banks-measuring-carbon-risk-as-clients-demand-more-analysis.

from the world investment community but is truly caught between a rock and hard place because of its business model.

Since mid-2017 the TCFD has changed the thinking and conversation about disclosure for carbon and climate change. But for a company that is 100% carbon, including "extreme fossil fuels" as framed by the ENGO movement, explaining how it is going to be commercially successful and attractive to investors in a world that increasingly believes its number one product is a menace is very difficult. But it is trying.

> The recommendations provide a useful framework to describe how businesses are managing climate risk and ensuring corporate strategies remain resilient in a low-carbon future. There are still many details to work out, particularly around the appropriate disclosure vehicles to ensure we can provide a transparent and fulsome discussion on our climate strategy over the long term while recognizing the challenges of providing forward-looking information within regulatory financial disclosure requirements.[260]

Which is where climate change crusade and publicly traded fossil fuel companies collide. The entire anti-carbon movement is based on "forward-looking statements," including excessive, outrageous, and frequently preposterous predictions of catastrophe. But nothing is more regulated and fraught with peril for public companies than "forward-looking statements." Companies can be sued if they predict things that don't happen. Conversely, climate change activists are most-often praised for their expressions of concern, no matter what they say.

The whole climate-money initiative is about is putting a value today on catastrophic economic damages that haven't happened yet. Using carefully selected words, Suncor discloses this challenge as well as anybody, writing, "We look forward to working with the task force on this journey to shape and evolve climate risk disclosure so it meets the needs of both companies and investors and leads to better understanding of what's required to transition to a low-carbon future."[261]

---

260 "Our Perspective and Engagement," *Suncor Report on Sustainability 2018* (2018),https://
    sustainability.suncor.com/en/climate-change/our-perspective-and-engagement.
261 Ibid.

Public-company carbon-disclosure risk re-emerged in New York State on October 24, 2018, when the attorney general sued ExxonMobil, charging with improperly advising investors of the future cost of the emissions of its Canadian oil sands investments. The litigation alleges ExxonMobil used one number publicly while using a lower figure internally. As was the case with previous litigation, ExxonMobil has denied wrongdoing.

Taken to the logical conclusion of those pushing for a zero-carbon world, Alberta's public forward- looking disclosure should read, "Sell you house today while it is still worth something and relocate to somewhere else with a future." The government has already published the future weather forecast: bad and getting worse.

The attack on carbon capital and investment appears to be working. At the January 2019 World Economic Forum in Davos, Switzerland, a separate gathering of representatives of oil producing countries and companies took place. According to a *Reuters* news report, the main subjects were climate change and the ongoing negative pressure from investors and capital providers.

Share prices for most of the world's publicly traded exploration-and-production and oil-service companies are significantly lower than a year ago, and many investors and analysts are starting to see sector valuations disconnected from the price of oil. More institutional investors are questioning whether carbon resources have a future and if so, what will it be?

Reuters wrote, "The conclusion of the discussions were worrying for those present – pressure is rising and the industry is losing a battle not to be seen as one of the world's biggest evils. The answer? Lure investors with higher returns and raise the PR game." OPEC's secretary general described the world oil business as, "Under siege."[262]

While this might be regarded as success for fossil fuel's critics, it also means less capital for future supplies. Considering the reality of growing demand, this could result in possible future price spikes and shortages.

How the fossil fuel capital squeeze and the forces behind it could jeopardize future oil supplies was highlighted again in a *Reuters* article February 26, 2019.

---

262 Dmitry Zhdannikov, "DAVOS—'Under siege', oil industry mulls raising returns and PR game," *Reuters* (January 24, 2019). https://www.reuters.com/article/davos-meeting-oil/davos-under-siege-oil-industry-mulls-raising-returns-and-pr-game-idUSL1N1ZO0MB.

Saudi Aramco CEO Amin Nassar told an industry event in London that, "Important stakeholders believe that the entire world will soon run on anything, but oil. These views are not based on logic and facts, and are formed mostly in response to pressure and hype." Nassar continued, "Our industry faces a crisis of perception, … Our traditional qualities of ample, reliable and affordable supply are not enough to meet society's expectations today."[263]

But as will be explained further in upcoming chapters, ensuring the world has cheap energy and all the comforts and convenience it delivers is certainly not what this is all about.

---

263 Rania El Gamal, "Aramco CEO says oil industry facing a 'crisis of perception'," *Reuters* (February 26, 2019) https://mobile.reuters.com/article/amp/idUSKCN1QF0YN.

4

THE MANY FACES OF CLIMATE CHANGE

"…whatever the apparent intentions are of those who recite the climate change mantra and tell everyone else what to do—as soon as possible—they are not transparent. Many have motives other than the size of the Antarctic ice cap in the year 2100."

Yet another old adage goes, "Do as I say, not as I do." Think of Neil Young, James Cameron, Jane Fonda, or Leonardo DiCaprio. They fly a long way—usually in a private jet from somewhere far away in the US—to Alberta's oil sands for a flash tour. Followed by TV cameras and reporters, they visit the area in a big, comfortable vehicle powered by gasoline or diesel.

They are there to trash the oil sands—and by default everyone associated with it—with a contempt normally reserved for totalitarian dictators or mass murderers. They tell everyone that mankind must use less fossil fuels and that oil sands are the most egregious example why. Their message is carried far and wide because they are really famous and, thanks to the internet, they have millions of followers on social media who hang on to their every utterance. ENGOs proudly add them to growing lists of important people who care.

They fly back home on their private jets, add a bunch more CO2 to the upper atmosphere, and resume their normal lives. All of these famous oil sands critics has a carbon footprint larger than a small, rural village without electricity in a Third-World developing country. The message? Everyone must use less carbon, particularly that dreadful oil-sands carbon.

Every year the United Nations has another COP climate change meeting somewhere. This is in addition to dozens of gatherings on various aspects of the same subject. The concerned fly from all over the world to agree, yet again, that fossil fuels must go. They issue yet another public statement about the disaster unfolding and plead for real political leadership. Their message calls for powerful, purposeful decision-making and the creation of regulations and legislation that, taken to their fullest extent, would ground the very airplanes that brought them there. The press release or report is generally issued after they go home, naturally.

These scenarios are but two of the many faces of climate change: a growing global movement to either champion the cause, benefit from it, or both. That you must burn a lot of fossil fuel to get rid of fossil fuel is overlooked by the believers.

This is not to say climate change is not a problem, or that significant reductions in fossil fuel consumption and CO2 emissions won't makes things better. It's just that whatever the apparent intentions are of those who recite the climate change mantra and tell everyone else what to do—as soon as possible—they are not transparent. Many have motives other than the size of the Antarctic ice cap in the year 2100.

Back in Stockholm in 1972 when the UN first acknowledged environmental problems must also be addressed, climate change was not the key issue. It was pollution. At that conference India's Prime Minister Indira Gandhi delivered a message that would affect all future global environmental discussions: since developed countries created the mess, they must bear the greatest financial burden to clean it up.

Gandhi declared, "The environmental problems of developing countries are not the side effects of excessive industrialization but reflect the inadequacy of development. The rich countries may look upon development as the cause of environmental destruction, but to us it is one of the primary means of improving the environment for living, or providing food, water, sanitation and shelter; of making the deserts green and the mountains habitable."[264]

Worried that the West was putting too much blame and burden for pollution on developing countries that were struggling to catch up economically, Gandhi stated, "There are grave misgivings that the discussion on ecology may be designed to distract attention from the problems of war and poverty. We have to prove to the disinherited majority of the world that ecology and conservation will not work against their interest but will bring an improvement in their lives. To withhold technology from them would deprive them of vast resources of energy and knowledge. This is no longer feasible nor will it be acceptable."[265]

---

264 "Indira Gandhi's Speech at the Stockholm Conference in 1972," *LASU-LAWS Environmental Blog* (July 18, 2012). http://lasulawsenvironmental.blogspot.com/2012/07/indira-gandhis-speech-at-stockholm.html.

265 Ibid.

The 1972 UN conference concluded with a resolution that greening the world would not put an undo economic burden on developing countries. This theme continued through all the UN environment climate change conferences and permeates the discussion today.

Meanwhile, since 1972, China has grown its economy to become the world's largest GHG emitter, due largely to coal power. The US is number two and India number three. These three countries account for 54% of global GHG emissions.

The UN's propensity to link controlling climate change with social and economic equality has caused many critics of proposed global-warming responses to develop conspiracy theories that the issue is more about global wealth redistribution than the environment. Much has been written about how climate change is a front for a master plan for global government.

There are two sides to this issue. The first is Westerners are energy pigs compared to developing countries. Data from WRTG Economics shows that this decade, the US is consuming about 23 b/o per person per year. Canada is a bit higher at 25. Germany is about the same as Canada. Estimates for China for 2018 are only 3.5 b/o per capita per annum; India 1.4.

However, since the issue of global environmental carbon equity emerged in 1972, US, Canadian, and German oil consumption per person has declined while that of China and India has skyrocketed. All forecasts indicate that the greatest growth in carbon energy demand and GHG emissions will come from Asia because that's where most of the people live and is where the economic growth rate is the highest.

Is it possible to reduce carbon emissions and ensure global economic and environmental equity at the same time? With the same policy set?

Another opportunity to use climate change as a front for another purpose is articulated by the socialists of the world, now renamed "progressives." They have never trusted, and often despised, capitalism and free markets. Their sentiment seems to be that now that the profit-first forces of darkness have altered the world's climate with devastating results, the imperative to lever climate change into a socialist revolution has never been stronger.

Nobody articulates this better than Naomi Klein, if not Canada's most famous socialist certainly the most widely read. She released a large book in 2015 titled, *This Changes Everything: Capitalism Vs. The Climate.* This follows earlier works with titles like *No Logo: Taking Aim at the Brand Bullies; Fences and Windows*

(1999); *Dispatches from the Front Lines of the Globalization Debate* (2002); and *The Shock Doctrine: The Rise of Disaster Capitalism* (2007). The theme of all of them is consistent. Most of the world's greatest problems are caused by free markets, free trade, free enterprise, and the big corporations that profit at the expense of the workers of the world.

This Changes Everything allows Klein to wrap everything she has ever believed is bad about capitalism into one narrative. Free enterprise is not only the cause of climate change, but impending climate disaster is the best reason yet to get rid of it. After providing yet another summary of the world's bleak future, she provides hope providing people understand the immensity of the challenge and are prepared to accept a completely different way of life and new world order. This will be accomplished by the transferring of economic power and influence away from corporations and to governments and the people. This includes international trade. The big bad oil companies are, of course, high on the list. (Klein 2015)

Klein's views that there is a single menace to mankind's current climate challenges is very popular among the political left. Look no further than the Leap Manifesto, a platform that has been put before the national New Democratic Party for consideration but has not yet been officially ratified. It starts with, "Canada's record on climate change is a crime against humanity's future. ... We could live in a country powered entirely by renewable energy, woven together by accessible public transit, in which the jobs and opportunities of this transition are designed to systematically eliminate racial and gender inequality. We know that the time for this great transition is short. Climate scientists have told us that this is the decade to take decisive action to prevent catastrophic global warming. That means small steps will no longer get us where we need to go."[266]

If the mass transit part reads a bit urban-centric, that's because it is. In 2018 Western Canada lost its regional bus service when Greyhound pulled out. The idea a new electric train fuelled by renewables will provide service from Edmonton to Slave Lake, Peace River, Manning and High Level and all towns, settlements and farm in-between is hard to conceive.

Climate justice and social justice are, to the left, the same thing. Canada can and must be rebuilt. This would include building and retrofitting energy-efficient

---

266 This Changes Everything team. *the leap manifesto* (2015). https://leapmanifesto.org/en/the-leap-manifesto/.

homes, job retraining, ending trade agreements that interfere with the master plan, and expansion of low-carbon sectors of the economy like government services.

The *Leap Manifesto* website reads, "The writing of the Leap Manifesto was initiated in the spring of 2015 at a two-day meeting in Toronto. ... The *This Changes Everything* team convened the meeting ..."[267] The website advertises Klein's book of the same name which was accompanied by a movie. The "team" includes Avi Lewis, Klein's husband and a broadcaster.

Climate change as a lever for restructuring society is not unique to Canada. In February 2019, Alexandria Ocasio-Cortez, a newly-elected Democratic Party member of the House of Representatives from New York, introduced a resolution titled, "Recognizing the duty of the Federal Government to create a Green New Deal" to combat climate change through a complete social and economic restructuring of the US.

For America to do its part in holding the global temperature increase to 1.5°C within 10 years, the country must ground all airplanes, retrofit all buildings to make them energy efficient, quit using internal combustion engines and everything they propel, and prevent cows from passing wind.

Concurrent with the rapid and massive reduction in US carbon emissions will be, "... a new national social, industrial and economic mobilization not seen since World War II and the New Deal." Government policy will, "... create millions of good, high wage jobs in the United States; ... provide unprecedented levels of prosperity and economic security; ... counteract systemic injustices (and) to secure for all people of the United States for generations to come – (i) clean air and water; (ii) climate and community resiliency; (iii) healthy food; (iv) access to nature; and (v) a sustainable environment."[268]

As a bonus, it will end exploitation and oppression of everyone including indigenous peoples, migrants, women, the poor, the young and the old, people with disabilities and those without housing.

Like Canada, transportation will be maintained by massive investment in high-speed rail powered by renewable energy. And, as it did in the Second World War, the government will borrow or print the money to pay for it all.

---

267 Ibid.

268 Resolution in the House of Representatives, Ms. Ocasio-Cortez, February 2019. https://www.govinfo.gov/content/pkg/BILLS-116hres109ih/pdf/BILLS-116hres109ih.pdf

While social justice and equality for all remain worthy objectives for any society, exactly how these wrongs can finally be righted in conjunction with changing the composition of the atmosphere is not intuitive.

Funding ENGOs to continue their ability to fight the good climate fight is another key element of climate change that is as important as reducing emissions. Measured by words written and books printed, the world's greatest ENGO leader is surely Bill McKibben, founder of *350.org*. He has written 16 books since 1989, most of them about the environment and climate change.

One, from 2011, is titled *eaarth; Making A Life on a Tough New Planet*. Others include *The End of Nature* (1989); *Fight Global Warming Now: The Handbook for Taking Action In Your Community (2007)*; and *GWR: The Global Warming Reader: A Century of Writing About Climate Change (2012)*. McKibben intentionally misspells "earth" because "It's a different place. A different planet. It needs a new name. Eaarth" (McKibben 2011).

McKibben tells the reader he wrote his book near his home in Ripton, Vermont, in the Green Mountains of that state.

The population of Vermont in 2017 was 623,627 according to *worldpopulation-review.com*. The US census bureau reports that the population of Ripton, VT in 2016 was 588. It is from here that McKibben, by his own admission, views the world. It is beautiful, tranquil, quiet, green and, compared to many parts of the world, empty.

Then he spends the rest of the book explaining how awful things are, and how wonderful things would be if more of the world could be like Ripton. And, of course, if enough people joined his crusade.

That's not exactly how McKibben tells the story, but his vision of the future seems difficult to export to where the people increasingly live in megacities like Tokyo (38 million), Delhi (26 million), Shanghai (24 million), Sao Paulo (21 million), Mumbai (21 million), Mexico City (21 million), Beijing (20 million), Osaka (20 million), Cairo (19 million), and Dhaka (18 million). These 10 cities alone have a combined population of 228 million, 70% of the entire US population of 327 million.

McKibben founded *350.org* with his friends in 2008. The name was chosen because 350 parts per million (ppm) is the concentration of CO_2 in the atmosphere that climate scientists agreed is the maximum to prevent global warming and devastating climate change. That number was exceeded early this decade.

The Earth Science Research Laboratory reports that the level CO_2 concentration in September 2018 was 405.51.[269] McKibben writes that before the Industrial Revolution, the figure was 275 ppm.

McKibben talks about the history of the global warming movement, the science, and the failed attempts by the UN to address the problem in a meaningful way. He was very disappointed by the 2009 climate change conference in Copenhagen where the attendees agreed to do something bold and constructive but instead settled for more expressions of great concern with no tangible action. Poorer and low-lying countries likely to be affected by rising oceans wanted to see significant emission reductions from the wealthier countries to ensure their survival. McKibben was disappointed no real action was forthcoming.

At Copenhagen, McKibben says the 350-ppm carbon dioxide ceiling was ratified by over half the attendees. But ultimately it failed. A major setback took place when the US senate failed to ratify the carbon cap and trade program introduced by the Obama government earlier in 2009.

McKibben's narrative is that all is not lost so long as people are prepared to make significant changes in their lifestyles and expectations, changes so significant that civilization will change. While most environmentalists want governments to raise the price of carbon energy to the point renewables or alternatives become economic and widely adopted, for McKibben it's back to the land. He writes about a world where we abandon factory farming and feedlots and get back to growing our own foods. McKibben describes how the Chinese fertilized the soil by recycling their own body waste instead of using massive amounts of ammonia and nitrogen-based fertilizer made from natural gas.

The solution is a return to small-scale agriculture, much smaller farms with no synthetic fertilizers.

The last chapter of *eaarth* is dedicated to reminding people one last time how awful things are, but that there is indeed hope. Just join the cause. "… ExxonMobil and BP and Peabody Coal are allowed to use the atmosphere, free of charge, as an open sewer for the inevitable waste from their products. They'll fight to the end to defend that business model, for its produces greater profits than any industry has ever known. We won't match them dollar for dollar; To fight back we need a different currency, our bodies and our spirit and out creativity. That's what a

---

269 "Trends in Atmospheric Carbon Dioxide," *Earth System Research Laboratory: Global Monitoring Division* (December 6, 2018). www.esrl.noaa.gov/gmd/ccgg/trends/.

movement looks like; let's hope we can rally one in time to make a difference" (McKibben 2011).

350.org has turned into an effective global ENGO and Bill McKibben is their disciple. Decide for yourself if his back-to-the land model that might work in Ripton, VT is transferable to the hundreds of millions of people who live in the largest urban centres of the world, or whether the land that surrounds these megacities is ready for their return to growing their own food by hand without fertilizer and living a simpler life.

The home page of *350.org*'s website reads, "350 uses online campaigns, grassroots organizing, and mass public actions to oppose new coal, oil and gas projects, take money out of the companies that are heating up the planet, and build 100% clean energy solutions that work for all. 350's network extends to 188 countries." The definition of success is to "keep carbon in the ground, … help build a more equitable zero-carbon society, … and "pressure governments to limiting emissions."[270]

350.org is not huge and refreshingly transparent. Its 2017 annual report disclosed US\$16.8 million in revenue, up from US\$12.9 million the prior year. Reported expenses were US\$12.9 million and US\$10.7 million respectively, meaning the organization is profitable. A big success (measured by the font size) in 2017 was under the heading "Resisting Fossil Fuel Development." It reads: "ENERGY EAST WIN – 1,100,000 BARRELS OF OIL PER DAY STOPPED." After years of fighting … the largest tar sands pipeline ever … is officially cancelled."[271]

Well into its second century, ENGO Sierra Club is a much bigger organization according to Activist Fact, which is "committed to providing detailed and up-to-date information about organizations and activists," all classified as "tax exempt non-profits."[272] For the Sierra Club, Activist Facts reports the organization had revenue in 2015 of US\$109 million. In its profile of the Sierra Club states, "… the organization advocates an end to obtaining energy from ANY fossil fuel sources, nuclear power, or large-scale hydroelectric projects even though leading climate change scientists note that wind and solar aren't enough to meet our energy needs."[273]

---

270 "About 350," *350.org* (n.d.). https://350.org/about/.

271 "Resisting Fossil Fuel Development," *350.org* (n.d.). https://350.org/2017-annual-report/.

272 "Sierra Club," *Activist Facts* (2015). www.activistfacts.com.

273 Ibid.

The proposed transition from, and ultimate demise of, fossil fuel consumption is perhaps the greatest economic undertaking and disruption in history. According to the IEA in its 2017 World Energy Outlook (revised, page 99) the world will spend about US$7 trillion annually early in the next decade on oil, electricity, coal, and natural gas. This is greater than the GDP of every country in the world except the US and China.

Whether or not this should be the case, of the 10 biggest companies in the world, six are in the energy business. Five produce oil and gas and the sixth electricity. According to the Forbes 500 report for 2017, the top 10 companies had revenues of $3.1 trillion. The five oil companies—Sinopec, China National Petroleum Corporation, Royal Dutch Shell, ExxonMobil and BP—had revenues of US$1.8 trillion or 47% of the total. The second-largest company was China's electricity giant State Grid with revenues of US$349 billion. State Grid burns coal to generate electricity but reported in late 2017 that, "… we will gradually reduce the share of fossil fuels in energy mix so that the installed capacity of clean energy will surpass that of coal-fired power in the next decade."[274] Include State Grid and the fossil fuel business grows to US$1.8 trillion or 59% of the top 10 total. Just from six companies.

The losers—the companies that must go—are well-known, the targets have been painted on their foreheads for decades so the assassins can't miss. Big Oil. Big Coal. Big Business. The anti-carbon propaganda is rife with vitriol written about the companies that supply the world with energy, fuel, and electricity.

But what is not written as often is how large the commercial opportunity to replace fossil fuels will be. This is the "$30-trillion cleantech" market opportunity federal Environment Minister Catherine McKenna speaks about. These are the good capitalists. Nothing wrong with them apparently.

On October 4, 2018, the IPCC released an updated report indicating that holding the increase in the earth's average temperature to 1.5°C by 2100 would require CO_2 emissions to fall by 45% by 2030 compared to 2010 levels and "net-zero" by 2050. The IPCC admitted this was ambitious but, "… possible within the laws of chemistry and physics."[275]

---

274 "Installed Capacity of Clean Energy or to Surpass Coal-fired Power in the Next Decade in China," State Grid (December 12, 2017). http://www.sgcc.com.cn/ywlm/mediacenter/industrynews/12/344022.shtml.

275 "Global Warming of 1.5°C," *IPCC* (October 4, 2018). https://www.ipcc.ch/sr15/.

Shortly thereafter, on October 7, 2018, *Bloomberg* newswire released a story titled "Climate Crisis Spurs UN Call for US$2.4 Trillion Fossil Fuel Shift." It started, "The world must invest $2.4 trillion in clean energy every year through 2035 and cut use of coal-fired power to almost nothing by 2050 to avoid catastrophic damage from climate change, according to scientists convened by the United Nations." It continued, "The US$2.4 trillion ... is also a sevenfold increase from the US$333.5 billion Bloomberg New Energy Finance (BNEF) estimated was invested in renewable energy last year."[276]

That is $US40.8 trillion. Where will the capital come from? What will it cost to service this investment once the new, carbon-free energy infrastructure is in place? What will energy cost?

Get used to seeing the name Bloomberg in every financial aspect of climate change. The BNEF website reads, "Bloomberg NEF is a leading provider of primary research on clean energy, advanced transport, digital industry, innovative materials, and commodities."[277] It was founded in 2008 because billionaire Michael Bloomberg figured the low-carbon-energy industry had a bright future.

Bloomberg is also chairman and a driving force behind the TCFD which is exerting pressure to persuade publicly traded fossil fuel companies to disclose their financial exposure to future climate change. When you're saving the world, there is nothing unethical about figuring out ways to weaken the competition and capitalize upon the new paradigm at the same time. That's why Michael Bloomberg is rich.

Another convert to making money by saving the world is Al Gore, of *An Inconvenient Truth* fame. In an interview with *Bloomberg Businessweek*, September 24, 2018, Gore reminded everyone about the challenge but, more importantly for a business magazine, the opportunity.

Gore co-founded Generation Investment Management in 2004, two years before he made his movie. The fund currently has US$21 billion under management targeted for investments in cleantech and renewables. Gore said, "We look for companies that don't borrow for the future, that help to create a more sustainable and healthier present without doing so at the expense of the opportunities that

276 Reed Landberg, Chisaki Watanabe, and Heesu Lee, "Climate Crisis Spurs UN Call for $2.4 Trillion Fossil Fuel Shift," *Bloomberg* (October 7, 2018). https://www.bloomberg.com/news/articles/2018-10-08/scientists-call-for-2-4-trillion-shift-from-coal-to-renewables.

277 "About: Overview," *BloombergNEF* (2018). https://about.bnef.com/about/.

should be available down the road. ... In the investor marketplace, there used to be this outdated view that fiduciary responsibility somehow prevented assets managers from taking the environment and social and governance factors into account. Now that's been turned on its head."[278]

While Gore did not disclose how the fund was doing—nor was he asked—he talked about some of the investments. One was the manufacturer of "the best electric bus in the entire world," the latest version of which Gore claims can go 1,000 miles without recharging. Another is an electric scooter with an interchangeable battery. The rider doesn't recharge the battery but swaps it out at a network of recharging stations. Another company in East Africa installs solar panels on houses off the grid as well as solar-powered TV sets. In the US, one of Gore's investee companies offers a digital system to manage restaurant food waste.

But Gore admits it won't be easy. "It's almost unimaginably difficult to make the transition that's now under way, because we still rely on fossil fuel for 80 per cent of the world's energy use." The issue is, will massive change occur "... in time to avoid crossing some dangerous threshold that would throw the climate balance out of kilter."[279]

Having helped identify the problem, he's working hard to solve it. "We can create tens of millions of new jobs retrofitting buildings with better insulation, better windows, LED lights, reducing the utility bills of the buildings' occupiers and owners, cleanup up the environment."[280]

If this fund were investing in fossil fuels, having a fund manager shamelessly pitch the bright future of his investments would be a conflict of interest. In saving the world it appears Gore is harkening back to the "enlightened self-interest" practiced by Alexander Galt in 1880. Gault, a developer of the CPR, made money in Britain promoting Canada's national railroad at the same time as he sold coal from his own mine to the CPR.

The carbon-trading market as envisioned at the Kyoto climate change conference in 1997 was also about big money. Kyoto was attended by John Palmisano, lobbyist for the disgraced and bankrupt Enron Corporation, a US energy company

---

278 "Al Gore is Still Optimistic," *Bloomberg Businessweek* (September 20, 2018). https://www.bloomberg.com/news/features/2018-09-20/al-gore-beholds-the-climate-apocalypse-and-is-still-optimistic.

279 Ibid.

280 Ibid.

that made a fortune in America's growing deregulated energy markets in the 1990s. Once called "the smartest guys in the room," Enron made a fortune buying, selling, and speculating on energy commodities such as natural gas and electricity.

The climate change agreement in Kyoto anticipated a global carbon-emissions trading system. Trading commodities and derivatives was Enron's core business. At Kyoto's conclusion, Palmisano wrote a memo to head office on what had just happened. It opened with, "If implemented, this agreement will do more to promote Enron's business than will any other regulatory initiative outside of restructuring of the energy and natural gas industries in Europe and United States. The potential to add incremental gas sales, and additional demand for renewable technology, is enormous. In addition, a carbon emissions trading system will be developed."[281]

While it would take some time, the memo concluded, "... I am sure that reductions will begin to trade within 1-2 years. Finally, Enron has immediate business opportunities which derive directly from this agreement. ... On the business front: During the next year there will be intense positioning of organizations to capture an early lead in a variety of carbon trading businesses."[282]

Enron had been lobbying for financial support for energy supplies with lower carbon emissions. Palmisano continued, "The endorsement of joint implementation within Annex-1 is exactly what I have been lobbying for and it seems like we won. The clean development will be a mechanism for funding renewable projects. Again, we won (We need to push for natural gas firing to be included among the technologies that get preferential treatment from the fund). A 'clean development fund' is included. The fund would allow for emission offsets from projects in developing countries. Joint implementation for Annex-1, developed countries and the transitional economies, is included. This means that Enron projects in Russia, Bulgaria, Romania or other eastern countries can be monetized, in part, by capturing carbon reductions for sale back in the US or other Western countries."[283]

The Kyoto strategy was to have the program working within a decade. After a couple more pages of details, he concluded, "This agreement will be good for Enron stock!!"

Enron eventually went broke in 2001 amid great scandal not because of its business model, but because of questionable and ultimately illegal activities by several

---

281 www.mastersource.org/kyoto-protocol/palmisano-kyoto-memo-10/.
282 Ibid.
283 Ibid.

of its key executives. Its reputation was so awful that in 2016, after Alberta's new NDP government discovered coal-fired electricity generators had wording built into contracts that allowed them to opt out of their agreements if the province's new coal-powered regulations were commercially punitive, Alberta called it the "Enron clause." Enron was, after all, run by crooks who cleverly erected barriers to Alberta's climate change progress.

Two things are illuminating about the Enron memo.

First, it describes in simple English how putting a price on carbon and creating an emissions trading market using cap and trade would work.

Second, it illustrates in the enormous financial opportunities that would be created by putting a price on carbon emissions. Producing carbon generated a lot of wealth. Recognizing that people would have to buy energy in some form, as carbon was penalized there was an equally large—if not larger—opportunity to generate wealth getting rid of it. At this point Enron's management was widely admired in the private sector as being clever, nimble and ahead of the curve. If Enron could see the opportunity, so could others.

The carbon market didn't go as far as conceived at Kyoto in 1997. The largest today in the EU. More on this next chapter.

An interesting twist to put a price on carbon to address climate change was a means by which carbon emitters could pay for their sins by purchasing or emission credits, or "carbon offsets".

Personal carbon offsets are ideal for virtue signalling. Many jet-setting carbon opponents proudly announce their travel is "carbon neutral." Air Canada sells offsets with every ticket through its association with Less Emissions, a "gold standard international offsets" vendor. Travellers can calculate their carbon emissions and determine how many credits would be needed to offset their travel. The calculation can be done right on the Air Canada website by entering the number of passengers, the starting point, and the destination.

At Less Emissions they sell "gold" level certified carbon offsets at $32 per tonne. Less Emissions Inc. is a unit of Bullfrog Power, one of the many new green and renewable energy companies. All offsets purchased are "certified," meaning they are not a scam. The website reads, "Gold Standard certified projects—A global standard for projects in developing countries that verifiably achieve GHG emissions reductions at the source and create positive impacts on social networks and their local economy."[284] The product is endorsed by the World Wildlife Fund, the David Suzuki Foundation, and the Pembina Institute. It is unknown whether ENGO endorsements are free for noble commercial undertakings or whether cash changes hands.

The industrial emissions purchasing market got off to a controversial start. An article in *The New York Times* on August 8, 2012 was titled, "Profits on Carbon Credits Drive Output of a Harmful Gas." It described how an Asian manufacturer of gases used in refrigeration and air conditioning discovered the money to be made "by simply destroying a ton of an obscure gas normally released in the manufacturing of a widely-used coolant gas." The gas was far more powerful than CO_2 for trapping heat. Destroying one ton of this gas was 11,000 times more valuable than eliminating one ton of CO_2, which made offset sales commensurately higher. Therefore, the company manufactured more coolant gas to generate more valuable and marketable offsets.[285]

This was in the early days of the international carbon market, and the rules have tightened considerably. There were also many questionable offset products in the early days of consumer offset sales. The notion of planting trees in the Third World sounded good, but the reality is that they would not actually remove much carbon from the atmosphere until they grew to an effective size years later.

To be inclusive on the many faces of climate change, one would be delinquent to not include what became known as "greenwashing," the rebranding of ordinary products to make them appear greener or less harmful to the environment.

According to an article in *The Guardian* in August of 2016, "The term greenwashing was coined by environmentalist Jay Westerveld in 1986, back when most consumers received their news from television, radio and print media – the same outlets that corporations regularly flooded with a wave of high-priced, slickly-produced commercials and print ads. The combination of limited public access to information and seemingly unlimited advertising enabled companies to

---

284 "About Less Offsets," *Less Emissions* (2018). https://www.less.ca/en-ca/lessoffsets.cfm.

285 Elisabeth Rosenthal and Andrew W. Lehren, "Profits on Carbon Credits Drive Output of a Harmful Gas," *The New York Times* (August 8, 2012). https://www.nytimes.com/2012/08/09/world/asia/incentive-to-slow-climate-change-drives-output-of-harmful-gases.html.

present themselves as caring environmental stewards, even as they were engaging in environmentally unsustainable practices."[286]

Marketing research revealed that by the 1990s the "green" movement had grown to the point where, all things being equal, people were more likely to buy a product if the manufacturer proclaimed its concern about the environment. This resulted in the massive rebranding or repackaging of just about everything in the past 25 years.

Think about it. What do you buy or see advertised any longer that doesn't claim to be as harmless or pure as possible. Most of this tremendous progress has been achieved will little or no harm done and without massive increase in costs.

Back in the 1950s toilet paper was available in all manner of stylish colors but in the 1980s white became western standard. This was because of the irritation of chemicals in the paper dyes on humans and the environmental impact. Nobody recycles used toilet paper and we all know where most of it goes.

Many products have been repackaged to make them easier to recycle and decompose. Leading the list today is the impending death of the plastic drink straw which time has proven to be virtually indestructible without heat. That more paper products are increasingly made from recycled paper is displayed prominently for customers to see. Plastic grocery bags are still available to stores, but more retailers and consumers prefer re-usable bags. Some stores charge for plastic bags to discourage their proliferation.

Even fossil fuel producers and the automobile industry have joined the parade to persuade the eco-wary that their products are in fact much more benign than critics have claimed.

Global oil giant BP (originally British Petroleum) became "Beyond Petroleum" long before its massive, unfortunate, and tragic Macondo oil blowout in the Gulf of Mexico in 2010. When Canada's Husky Oil added ethanol to its gasoline—mandated by federal law—the new product became "Mother Nature's Fuel." An engine installed in a Ford 150 pickup truck that used the combination of turbocharging and direction injection to improve performance and fuel economy was called "EcoBoost."

It's everywhere. If you're 40 or under, you won't remember "Tony The Tiger," Esso's slogan that its gasoline would "Put a Tiger in Your Tank." That was when

---

286 Bruce Watson, "The Troubling Evolution of Corporate Greenwashing," *The Guardian* (August 20, 2016). https://www.theguardian.com/sustainable-business/2016/aug/20/greenwashing-environmentalism-lies-companies.

gasoline was about power and speed, not the end of the world. Imagine the reaction today if the Canadian subsidiary of the world's most denigrated oil company used an endangered species to flog its climate-destroying fuel.

There are no surprises in the many faces of climate change. But if you're trying to make your own decision about what to believe and the best course forward, it is considerably more complicated than finding the information you agree with and shutting your brain off.

5
THE CARBON TAX CONUNDRUM

"What is remarkable about the carbon tax debate in Canada is that it no longer has any relationship to its original intent, which was to save the world by discouraging the consumption of fossil fuels. With political parties at war within Alberta and several provinces at war with federal government, the discussion is more about political will and electoral success and how few people will actually have to pay it."

C arbon taxes have been part of the global-warming and climate change vernacular for over 20 years. They are widely hailed as one of the most effective ways to use market forces to reduce fossil fuel consumption and emissions. How? This is not complicated. If something costs more, people will buy less. This is one of the fundamental tenets of economics.

But the simplicity of the solution is also the root of the problem. Given a choice, people don't want to spend more than they must for anything without a compelling reason. This is complicated by the timing of the expenditures. We're told we must pay higher costs right now for the necessities of life to ensure a lower average temperature for the earth 20, 50, or 100 years into the future.

In today's world you pay now for most things now and get the product—fast food, clothing, fuel—immediately. The advent of credit created a situation where you get the goods or service today and pay later. You can have up to 25 years to pay for your house but you still get to live in it.

While climate change crusaders continually stress that people must pay today to ensure a better future for themselves, their children, their grandchildren, and the planet, why people aren't embracing carbon taxes is hardly complicated.

Persuading people to accept carbon taxes requires: 1) agreement on the problem; 2) agreement that higher costs are the solution even without immediate tangible benefits; 3) trust that policy-makers introducing the taxes know what they are doing, and that the revenue is truly earmarked for the greater good, not short-term political gain.

Carbon taxes have been headline news almost as long as climate change. They may be forever. A Google search for the words "carbon tax" brings up 180 million hits. Searching "Premier Rachel Notley," the creator of Alberta's consumer carbon tax, only gets 344,000.

The first discussions about turning carbon taxes into policy began among European Economic Community members in 1991 and gained more traction at the United National Conference on Environment and Development in Rio de Janeiro, Brazil in 1992. What emerged from the "Earth Summit" was the UNFCCC, a plan to stabilize then reduce global GHG emissions. This was formally ratified in Kyoto, Japan, in 1997. Part of the so-called "Kyoto Protocol" was a "cap and trade system," which would put a price on carbon globally.

Treating carbon as a sin by taxing it was an attractive solution because history has proven if you want less of something, tax it more. This applies to virtually everything. You want fewer rich people, tax them more. If you want people to consume less tobacco or alcohol for their own good, tax both more. You want people to use less water, tax it more. Nothing to it.

If you want more of something, the opposite is true. While modern society has little experience adapting to tax cuts, it has happened.

Reductions in income taxes are stimulative because they increase the disposable income of the populace, thereby putting more money into the economy. Corporate tax cuts assist businesses in expansion and attract new investment. Alberta's corporate tax cuts in the 1990s helped persuade major corporations to move their head offices to the province. Modern society has limited experience with tax cuts.

When governments raise taxes too high, to the point that it affects the economy in ways not planned, the result is "unintended consequences." In the 1970s Sweden raised personal tax rates on its wealthiest residents to 85%. Tennis star Bjorn Borg, filmmaker Ingmar Bergman, and pop group ABBA all left the country. Closer to home, in the 1970s and 1980s high taxes and royalties on oil and gas production introduced by the federal and provincial governments chased capital out of the country at the same time Canada expressed concern about insufficient domestic oil supplies. When the Alberta government under Premier Ed Stelmach raised royalties in 2007 through the New Royalty Framework, investment fled Alberta.

Thanks to higher taxes from Edmonton and Ottawa for almost everything, this is what's happening in Canada today. Because US President Donald Trump has cut taxes, all Canadian businesses are worried about Canada's growing non-competitiveness caused by higher taxes. Powerful tools, those taxes.

Carbon taxes were conceived to discourage and reduce the consumption of fossil fuels—coal, oil, and natural gas. The application of such a tax is technically referred to as a "negative externality," which is described as follows:

> A negative externality occurs when an individual or firm making a decision does not have to pay the full cost of the decision. If a good has a negative externality, then the cost to society is greater than the cost the consumer is paying for it. Since consumers make decisions based on where their marginal cost equals their marginal benefit, and since they don't take into account the cost of the negative externality, negative externalities result in market inefficiencies unless proper action is taken.[287]

And so it goes for taxing carbon. Cheap fuel or energy is of great benefit to an individual but according to climate science it has a negative impact on everyone else. Another term is "tragedy of the commons." "The tragedy of the commons is a term used in social science to describe a situation in a shared-resource system where individual users acting independently according to their own self-interest behave contrary to the common good of all users by depleting or spoiling that resource through their collective action."[288]

Carbon taxes were immediately attractive to some politicians. Raising taxes to save the world was an intriguing opportunity for legislators and regulators. The first country to introduce a carbon tax was Finland in 1990. Website *carbontax.org* says in the remainder of the 20th century Finland was joined by, in chronological order, Poland, Sweden, Norway, Denmark, Latvia, Slovenia and Estonia.

From 2001 to 2017 another 36 jurisdictions would join the first eight countries in "carbon-pricing initiatives." By 2017, 43 countries and subnational jurisdictions like Alberta, BC, and California were doing something in this area. The participating jurisdictions accounted for about 13% of global GHG emissions.[289] Notably absent were the United States, Russia, and India. In late 2017, China announced

---

287 "Economics," *Fundamental Finance* (n.d.). http://www.economics.fundamentalfinance.com/.

288 "Tragedy of the Commons," *Wikipedia* (December 17, 2018). https://en.wikipedia.org/wiki/Tragedy_of_the_commons.

289 *Carbon Tax Center* (2018). https://www.carbontax.org/.

that it would be introducing a "carbon market." If and when it is fully functional, this would raise the percentage of global GHSs being priced and taxed in some way closer to 25%.

That people are unlikely to voluntarily use less carbon to avoid negative externalities is well understood; hence the call for government action. The failure of politicians to do what should have been done with carbon emissions years ago is named by commentators as the primary cause of the undesirable weather events of the summer of 2018.

What's the problem? How can something this simple be so complicated? The reality is that telling people what needs to be done is infinitely simpler than putting in place the processes to make it happen.

There is broad agreement among economists that the most effective way to reduce carbon consumption and subsequent emissions is to price it. Where and how remains the subject of great debate after a quarter century of discussion, analysis and attempts. Because for any form or method of carbon pricing to be effective, the cost of one of life's most essential commodities—energy—must rise. It follows that the cost of all products made from carbon and delivered with the assistance of carbon must also rise.

The critics of a carbon tax have declared it "a tax on everything." Which is precisely what it is.

You don't need to be a political genius to figure out what happens when governments decide to tax everything. Going back to James Buchanan's 1986 Nobel Prize winning *Public Choice Theory*, politicians—democratically elected or not—are not likely to make decisions that seriously imperil their own futures. Governments that intentionally impoverish their citizens are eventually replaced by either the bullet or the ballot. Dictatorships are overthrown through civil disobedience or insurrection. Democratic governments are replaced on election day.

Even authoritarian governments face dissent such as riots when costs go too high. Countries that subsidize gasoline experience mass protests when they try to raise the price. Higher food prices were partially responsible for the insurrections that led to the so-called Arab Spring. In France, a democracy, farmers routinely take to the streets when the government tries to *reduce* the cost of food by reducing subsidies to farmers. Riots over rising fuel taxes rocked that country in late 2018.

The first form of carbon tax, the one endorsed by the Kyoto Protocol, was "cap and trade." This is primarily industrial and applied to companies or industries.

The policy was directed at companies that produce carbon energy and products or consume a lot of carbon in the production or manufacture of goods or delivery of services.

The cap—the ceiling of permitted emission levels—is determined by analyzing carbon emissions and setting an upper limit. The amount of carbon emissions subject to taxation is called the "allowance." Government-issued credits to specific industries or companies that exceed the cap are also called allowances but are actually exemptions. Carbon has a value when emissions are either above or below the cap. Emissions exceeding the cap are taxed as a penalty. This provides an economic incentive to stay under the cap. Carbon emissions below the cap also have value because they can be sold or traded. The entire process is called a carbon market.

For example, if a company exceeds its emission cap and is facing additional levies, it can purchase carbon credits or offsets from another company below its cap. Those exceeding their emission caps are punished. Those reducing emissions below the cap are rewarded by being able to monetize the carbon reductions.

The problem was implementation and how to enforce emission caps. This is how it was meant to work on a country-to-country basis as conceived at Kyoto:

> Carbon trading began in response to the Kyoto Protocol, signed by 180 countries in 1997. The Kyoto Protocol ... called for 37 industrialized countries to reduce their greenhouse gas emissions between the years 2008 to 2012 to levels that are 5% lower than those of 1990. Article 17 of the Kyoto Protocol established emissions trading by allowing countries that have emission units to spare (emissions permitted to them but unused) to sell this excess capacity to countries that are over their emissions limits. In effect, this created a new commodity in the form of emissions and created a carbon market. Since CO2 is the principal greenhouse gas, emissions trading effectively became carbon trading.[290]

By 2005, dozens of the world's largest companies wanted clarity on the rules by which the global carbon-trading market would operate. They sought, "... 'clear,

---

290 "Carbon Trading," *Source Watch* (March 19, 2015). https://www.sourcewatch.org/index.php/Carbon_trading.

transparent, and consistent price signals' through 'creation of a long-term policy framework' that would include all major producers of greenhouse gases. By December 2007 this group had grown to encompass 150 global businesses."[291] By 2008 the market was said to be US$47 billion, and another source said US$129 billion only a year later.

While sound in theory, the problem was the inability of countries to set "clear, transparent and consistent price signals." As caps were set, so were allowances, particularly when companies or industries complained they deserved preferential treatment by politicians or policy makers. By 2008 the ENGO movement was calling the whole process flawed because some companies had, of course, figured out how to make money. That's what companies do. There was a profit incentive built into the process as a reason to change carbon emission behaviour.

Individual environmentalists attacked established ENGOs like Environmental Defense Fund for supporting the carbon-trading market. By 2011 there were reports of hackers stealing emissions trading credits and selling them. Farmers marched on the Chicago Mercantile Exchange, a major commodity trader, in 2011 to protest carbon trading because by then it also involved food.

As the first adopters of carbon taxes in the 1990s and being primarily fossil fuel consumers (not producers), Europe led the way in adopting the Kyoto cap and trade model. The European Union Emissions Trading System was implemented in 2005 and continues today. Its website describes itself as accounting for over three-quarters of global carbon-emissions trading activity. It operates in all 28 EU countries plus Iceland, Norway, and Liechtenstein. It "limits emissions from more than 11,000 heavy energy using installations (power station and industrial plants) and airlines operating between these countries." It involves about 45% of the EU's GHG emissions.[292]

Its own review of how the trading system works, or doesn't work, is illuminating. In the first trading period from 2005 to 2007, the EU admits participating countries granted too many allowances, so the price of carbon fell to zero. The second trading period from 2008 to 2012 saw allowances reduced by 6.5% but the recession had a significant impact on cutting emissions, resulting in more credits for sale than buyers. The third trading period from 2013 to 2030 is predicted to see a further

---

291 Ibid.

292 "EU Emissions Trading System (EU ETS)," *European Commission* (n.d.). https://ec.europa.eu/clima/policies/ets_en.

reduction in the total EU emission cap of 1.74% and "a progressive shift toward auctioning of allowances in place of cost-free allocation." The EU is optimistic that by 2020 GHG emissions among participating countries and companies will be 21% below that of 2005. The EU claims its GHG emissions in 2016 were 22% lower than in 1990, although they increased slightly in 2017.[293]

Carbon-emissions trading never had the impact envisioned at Kyoto in 1997. There was some optimism the US was going to create a carbon-trading market following the Copenhagen Accord of 2009. Legislation was introduced by the Obama administration in 2009 and passed in the House of Representatives. However, it was defeated in the senate. Outside of Europe, California introduced an emission-trading system in 2011. Quebec did the same in 2011 then joined with California in 2013. Ontario connected with California and Quebec in 2017 but that was all undone by new PC government after the 2018 Ontario election.

The political attraction of a corporate emission-trading market using cap and trade is the cost of carbon to the consumer is hidden, buried in the final cost. The drawback is that it does not capture any of the billions of smaller carbon reduction opportunities available globally at the personal consumption level.

When Alberta introduced a large-emitter carbon tax in 2007 it was a hybrid; a direct levy on carbon emitters but not on consumers. In this case the funds were earmarked for carbon-reduction research and found their way back to some of the same companies that paid them. BC's 2008 carbon tax went further including direct consumer expenses for products like gasoline, but it also involved a concurrent reduction in other taxes. The final outcome was revenue neutral; a change in behaviour by taxing carbon but without a net financial penalty.

The "perfect world" carbon tax, at least to economists, is a tax on the carbon component of everything at the end-user consumption level. This would result in billions of individual pricing decisions every day all over the world. Since the intent is to reduce carbon consumption—not the ability of consumers and households to buy food clothing and shelter—the perfect consumer carbon tax is also revenue neutral. Carbon taxes should be designed to be offset by tax reductions in some other area.

Companies would pay for the carbon content of all industrial inputs. However, carbon taxes would be accompanied by corporate tax decreases. Individuals

---

293 Ibid.

would be treated the same way, with reductions in income taxes or other federal or provincial sales taxes. By taxing carbon but offsetting the cost by personal tax cuts or corporate tax cuts, individuals and businesses could be more profitable than before if they exhibited the desired behaviour of consuming less carbon. Somebody with income who figured out how to use no little or no carbon would be far better off financially.

Unfortunately, in Canada the new NDP government in Alberta and the new Liberal government in Ottawa never got the memo from economists. Or if they did, they ignored it.

In Alberta, the NDP government's 2015 Climate Change Leadership Plan included broad-based carbon taxes on most things with some exemptions. This came into effect January 1, 2018. The carbon tax followed the mid-2015 corporate tax increases and personal income tax increases on higher income earners. Alberta's carbon tax was highly visible and appeared as a line item on electricity and natural gas bills. At one point, when natural gas prices were very low, on a gas bill the carbon tax was greater than the carbon-based energy it was penalizing.

But the NDP plan included rebates, not tax cuts. Accompanying the carbon tax was a series of payments to lower income earners. Below a certain income threshold, individuals would get a $300 annual rebate, couples $450, and families $540. The Alberta government website reads, "Carbon levy rebates protect low- and middle-income Albertans who spend a higher percentage of their income on energy costs and have fewer financial resources to invest in energy efficiency products. An estimated 60% of households will get a full or partial rebate."[294] Businesses and higher income earners would bear the burden of the carbon tax.

In the 2015 federal election campaign, the Liberal Party telegraphed its intentions to take some sort of action on carbon taxes if elected. The policy platform released prior to the October 2015 federal election stated, "The provinces and territories recognize the need to act now, and have already begun to price carbon and take action to reduce greenhouse gas emissions. ... We will end the cycle of federal parties – of all stripes – setting arbitrary targets without a real federal/provincial/territorial plan in place. We will work together to establish national emissions-reduction targets, and ensure that the provinces and territories have

---

294 "Carbon Levy and Rebates," *Alberta.ca* (2018). https://www.alberta.ca/climate-carbon-pricing.aspx.

targeted federal funding and the flexibility to design their own policies to meet these commitments, including their own carbon pricing policies."[295]

Once elected, the federal government announcement that effective January 1, 2019, Ottawa would impose a carbon tax on provinces that had not yet done so. This isn't exactly what had been promised during the campaign, but it has been the stated law of the land for the past three years. It is scheduled to be implemented April 1, 2019.

On October 24, 2018, the Liberals announced how it would handle carbon taxes in provinces without one. Rebate cheques would be sent to individuals and households that in some cases would exceed the anticipated carbon taxes paid. If this looks like pure politics, it is. The *Canadian Press* version of the announcement read, "Trudeau said there is both a 'moral and economic imperative' to manage climate change, so the next generations of Canadians aren't left in a world where monster storms and massive droughts burn up or flood out major portions of the planet."[296]

Trudeau continued, "Will we kick this can down the road yet again to be dealt with in another place or another time or will we show some courage and do what must be done for this generation or the next?" The article stated, "Officials said more than 70 per cent of households will end up receiving more than they pay."[297]

Inspiring words and politically attractive financially. But will it accomplish anything?

In Alberta in 2017, on the day Jason Kenney entered the race for the UCP partly leadership, he declared war on the governing NDP's carbon taxes. Kenney was later elected party leader and then became an MLA. The Ontario election of June saw an upset over the same issue. The Liberal government, which had been in power since 2003, had embraced carbon taxation through cap and trade as one of many tools to combat climate change. The party was humiliated at the ballot box by Doug Ford's Progressive Conservative Party and its campaign pledge to eliminate them.

---

295 "Climate Change," *The Liberal Party of Canada* (2018). https://www.liberal.ca/realchange/climate-change/.

296 Mia Robertson, The Canadian Press, "Federal Carbon Tax Rebates will Exceed the Cost for Most People Affected," *CTV News* (October 23, 2018). https://www.ctvnews.ca/politics/federal-carbon-tax-rebates-will-exceed-the-cost-for-most-people-affected-1.4145915.

297 Ibid.

All taxes are regressive. This means the lower the income, the greater impact a certain tax has on disposable income. Which is fine. But if the rebate offsets the carbon taxes paid, the net effect could be no reduction in carbon consumption or emissions. Everyone understands the concept of tax fairness in an egalitarian society like Canada. But we're supposed to be saving the world. How do you do that if 60% or 70% the people paying carbon taxes are fully or partially exempt?

On October 5, 2018, UCP leader Jason Kenney and Ontario Premier Doug Ford held a boisterous rally in Calgary during which Ford reiterated his determination to challenge a federal plan for a national carbon tax network and Kenney reinforced his commitment to end Alberta's.

As this book goes to press, Saskatchewan, New Brunswick and Ontario are undertaking legal challenges against Ottawa's right to impose a carbon tax if a province fails to introduce one satisfactory to Ottawa. Manitoba has objected to a carbon tax, supported it, then changed its mind again and is now opposed. Prince Edward Island is not on side.

This opposition is to taxes aimed at consumers. The same provinces that object to Ottawa's broad consumer carbon tax have all introduced large-emitter carbon tax programs.

Following the 2017 election, BC's new NDP government increased carbon taxes. After Ottawa's failure to deliver on the expansion of the Trans Mountain pipeline to Burnaby in August 2018, Alberta withdrew from the federal plan that it once endorsed and postponed a scheduled future provincial carbon tax increase.

On January 21, 2019 in a speech to the Economic Club of Canada in Toronto, Ontario Premier Doug Ford told the audience, "I'm here today to ring the warning bell that the risk of a carbon tax recession is very, very real", assuming that the federal government proceeds with forcing at tax on the provinces that don't have one in place.[298]

What is remarkable about the carbon tax debate in Canada is that it no longer has any relationship to its original intent, which was to save the world by discouraging the consumption of fossil fuels. With political parties at war within Alberta and several provinces in court with federal government, the discussion is more about political will and electoral success and how few people will actually have to pay it.

---

298 The Canadian Press, "'A total economic disaster': Doug Ford says the Liberals' carbon tax will plunge Canada into recession," *The National Post* (January 21, 2019). https://nationalpost.com/news/canada/doug-ford-says-the-liberals-carbon-tax-will-plunge-canada-into-recession.

Secondary is if or how Canadians can or should contribute to the global commitment to mitigate the impacts of climate change, or the fact that Canada is a tiny carbon emitter compared to the US, China and India which are not following Canada's lead. Which means Canada's actions, whatever they may be, will not move the needle on global GHG emissions.

There is the economic purity of carbon taxes and the political debate, which are not the same thing. Trevor Tombe is an associate professor of economics at the University of Calgary and is in the centre of a debate that is increasingly about politics and not climate change. Tombe was interviewed for this book in July 2018. He criticizes both sides, the climate change alarmists who want to "decarbonize whatever the cost" and Alberta politicians with "communications strategies and campaigns which force polarization." He expresses frustration with the absolute black and white views of both sides.

As an economist, Tombe takes a classic view that carbon should be priced "at the margin" to force the right decision one way or another. He explains how "certain uses of carbon have benefits that far outweigh the cost," and when that is the case then carbon is the answer. He uses the example of vehicles. A gasoline-powered ambulance rushing a sick or injured person to hospital is an excellent use of carbon. The benefits clearly outweigh the cost. According to Tombe a poor use of carbon is a car left idling in the winter.

Tombe asks the key question, "How do we know when emissions are less valuable?" He then answers it by bringing in the negative externalities caused by carbon emissions, measured by the tonne. The answer is in the cost, which should be "set at the price we think the damage is of that one tonne of emissions." He also believes corporate taxes should be reduced to "lower the cost of investing" in technologies and processes to consume less carbon and reduce emissions. To this end he is in favour of 100% tax write-offs for investments in carbon reduction. Because in the end, when carbon taxes are designed, Tombe reminds everyone, "What is the goal? What are we trying to achieve?"

As an economist, Tombe believes that the market has the best chance of dealing with this issue in a meaningful way by pricing carbon at the margin so every purchasing decision has the appropriate cost/benefit, risk/reward decision behind it. "That's the beauty of market-based policy. Where we are today is a direct result of how we are thinking about the problem – it is black and white on both sides."

The NDP claims its carbon taxes are aiding the climate change battle even though the net cost to 60% of Albertans reflects a part of, or is less than, the taxes themselves. Ottawa has now upped the ante by offering rebates, which for many are promised to be higher than the taxes paid. The financial incentives of reducing carbon use by offsetting lower taxes for high-income earners and corporations are gone. In Alberta, those taxes were increased long before the carbon tax was introduced.

On the other side, the UCP claims this is a tax on being alive in Alberta, a tax that will have no meaningful impact on global emissions. Both facts are technically correct. Which is why the subject is so complicated.

Economist Jack Mintz is president's fellow at the School of Public Policy at the University of Calgary. He was interviewed in August 2018. On climate change he says, "You need to know where you're going. The real test is can you get your policy turned into politics?" Writing policy, which is really coming up with ideas, is much easier that implementation, particularly on a broad scale. Mintz says, "A lot of the policy forecasts don't really do very good job of judging how people react to these things." The reality is, says Mintz, "paying today for a problem that won't come for 100 years is a tough sell."

While change is inevitable, it won't happen overnight. He says, "There is a still room to take advantage of the resources we have right now. We must be careful with our policies if they undermine what we should be doing; false policies which create a lot of costs which undermine the objectives. Policies should put the minimum cost on the economy. I support putting a cost on carbon but tax carbon alone. That is not what is happening."

Mintz points out there have been many programs aimed at mitigating climate change, but they ended up being very expensive with minimal emission reductions. He cites the phasing out of coal-fired electricity generation in Ontario, which ended up costs $100 per tonne of carbon eliminated, and electrical vehicle subsidies that cost $200 tonne. On the matter of carbon reduction, "No one really knows what will work best so just put a price on it," which will allow the market to decide. "We should get rid of all the other policies that compete [with market decisions] like subsidizing renewables like solar and wind."

He figures another approach would be to put money into research and development tax credits. Mintz used as an example the federal IRAP (Industrial Research Assistance Program), which works well, mainly because it is depoliticized. Alberta's

Climate Leadership Plan was very politicized because at the same time carbon taxes were introduced, so was the accelerated phasing out of coal-fired electricity generation (also for health reasons) and the introduction of subsidies for renewable energy. In this manner, "Governments are undermining the effectiveness of carbon taxes, which is why carbon taxes are so unpopular. I don't like the direction of carbon policy right now."

Mintz is on the record through his regular columns in the *National Post*, where he explains how federal and provincial personal and corporate tax increases, combined with the increasing cost of climate change initiatives, are affecting the competitiveness of the entire country. Personal tax rates are now at 53% in Canada and only 44% in the US. Young people who can move south will move. When corporate tax rates were increased, "We lost one of the major advantages we have."

In a *Financial Post* column in April 2018, Mintz reiterated how the Canadian approach to carbon taxes was a failure, writing, "… and economists have touted a potential 'double dividend' in carbon tax revenues are used to reduce the most harmful taxes on the economy, such a corporate and personal income taxes." He continues, "Therefore, regulations to phase out coal, promote ethanol, or subsidize electric cars, among other policies, are unnecessary. Wherever the cost to reduce emissions is less than the carbon tax, consumers and businesses will naturally adopt that technology without needing any further incentive."[299]

Reacting to Ottawa's new plan to issue rebates that in some cases exceed carbon taxes paid, on October 25, 2018 Mintz wrote a *Financial Post* commentary titled, "The Liberals Ignored My Carbon Plan. Theirs is Much Worse."[300] Mintz recalled a 2008 paper he co-authored that proposed converting the federal fuel excise tax into a carbon tax, which is already $0.10 a litre.

Total fuel taxes on gasoline—federal, provincial, GST, and PST—are already significant. If taxes actually force demand destruction, surely it would have happened by now. The Canadian Taxpayer Federation reported that in May 2018 gasoline taxes ranged from a low of $0.30 a litre in Manitoba to $0.55 a litre in

299 Jack Mintz, "Jack Mintz: Carbon Pricing has been Fully Exposed as Just Another Tax Grab," *Financial Post* (April 10, 2018). https://business.financialpost.com/opinion/jack-mintz-carbon-pricing-has-been-fully-exposed-as-just-another-tax-grab.

300 https://business.financialpost.com/opinion/jack-mintz-the-liberals-ignored-my-carbon-tax-plan-theirs-is-much-worse.

Montreal. The tax portion ranged from 25% to 40% of the retail price. The latest federal plan will supposedly add $2 to a tank of gas. So what?

Mintz wrote, "I stand behind the points of the 2008 paper. ... And we should integrate existing federal-provincial fuel taxes ... by transforming them into general carbon taxes. With a single carbon tax rate, businesses and consumers can determine the best way to reduce emissions."[301]

He continues with his argument that carbon taxes will work best if left alone. "If producers see that they can save more carbon tax by investing in a technology or process, they will adopt it. If the technology is prohibitively expensive (such as carbon capture and storage) they won't. This is how a carbon tax works: Not governments trying to make decisions for us, without any idea what new technologies might arrive tomorrow, lacking the wisdom to assign differential carbon prices. Yet that's what the government will continue doing with a plethora of taxes, regulations and subsidies, despite politicians attempts to argue that the carbon tax is a 'market based' solution to climate control."[302]

Calgary's Jim Dinning was Alberta's former PC Minister of Finance in the Ralph Klein era and a former chancellor of the University of Calgary. He now sits on an advisory board of Canada's Ecofiscal Commission and was interviewed for this book in August 2018. The website of the Ecofiscal Commission summarizes its purpose: "Ecofiscal policies put a price on pollution. They offer real incentives for investment in innovative technologies so that we can continue benefiting economically from our natural wealth while also providing better protection to the environment. The revenue generated from ecofiscal policies can create further economic benefits; for example, by reducing income and payroll taxes or investing in new technologies or critical infrastructure."[303]

This is carbon tax purity because it prices what politicians have labelled pollution, yet doesn't remove money from the economy to ensure there are adequate funds to use technology and innovation to accelerate the process.

Dinning says to makes progress you "need to start with the end in mind and work yourselves backward from there. If you make carbon taxes revenue neutral, then carbon taxes are the most conservative public policy." Dinning sees real opportunity in Alberta. "We have to do something and we can because we are so

---

301 Ibid.
302 Ibid.
303 "What is Ecofiscal?" *Canada's Ecofiscal Commission* (n.d.). https://ecofiscal.ca/what-is-ecofiscal/.

knowledgeable about energy. We know a lot about other things about energy; we shouldn't just talk about oil and gas." And, most importantly, "We should not be on the bleeding edge of climate policy."

"Bleeding edge" is business slang for the money-wasting, exact opposite of leading-edge, commercially valuable innovation.

Having spent much of his working career in politics at the most senior levels, Dinning is not a fan of the ability of governments to tackle major problems. What governments say and what they do is contradictory. "Governments are horrible at saying we are not out to change your behaviour but we're not going to change anything until we change the behaviour of people and companies." Dinning repeats the Ecofiscal Commission mantra that to be effective, carbon taxes should be accompanied by reductions in personal income tax rates and corporate income tax rates.

And he reminds people what success looks like with carbon taxes. "If the policy is successful, carbon taxes will be a declining source of revenue" because carbon will gradually disappear from the economy as alternatives are developed. Putting carbon taxes into general revenue without clear signs it is either neutral, or intended to fund other benefits and technologies, is certain to invite greater attacks from political opponents. Thinking about governments in general, Dinning adds, "There is no problem we [governments] can't make bigger."

In Ottawa growing complaints are being heard about how simultaneously solving climate change and raising taxes affects Canadian competitiveness. In August 2018 the federal government announced it was reducing the scope of a carbon tax (that didn't yet exist) to keep five major industries—steel, iron, lime, nitrogen fertilizers, and cement—competitive with international competitors.

One of the godfathers of carbon taxes is Yale University economics professor William Nordhaus, who has been publishing papers on climate change and economics since 1977. He was recently awarded a Nobel Prize in Economics for his contributions in this field. He has written, "Climate change is a member of a special kind of economic activity known as global public goods. At a minimum, all countries should agree to penalize carbon and other GHG emissions by the agreed upon minimum price."[304]

---

304 William Nordhaus, *Carbon-Price* (2005). https://carbon-price.com/climate/experts/william-nordhaus/.

In 2005 Nordhaus summarized the global political reality of solving the problem stating, "... there is no legal mechanism by which disinterested majorities of countries can coerce free-riding countries into mechanisms that provide for global public goods. For economic public goods, there are three potential approaches; command-and-control regulation, quantity oriented market approaches, and tax or price-based regimes. Of these, only the tradeable-quantity and the tax regimes have any hope of being reasonably efficient."[305]

Nordhaus suggests that if enough countries with sufficient economic clout could work cooperatively and adopt a common carbon tax platform, then their collective might would then persuade other countries to join through punitive tariffs and penalties. While it would work if it existed, it remains theoretical and aspirational for now.

The approach of Stephen Harper's CPC administration was that carbon taxes had to be global to be effective. Harper was not publicly opposed to carbon taxes in Canada but saw that if Canada alone introduced them while other major economies like the US did not, Canada would be placed at a severe economic disadvantage.

A paper titled "Overcoming Public Resistance to Carbon Taxes" was posted on *WIREs Climate Change*, published by Wiley Periodical Inc. in May 2018. Reviewing studies and papers from all over the world, the authors concluded that there were five main sources of resistance to carbon taxes. Carbon taxes had a greater impact on low-income earners; direct subsidies into low-carbon energy sources like renewables were more effective; they would damage the greater economy (see Harper above); and governments just wanted more money, which is fundamentally a trust issue.

While the authors repeated the theoretical economic genius of carbon taxes to reduce carbon emissions, the data shows governments prefer direct subsidies. The paper states, "Despite these advantages, carbon taxes are one of the least used climate policy instruments. In 2016, 18 countries and two Canadian provinces have implemented a carbon tax. In comparison, 176 countries had policy targets for renewable energy and/or energy efficiency, and 110 national and subnational jurisdictions had a feed-in tariff (pay more for renewables from small generators). Carbon tax proposals have been undone, sometimes at an advanced political stage,

---

305 Ibid.

for example in Australia (2014), France (2000), Switzerland (2000, 2015), and most recently in the United States in Washington State (2016)."[306]

They concluded the best way to gain greater acceptance would be phase carbon taxes in gradually, "earmark" (make clear the use of the revenue that will go to emission reduction), compensate lower income households, and cut other taxes.

Their summary of what the people who are expected to pay carbon taxes believe deserves repeating because this is a global view, not just local. "Voters are instinctively against new taxes, even if they are explicitly aimed at preventing environmental harm. They are doubtful about the effectiveness of a carbon tax, dislike its coercive nature, and are concerned about its impact on low-income households. These perceptions are not necessarily all correct, but they matter, and they have made it difficult to pass carbon tax proposals in the past."[307]

The unpopularity of carbon taxes erupted onto the global stage in late 2018 when the so-called "Yellow Vest" movement in France staged violent demonstrations across the country. This coincided with another UN climate change conference in Poland where the prevailing theme once again was the world isn't moving quickly enough to save itself from carbon-induced climate catastrophe, the problem carbon taxes are supposed to help fix.

The French are notably obstreperous as a species, often taking to the streets to express their dissatisfaction with anything that disrupts their lifestyles or costs them money. The issue of another round of carbon tax increases was in addition to others such as illegal immigration, declining real incomes, and diminishing job opportunities. The cause of the riots and the characteristics of the rioters was well analysed by the world's media. But ultimately it was the working French rebelling against well-paid government officials—the ones with job security, indexed pay raises, and pensions—telling them what sacrifices they needed to make for their own good. A headline in *The Globe and Mail* December 7, 2018, read, "France's Fuel Protests Reveal Dangers of Heaping Carbon-Reduction Costs on the Poor." Which would be fine if the world didn't have billions and billions of poor people who need fossil fuels to get through the day, or who aspire for a time when they

---

306 Stefano Carattini, Maria Carvalho, and Sam Fankhauser, "Overcoming public resistance to carbon taxes," *WIREs Climate Change* (June 6, 2018). http://wires.wiley.com/WileyCDA/WiresArticle/wisId-WCC531.html.

307 Ibid.

will be wealthy enough to afford the most basic trappings of modern society such as electricity or mechanized transportation.[308]

Since the riots in France and the recognition by Canadian political parties that carbon taxes have become politically attractive wedge issues to replace governments in Edmonton and Ottawa, more commentators are wondering about the usefulness of the climate change mitigation tool. Will carbon taxes solve anything if the people who must pay them for the plan refuse to work?

Mark Jaccard, a professor at the School of Resource and Environmental Management at Vancouver's Simon Fraser University, has written extensively about the climate and the economy. In a blog on his website on December 15, 2018, Jaccard lamented about the current highly politicized state of carbon tax policy. He wrote, "Carbon pricing is not essential to stop burning coal and gasoline. We economists only say it is because we prefer it. If we were honest, we would explain that decarbonization can be achieved entirely with regulations. These will cost more, but not a great deal more if policy-makers use flexible regulations, or 'flex-regs,' that allow companies and individuals to determine their cheapest way to decarbonize." This also appeared in a column in *The Globe and Mail*.[309]

People are asking the same question in the US. An article on the website *Politico* on December 9, 2018 speculated about the success of policy tools that so many clearly despise. The headline was, "Why Greens are Turning Away From a Carbon Tax." The subtitle continued, "Putting an economic cost on greenhouse gases is proving a hard sell with the public, even as time to head off climate change shrinks."[310]

Sensing that taxing Canadians for their own good is not of much a vote-getter once people actually have to pay them, federal Environment Minister Catherine McKenna repeats the new official government line that this is a tax on pollution, not on the necessities of life.

---

308 Eric Reguly, "France's Fuel Protests Reveal Dangers of Heaping Carbon-Reduction Costs on the Poor," *The Globe and Mail* (December 7, 2018). https://www.theglobeandmail.com/business/commentary/article-frances-fuel-protests-reveal-dangers-of-heaping-carbon-reduction/.

309 Mark Jaccard, "Carbon Pricing: Wasting Time We Cannot Spare on the Optimal Steering Mechanism for the Titanic," *Sustainability Suspicions* (December 15, 2018). http://markjaccard.blogspot.com/2018/12/carbon-pricing-wasting-time-we-cannot.html.

310 Zack Colman, "Why greens are turning away from a carbon tax" https://www.politico.com/story/2018/12/09/carbon-tax-climate-change-environmentalists-1052210.

On Twitter on January 3, 2019, McKenna wrote, "Fact: Unlike Conservatives, Canadians understand pollution shouldn't be free. Under our plan, we're pricing pollution, emissions & putting more money back into the pockets of families. With Climate Action Incentive, 8 in 10 families will get more money back than they pay."

Problem solved. Don't actually pay carbon taxes but pay more for a few things today that include carbon, and then get a bigger cheque back from the government for cooperating. Or, put differently, under the Liberal plan hopefully you won't riot in the streets like the French.

Small business lobby group Canadian Federation of Independent Business (CFIB) joined the debate February 12, 2019 when it released a survey indicating 80% of its members didn't believe they would be able to pass along more than 25% of the federal carbon taxes to customers. Widely publicized federal rebates are focused on voters, not businesses.

CFIB wrote on its website, "The majority of small businesses in Saskatchewan, Manitoba, Ontario and New Brunswick oppose the federal carbon backstop plan, which is set to come into effect this April in the four provinces, says the Canadian Federation of Independent Business (CFIB). The plan unfairly burdens small businesses, which will contribute almost 50 per cent of carbon tax revenues along with municipalities, universities, school boards and hospitals, but will receive just 7 per cent back in the form of rebates and grants."[311]

The CFIB continues, "'With consumers receiving 90 per cent of the 'Climate Action Incentive' payments and many large emitters receiving carbon tax exemptions, small businesses are left holding the bag,' said CFIB president Dan Kelly. 'The government's technical documents state that small businesses are expected to simply pass their added costs down to consumers, but 80 per cent of small firms in the four provinces report they will be able to pass on less than one quarter of the new costs.'"[312]

A terrible accident in Mexico on January 18, 2019 was a reminder that the ideal price for carbon-based fuel for many remains zero. Trying to scoop up gasoline leaking from a pipeline, hundreds were collecting fuel with pails when it caught fire, resulting in a massive explosion. This section of pipe leaked because it had been tapped to steal gasoline before and patched. At last count the official death

---

311 Canadian Federation of Independent Business, https://www.cfib-fcei.ca/en/media/small-firms-pay-almost-half-federal-carbon tax-get-7-cent-back.
312 Ibid.

toll was 85 with more people still missing. A similar event in Nigeria in 2006 killed some 200 people, except in this case the line was tapped into.

The fundamental problem with carbon taxes of any kind remains that nobody wants to pay them. While there remains broad academic consensus carbon taxes are the most effective way of reducing emissions, politics and politicians are in the middle. Which is what Jack Mintz observed by saying, "The real test is can you turn your policy into politics?"

6
ALBERTA WITHOUT CARBON

"The world has changed much more rapidly than Alberta has been able to adapt. All the key external drivers that made Alberta what it is, or more romantically what it was until 2014, are different... how much longer can the province and its governments afford to continue down the current path?"

Today, Alberta is confronted by challenges from domestic and international forces that have never before existed, including obstruction to its success and its future from other governments that were once supportive partners in Confederation. Directly and indirectly, governments in Vancouver, Burnaby, Montreal, Quebec, Ontario, British Columbia, and Ottawa have been able to impose their views of what four million Albertans should be allowed to do next, and not care a whit about the human and economic impacts of their actions.

It is clear that many who opine about the future of Alberta do not understand how dramatically the province's economic prospects have changed since the start of the twenty-first century. Or they know and don't care, which in some ways is even worse. The future looks so significantly different from the past that you'd think somebody would write a book about it.

For more than a century Alberta's almost continual growth—economic and lifestyle, personal and corporate—was encouraged and sustained by the rest of Canada and the rest of the world. The federal government and other provincial governments were for the most part allies; Alberta's success was Canada's success. Internationally Alberta's unique carbon treasure trove was envied, and the province was seen as a fair, equitable, stable, and predictable jurisdiction in which to invest. The US would buy all the oil and gas Alberta would sell. American, European, and Asian investors were pleased put significant capital into Alberta's growing and unlimited carbon resources.

Hundreds of thousands of people from all over the world moved to Alberta, attracted by economic freedom and unlimited opportunity. Alberta was place where newcomers were welcome and could achieve personal growth, fulfillment and prosperity.

But that all ended. The world price of oil collapsed, the climate change crisis gained greater momentum, and Alberta elected an NDP government.

Perhaps the first sign of trouble with the new NDP government was when the party's leader, Rachael Notley, referred to her home province as the "embarrassing cousin" of Confederation, primarily because of its environmental record. Things went downhill from there. In one year—from the OPEC meeting that crashed oil prices on November 27, 2014 to the announcement of the NDP's Climate Leadership Plan on November 22, 2015—Alberta was set on a new path that would seriously damage the province's economy for years to come, possibly forever.

To be fair, not all of Alberta's current challenges are the fault of a single provincial government. The NDP didn't invent climate change, collapse the price of natural gas, or build the bloated and expensive provincial administration it inherited. Big government, big spending, and big deficits in Alberta took decades to develop. Nor was the Notley administration responsible for the federal election of Justin Trudeau's Liberal government in October of 2015. No one anticipated that by February 2016 the price of oil would plummet to only 25% of its 2014 high. And certainly the NDP didn't know that after years of loud and growing opposition from US environmentalists, President Obama would officially kill the Keystone XL pipeline in late 2015.

But the NDP introduced broad, interventionist, and expensive policies after forming government in May 2015, only five months after oil prices had begun to collapse. And these policies were stunning in their indifference to economic reality. Like kids in a candy store, Alberta's new NDP government had a list of social wrongs to right and a majority government to do it with. Reshaping Alberta in its own image was more important to the NDP than paying attention the economic carnage being wrought on the province's most important industry.

In business you plan for the worst and hope for the best. But the NDP caucus was not oversupplied with MLAs with a lot of business or financial management experience. Everything that could go wrong with pipelines and oil prices did go wrong.

When Alberta's NDP assumed power, the full impact of the oil price collapse was not yet known. WTI was still fetching over US$59 a barrel. It would be less than half that value nine months later. US President Barack Obama had not officially killed the Keystone XL pipeline, nor had the Federal Court of Appeal yet overturned the previous NEB and federal government decision on Northern Gateway, as was the case in June of 2016.

As the economy and commodity prices collapsed in 2015, Alberta's punch-drunk NDP government raised taxes on virtually everything and everyone that made money, capped oil sands emissions, moved to phase out coal for electricity generation, increased subsidies for renewables, and raised the minimum wage to the highest in the country. Only small businesses got a much-publicized tax break. To act as a "shock absorber" for a collapsing economy, the government intentionally and proudly cranked up public spending and borrowed billions to pay for it. Deficits unseen in Alberta's history began to pile up.

With the Climate Leadership Plan in hand, the NDP sent a large delegation to the COP 21 climate change conference in Paris in late November 2015. They basked in their progressive, interventionist glory, proclaiming that the internationally demonized "tar sands" were now under the direct supervision of enlightened climate change believers. Coal phased out. Carbon taxes. Renewables. Oil sands emission cap. Big oil sands producers in agreement with the new plan, at least on television. If only the rest of the world could get with the program like Alberta. Then mankind might have a chance.

The Trudeau administration officially buried Northern Gateway in November of 2016 when it approved Trans Mountain and the Enbridge Line 3 expansion with great fanfare. The federal Liberals and provincial New Democrats bragged about how their combined progressive administrations and carbon taxes would get offshore export pipe built when previous backwards conservative politicians could not.

Through aggressive climate change policies, the two governments claimed to have purchased a "social license" that would move oil sands and pipeline opponents to drop their resistance. Why? Because different politicians were making the ask.

The ultimate death of Northern Gateway was not regarded by the NDP as a problem for Alberta but as another example of conservative incompetence. The Energy East pipeline was still alive plus, in early 2017, new US President Donald Trump had revived Keystone XL, killed by his predecessor. The sentiment was Alberta had three pipelines; one east, one south, and one west.

But later in 2017, a combination of events caused TransCanada Corporation to abandon Energy East. For Canada, all hopes were placed on the federally sanctioned Trans Mountain pipeline, the one that was seen as Premier Notley's very public ace in the hole. Multiple photo ops of shovels in the ground and pipe stacked in the background appeared on TV. Any day now. The Trans Mountain court challenge scorecard to date was Trans Mountain 16, opponents 0.

But in August 2018 relentless opposition by the BC government and court challenges by Indigenous and environment groups would kill that too.

As the export pipelines dropped off one by one, so did investor interest in Alberta. The costs were rising thanks to higher federal and provincial taxes, while the returns were shrinking as low-cost, low-risk market access through pipeline disappeared. Multiple oil sands producers with other places to invest voted with their feet.

What followed the policies of 2015's new Alberta and federal governments and the pipeline cancellations was the greatest financial exodus of foreign carbon-resource investment capital in the shortest period time in Alberta's history. Over $30 billion changed hands. Following decades of massive capital inflows, in 2016 and 2017 companies from Europe and the US sold their oil sands operations to Canadian buyers. One of the sellers was Shell Canada, whose CEO had appeared on TV with Premier Notley, endorsing Alberta's Climate Leadership Plan only 15 months earlier.

In early 2017 the rest of the oil and gas producing world began to recover thanks to a decision in late 2016 by OPEC and other producers like Russia to withdraw 1.8 million b/d of production from the market. After hitting record lows in February of 2016, by February of 2017 WTI had doubled to over US$53. By February 2018 it exceeded US$62. The world's oil industry was back on its feet. The oil surplus that had helped collapse prices was replaced again by concerns about future supplies.

Alberta, however, did not participate in the recovery. Due to all the higher taxes, the pipeline cancellations, and the complete destruction of investor confidence in Alberta as a sound place to invest, by mid-2018 the province was going in the exact opposite direction. This was the only major oil-producing region in the world where activity and investment was going down, not up.

While nobody in Ottawa or Edmonton would admit it, foreign capital wasn't going to return to Alberta's oil and gas industry until something big changed.

People who work in Calgary's oilpatch finance community have never seen a time when corporate valuations were so low and investor interest so scarce. The only activity in capital markets was among companies that were stuck in Canada, attempting to buy their competitors at rock-bottom prices.

The default group to blame for the stagnation of Alberta's primary industry is US ENGOs, which had so successfully declared war on Alberta's oil sands and export pipelines. However, these activists were assisted by Canadian municipal and provincial governments in British Columbia and Quebec and the Liberal government in Ottawa. All put their own political agendas ahead of whatever economic damage their actions might inflict on Alberta, its industries, and its citizens, by interfering in the construction of crude oil pipelines to the Pacific and Atlantic Oceans.

However unhappy Albertans may be with the NDP government BC elected in 2017, it was the BC Liberals under Premier Christie Clark that helped foster resistance to Northern Gateway by introducing the infamous "five conditions" for BC approval in 2013. The mayor of Montreal and premier of Quebec, who opposed Energy East and contributed to its ultimate abandonment by sponsor TransCanada, are both gone. But the voters they appeased have not. It was in 2014 when Quebec Premier Philippe Couillard and Ontario Premier Kathleen Wynne teamed up to announce to the world their concerns about Energy East.

There was jubilation in Montreal when TransCanada abandoned Energy East in 2017 and a similar reaction in the Lower Mainland of BC when the courts killed Trans Mountain in 2018. While American radical environmentalists helped create many of Alberta's pipeline problems, they got lots of support among Alberta's fellow Canadians.

In the fall of 2018 the market spot prices for Alberta's core crude oil and natural gas resources were the lowest in history relative to international prices. The discount at which Canadian crude was priced was exacerbated by the simultaneous shutdown of several US refineries that process Alberta crude. Because of the significant drop in short-term cash flow, an industry that has been downsizing for the previous four years returned to laying off workers as spending programs were cancelled or suspended to reflect unplanned cash flow declines.

Things were so awful some major oil producers and captains of industry were calling on the province to shut-in oil to mop up the surplus and raise prices. While having the province control production for reservoir conservation and to ration pipeline space was not a new concept, it had never before been done with the stated purpose of raising prices. This raised a lot of issues surrounding collusion and price-fixing. This was beyond desperation. How the mighty had fallen.

Finally, in December the Alberta government announced that effective January 1, 2019 the province would enforce the withdrawal of 325,000 b/d of oil production from markets until prices stabilized. Called "curtailment", it was Alberta's response to the forces that had helped create a financial crisis and was probably not what most anticipated. Unless you knew the province hadn't actually attended the latest OPEC meeting in Vienna, you might have though Alberta had joined the cartel.

Prices recovered and this move saved many smaller companies from bankruptcies and allowed thousands of oil workers to keep their jobs.

Can't get worse you say? Sure it can. The increasingly frequent and alarming warnings that the world must decarbonize or face certain climate disaster continue. In October 2018 the IPCC stated the only way to preserve the earth's climate in more or less the same state as today is to reduce CO_2 emissions by 45% by 2030. Thereafter, the world would achieve "net zero" carbon emissions by 2050. Clearly, the future of the world and the future of Alberta are irrevocably conflicted. Success for mankind rests on the economic failure of Alberta.

As this book is completed in early 2019, virtually everyone who works in what has always been Alberta's number one industry and engine of economic success is discouraged. Some are in despair. Billions and billions of dollars of personal wealth have been obliterated. Tens of thousands of people—particularly formerly well-paid technical professionals such as engineers, geologists, and geoscientists—remain unemployed. As new office towers conceived in better times are completed, the downtown office vacancy rate is as high as 30%, the highest of any major Canadian city. Calgary City Hall is worried about the diminishing property tax revenue from the city's core.

The problem was highlighted in a *Calgary Herald* article January 5, 2019 titled, "Plunging downtown values trigger huge tax hikes for businesses." Calgary's robust downtown core has helped fund city operations for years but property values are a factor of rental rates and vacancy rates. Both are under pressure. A chart in the article shows the assessed value of 16 of Calgary's landmark office towers for 2019 compared to 2018. Their assessed value has plunged an average 22% in one year from $8.1 billion to $6.2 billion, a $1.9 billion decline. This has been going on since 2014.

According to the article, the loss of property tax revenue from downtown will have to spread among other businesses to keep funds flowing to City Hall. "If nothing else changes, 64 per cent of non-residential property owners – more than 8,000 tax accounts – will see increases of 10 per cent or more this year…"[313]

Certainly there are multiple world economic and environmental realities beyond the control of Alberta's NDP government, but there are important aspects of Alberta's future that are entirely within its purview. These are the size, cost, role, and direction of the provincial government. Seventy years of oil money have been kind to Alberta in many ways, cruel in others. As was explained in the previous section, under the boring and "moribund" Social Credit government, most of the carbon-resource money was spent expanding the infrastructure to keep up with the growing population and industry. Under Peter Lougheed's Progressive Conservative administration, spending expanded well beyond the historical obligations of government. But OPEC provided the cash to pay for it.

Because it had the money—or until recently had the money—Alberta has built up the largest provincial public administration on a dollars per-capita basis of any of the other nine provinces. Regrettably, there is no indication that the healthcare system, education system, or public infrastructure in Alberta is materially better than those of the other nine provinces. Where is all the money going? Where has it gone?

On February 28, 2018, RBC Economics/Research released its Canadian and Provincial Fiscal Tables, which track government spending of each administration from 1980/81 to 2017/18, the last figures available. One chart is titled "Program expenses per capita," which divides jurisdictional spending by the population. For analysis, Alberta is compared to the other nine provinces every five years, commencing in 1980/81.

Because world oil prices in the early '80s were ten times what they had been seven years earlier, Premier Lougheed's administration had lots of cash. As promised, they spread the money around. In the 1980/81 fiscal year RBC reports Alberta spent $3,744 per capita, 43% above the national average of $2,645 and

---

313 Chris Varcoe, "Varcoe: Plunging Downtown Values Trigger Huge Tax Hikes for Businesses," *Calgary Herald* (January 4, 2019). https://calgaryherald.com/business/energy/varcoe-1300-city-businesses-face-threat-of-30-tax-hike-under-property-reassessment.

26% above that of BC, Alberta, Ontario and Quebec, hereinafter referred to as the "Big Four."[314]

By 1986/86, the last fiscal year of Lougheed's premiership, spending hit an all-time comparative record of $5,770 per capita compared to $3,520 nationally, 64% higher than the 10-province average. Spending was 33% higher than the average of the Big Four, which includes Alberta. Remove Alberta from the average and its spending rises to 44% more per person than the average of BC, Ontario, and Quebec.[315]

In the 1990/91 fiscal year some of the spending cuts of the Don Getty administration began to materialize but slowing down the big machine of Alberta government while spending big to diversify the economy did not yield significant cuts comparatively. That year Alberta spent $5,783 per capita, 26% above the national average of $4,593 and 19% more than the Big Four average of $4,701.[316]

Ralph Klein won the PC leadership in 1993 and began the provincial spending cuts, which remain legendary to this day. For the 1995/96 fiscal year, when the full impact of the reductions and rollbacks was measurable, Alberta joined the rest the provinces in spending for the only time in the past four decades. That year spending per capita was $4,637, 6% higher than the national average of $4,382 and 8% below the average for the Big Four of $5,055. BC, Ontario, and Quebec all had higher spending than Alberta that year.[317]

By 2000/01, prosperity was returning to Alberta and things were back to normal—Alberta normal. Per capita spending of $5,984 was 7% higher than the national average of $5,615 and 6% higher than the Big Four, which came in at $5,649.[318]

But by 2005/06, Ralph Klein's last year, Alberta was back in the chips and it was time to once again share the wealth. That year spending was up to $7,998 per person, which was 9% above the national average of $7,410. However, by the twenty-first century, Big Four provincial governments were starting to demonstrate economies of scale whereby the smaller provinces were spending more per

---

314 *RBC Economics/Research - Canadian and Provincial Fiscal Tables* (February 28, 2018). http://www.rbc.com/economics/economic-reports/canadian-fiscal-reports.html.

315 Ibid.

316 Ibid.

317 Ibid.

318 Ibid.

capita than the larger ones. Except Alberta. Of interest that year is Alberta spent 13% more per person than BC, Ontario, and Quebec, which managed to get by with an average of only $6,644.[319]

For the 2010/11 fiscal year, Alberta's spending of $11,530 per person was 8% above the national average of $10,864 but 15% more than the Big Four, which average only $8,691. That year Alberta spent $1,513 more on each citizen than BC, $2,242 more than Ontario, and $164 more than Quebec. [320]

By 2015/16, the first year of the NDP, the spread increased as Alberta's government spent $11,530 per capita, 8% higher than the national average of $10,864 and 17% more than the average of the Big Four. That year BC only spent $9,373, Ontario $9,288, and Quebec had provincial spending well in hand at only $7,943.[321]

For the 2017/18 fiscal year RBC reported Alberta was back on top as the undisputed leader in terms of provincial spending at $12,717 per person. This was 14% above the national average of only $11,172 and 19% more than the Big Four's average. The only province that came close was Saskatchewan, but in its 2018 budget it promised big spending cuts. Alberta's did not.[322]

In the spring of 2018 Alberta was on the same spend-now/pay-later program. The title of the budget link on the government website reads, "Budget 2018 builds a diversified economy, protects vital public services and sets a path to return to balance – a recovery built to last."

Obviously, the budget presentation was political. It always is. All the hot buttons were pushed, including a commitment to balance the budget in only six years. This is seven years later than the NDP's original pledge during the 2015 election. But first there would be another $29.4 billion in deficit spending.

The province only provided granular financial details three years out to the 2020/21 fiscal year. By then spending would be 6.9% higher than the estimates for the 2017/18 fiscal year, bitumen production would be up about 600,000 b/d, and

---

319 Ibid.

320 Ibid.

321 *RBC Economics/Research - Canadian and Provincial Fiscal Tables* (February 28, 2018). http:// www.rbc.com/economics/economic-reports/canadian-fiscal-reports.html

322 Ibid.

western Canada Select would fetch $50.90 per barrel, nearly double the terrible numbers of the fall of 2018.[323]

By 2021 the deterioration of the province's balance sheet will be significant. The "net financial assets" (financial assets such as cash, accounts receivable, savings, and loans to government agencies less borrowings, pension obligations and accounts payable) would be -$48.2 billion. Total debt would be $77.2 billion. Interest on the debt by then was forecast to be $2.9 billion per year, almost 5% of total government spending.[324]

The proliferation of red ink and growing debt in the past decade has been breathtaking. What should be noted is how non-renewable resource revenue has declined while spending has steadily marched upwards. Alberta's government is ill-prepared for a non-carbon world where protecting the climate will require continued decarbonization. The province requires a properly planned transition to a global market where its key carbon products will have lower demand, and hence less value.

The natural gas and oil sands leasing boom of the first decade of this century, combined with tightly controlled spending under the Ralph Klein PC administration, worked wonders on the province's balance sheet. In 2004 the province reported net financial assets of $10.5 billion. But the cash was pouring in faster than anybody could figure out how to spend it. A year later in 2005 it was $15.2 billion, $26.0 billion in 2006, $34.1 billion in 2007, and it reached its all-time peak of $35.0 billion in 2008. The largest single growth area was called "portfolio investments," the Heritage Fund and the new Sustainability Fund. From 2004 to 2008 this figure increased from $16.9 billion to $36.1 billion, or 114%. During the same period, liabilities only grew from $18.7 billion to $24.4 billion, or $30%.[325]

Alberta's financial problems started in 2009 with a combination of continued high spending and falling revenue. By that point Alberta's "infrastructure deficit" was a major public policy issue. The recession of 2008/09 had clobbered non-renewable resource revenue and the development of US shale gas had a dampening impact on natural gas prices and Crown gas royalties.

---

323 Government of Alberta, Ministry of Treasury Board and Finance Annual Reports, https://www.alberta.ca/government-and-ministry-annual-reports.aspx#toc-0.

324 Ibid.

325 Government of Alberta Ministry of Treasury Board and Finance Annual Reports, https://www.alberta.ca/government-and-ministry-annual-reports.aspx#toc-0.

For now that appears permanent.

While oil prices and land sales increased non-renewable resource revenues between 2010 to 2014, there was no adjustment to the spending spree. By 2015, when the PC dynasty was replaced, net financial assets had declined to $13.1 billion, a staggering $21.9 billion reduction in only seven years. They were all good years for oil and gas, except 2009. While portfolio investments like the Heritage Fund and Sustainability Fund were left alone, the line item for "unmatured debt" had risen from $2.5 billion in 2008 to $12.4 billion by 2015. Although the asset side of the balance sheet had grown, liabilities went from $24.4 billion in 2008 to $53.0 billion only seven years later.[326]

In the first three years of the NDP administration, the spending continued. By 2018 the net finance assets had deteriorated to -$19.3 billion, a $54.3 billion negative swing in only 10 years. The 2018 budget sees the net financial asset position deteriorating by almost $30 billion by 2020/21.[327]

Over the past 40 years, the existence and use of the Heritage Fund has been widely discussed among Albertans. It was capped in 1987 at $12.7 million. No additional funds would be contributed. Part of the fund's investment income was used for general revenue in the 1982/83 fiscal year. By 1983/84 all the income generated went into general revenue. Corrected for inflation, the Heritage Fund today is much smaller than it was in 1987.

The Heritage Savings Trust Fund was set up to help Alberta get through a "rainy" day. Obviously, the government saw a severe storm out the legislature windows 35 years ago and withdrawals began only seven years after it was created in 1976. Had the capped fund been left alone and its internal returns reinvested it would be worth considerably more today. Using the annual CPI rate of inflation for each year as a rate of return, the compounded value today would be $25.8 billion. At an average compounded rate of return of 4%, the fund would total $42.8 billion in 2018, enough to retire the province's long-term debt, which was $43.3 billion according to the first quarter fiscal report, dated June 30, 2018. Had the fund generated a 6% annual average compounded rate of return, the value today would be $82.0 billion.

---

326 Government of Alberta Ministry of Treasury Board and Finance Annual Reports, https:// www.alberta.ca/government-and-ministry-annual-reports.aspx#toc-0.

327 Government of Alberta Ministry of Treasury Board and Finance, https://www.alberta.ca/ budget.aspx.

Since cash began to flow from the Heritage Fund 35 years ago, the capital has kept spending and taxes down. To say it could be worth $82 billion is interesting, but not particularly relevant. Assuming it has earned an average of $1 billion per year, either Albertans would be $35 billion poorer because they had to pay higher taxes, or the province's balance sheet would be that much worse. Without cutting spending, the government would have borrowed the money.

Alberta's fiscal balance sheet also includes "tangible capital assets," public infrastructure owned by the province such as land, buildings, equipment, roads, bridges, and dams. That figure has risen considerably in the 10 years from 2008 to 2018. Assuming the accounting methods are the same, this is where a bunch of the money has gone. The 2008 balance sheet reported tangible capital assets of $14.1 billion. The 2018 figure was $51.4 billion. Since the assets are depreciated for financial reporting purposes, the significantly higher figure indicated a lot of assets are new, which they are. Capital spending by the province has been $6 billion per year or higher for recent years.[328]

While the infrastructure in the form of schools and roads is always welcome, these investments also ensure higher structural operating costs for staffing, operations, repairs, and maintenance. Government infrastructure provides no revenue, only expenses. The Notley administration has regularly criticized the Ralph Klein administration as the architects of modern infrastructure deficits by slashing public spending too far.

But as the balance sheets reveal, the post-Klein PC administration started investing heavily in infrastructure long after Klein was gone and before the NDP were elected. By 2015 the tangible capital assets on Alberta's balance sheet were already up to $44.3 billion.[329]

Hopefully, at some point the current government and vested public-sector interests will quit blaming today's problems on a premier who was replaced in 2006 and passed away in 2013.

During the "dark years" of the Klein administration, the news was not entirely as bad as the NDP would have us believe. Capital was flowing in from all over the world. Cash was piling up in the provincial treasury. House prices were rising,

---

328 Government of Alberta Ministry of Treasury Board and Finance Annual Reports, https://www.alberta.ca/government-and-ministry-annual-reports.aspx.

329 Government of Alberta Ministry of Treasury Board and Finance Annual Reports, https://www.alberta.ca/government-and-ministry-annual-reports.aspx.

unemployment was falling, and head offices of major Canadian corporations—like Canadian Pacific, Imperial Oil, and TransCanada Corporation—were relocating to Alberta.

In their determination to fix things, the NDP have apparently solved one problem to their satisfaction. Heavy spending on infrastructure and government employment have indeed acted as an economic shock absorber.

But on the other side of the ledger, no more head offices are moving to Alberta, capital is leaving, house prices are falling or flat, and thousands of oil and gas workers are still unemployed.

The government claims there is a recovery underway, and by some measures there is. The primary drivers have been higher oil prices and government spending. The *Fraser Institute* released a report in August of 2018 titled, "The Illusion of Alberta's Jobs Recovery: Government vs. Private Sector Employment." From 2014 to 2018 the report claims government employment has increased by 78,733 jobs, or 21.5%, while over the same period private sector (excluding self-employed) fell by 46,267, or 3%. The report reads, "As a result, the government's share of total employment (excluding self-employment) increased from 19.5 percent to 23.2 percent – a level not observed in Alberta since 1994."[330]

Another way to look at where the jobs are in Alberta is by sector. A monthly government report is issued titled Alberta Labour Statistics and Annual Reviews. It relies on data from Statistics Canada, which is seasonally adjusted.

In October 2014 Alberta reported there were 2.396 million workers in the province's labour force. A total of 431,000 worked in resources (forestry, fishing, mining, oil and gas) and construction while 467,000 worked in education, healthcare, and public administration.

Four years later in October of 2018, the total workforce had expanded by only 0.5%, or 123,000 jobs to 2.519 million. Resources and construction had shrunk by 30,000 jobs or 7% to total 401,000. Education, healthcare and public administration had added 84,000 jobs to expand by 18% to 551,000.[331]

330 Charles Lammam and Hugh MacIntyre, "The Illusion of Alberta's Jobs Recovery: Government vs. Private Sector Employment," *Fraser Institute* (August 30, 2018). https://www.fraserinstitute. org/studies/illusion-of-albertas-jobs-recovery-government-vs-private-sector-employment.

331 Government of Alberta, "Alberta Labour Force Statistics: Seasonally Adjusted," *Open Government* (). https://open.alberta.ca/publications/2727158.

Funded in part by growing public debt, this too must be part of the shock absorber Premier Notley deemed Albertans essential to help cope with the oil price collapse.

As significant, the government pays its employees more. Another 2018 analysis by the *Fraser Institute* revealed that after normalizing the data to include factors like gender, age, education, company size, industry, occupation and other variables between the private and public sector, "... Alberta's government sector workers (federal, provincial, and local) enjoyed a 9.6% wage premium, on average, over private-sector counterparts in 2017."[332]

The already legendary government employee benefits are also quantifiable. Using what information it could find, the report reads, "... the government sector enjoys an advantage over the private sector. For example, 72% of government workers in Alberta are covered by a registered pension plan, compared to 24.2% of private-sector workers. Of those covered by a registered pension plan, 95.3% of government workers enjoyed a defined benefit pension compared to 29.3% of private-sector workers." They also retire, on average, 1.7 years sooner and are much less likely to lose their jobs (4.2% in the private sector compared to 0.7% in the public sector).[333]

Now what? The NDP government sees non-renewable resource income rising in the next few years but not approaching anywhere near the levels of the past in terms of percentage of total revenue. Whatever has been said and done about industrial diversification over the years, the Alberta treasury was a one-trick pony from the 1970/71 fiscal year to 2014/15. Over this 44-year period, non-renewable resource revenue as a percentage of total revenue averaged 29.8%. The best year was 1979/80 when it reached 77.4% and the worst 1998/99 at only 14.1%.[334]

332 Milagros Palacios, David Jacques, Charles Lammam, and Steve Lafleur, "Comparing Government and Private Sector Compensation in Alberta, 2018," *Fraser Institute* (2018). https://www.fraserinstitute.org/sites/default/files/comparing-government-and-private-sector-compensation-in-alberta-2018.pdf.

333 Milagros Palacios, David Jacques, Charles Lammam, and Steve Lafleur, "Comparing Government and Private Sector Compensation in Alberta, 2018," *Fraser Institute* (2018). https://www.fraserinstitute.org/sites/default/files/comparing-government-and-private-sector-compensation-in-alberta-2018.pdf.

334 Government of Alberta, Ministry of Energy, https://open.alberta.ca/opendata/historical-royalty-revenue.

After the gas price crash of 2009 and the oil price collapse of 2014, the percentage fell to only 6.5% in 2015/16 and 7.3% the following year. Looking forward, for the next three years to 2020/21, government estimates for non-renewable resource revenue as a percentage of total revenue are only 8.0%, 8.3%, and 9.3%. Based on what the government knew in the spring of 2018, the oil and gas business is not going to work its magic on the province's finances again anytime soon.[335]

Exactly what the Ed Stelmach PC administration was thinking when it raised royalties through the New Royalty Framework in 2007 remains a mystery to those who understand the fluctuations of commodity prices, export markets, production volumes, and sources of capital. In the 2006/07 fiscal year, Alberta collected $12.3 billion, which was 32.2% of all funds collected. The prior year, 2005/06, the values were $14.3 billion and 40.4% respectively.[336]

On the subject of cutting government spending to bring Alberta's growing government in line with the province's shrinking economy, the NDP and public service unions are quick to play the fear card. They maintain the only possible way to balance the books is through draconian reductions in core services such as healthcare and education. Sure, spending can be cut. But it will probably cost you your health, your life, your children's education, or all three.

Meanwhile, in the private sector the adjustment to the new realities has been devastating. To demonstrate how much wealth has been eradicated, compare the market value of 20 publicly traded Canadian oil and gas exploration and production companies today to 2014. Twenty Canadian-founded, Calgary-headquartered medium and larger upstream (no refining) companies were chosen for analysis because Calgary Economic Development reported they all had revenue exceeding $300 million in 2017, the last full year for which figures are available. The total revenue of these 20 companies that year was $25.5 billion, not all of it in Alberta. The largest, EnCana Corporation, had revenue of $5.8 billion. The smallest, Bellatrix Exploration, reported revenue of $302 million. Eight of these companies had revenue exceeding $1 billion in 2017.

Based on the closing price on January 17, 2019 on the Toronto Stock Exchange, when compared to their 2014 market high, their share price values had declined

335 Government of Alberta, Ministry of Energy, https://open.alberta.ca/opendata/historical-royalty-revenue.

336 Government of Alberta, Ministry of Energy, https://open.alberta.ca/opendata/historical-royalty-revenue.

an average of 76%. Their combined market capitalization on January 17 was $47.7 billion. Based on their 2014 share price, the total loss in investor equity value can be estimated at $160 billion (this methodology assumes the 20 companies had the same number of shares issued and outstanding in 2014 as 2019, which will not be the case for all of them). Although they are broadly held by investors outside Alberta, much of the personal wealth of the founders, executives and employees—all Albertans—has been destroyed by the real-world economy.

And, of course, worker pay cuts in the carbon-resource industries to adjust to market realities are well understood. Many in this business, particularly in oilfield service, are making less today than they did in 2014.

But where it really hurts is in the individuals' residential properties, one of the biggest assets and sources of wealth for most homeowners. In its energy-use tables, Natural Resource Canada reports that in 2016 Alberta's total "housing stock" was 1.68 million dwellings consisting of 1.01 million single detached homes, 226,000 single attached homes (duplex, fourplex, private entrance condos), 363,000 apartments, and 72,000 mobile homes. For analysis purposes, assume the detached houses have an average value of $400,000, the attached dwellings $250,000, the apartments $200,000, and the mobile homes $100,000. This would give Alberta's residential housing stock a total value of $542 billion. A 5% decline, certainly not an unreasonable number in many communities in the past few years, is $27 billion. This is the equity portion, not the mortgage.[337]

Obviously public-sector workers own houses, condo, mobile homes, and oil and gas stocks and are therefore not immune to the economic corrections being endured by all Albertans.

During the next provincial election and thereafter there will be a debate on what can and cannot be done with Alberta public-sector spending. Albertans can decide for themselves if it reasonable for almost everyone in Alberta to undergo such a massive adjustment to their personal net worth while at the same time the lucky ones who work for the government can refuse to participate. Plus warn their fellow Albertans of nothing but dire consequences if adjustments are made. Alberta runs the most expensive provincial government per capita in the country

---

337 "Comprehensive Energy use Database, Residential Sector, Table 15," *Natural Resources Canada* (2016). http://oee.nrcan.gc.ca/corporate/statistics/neud/dpa/showTable.cfm?type=CP&se ctor=res&juris=ab&rn=15&page=4

and has for years. Why? How? Is it impossible in light of the province's vastly changed fortunes to even have this conversation?

Luckily, there are several glimmers of hope for relief from some of the market access problems that are crushing oil and gas prices. But that break won't come soon. The Enbridge Line 3 pipeline replacement to the US is under construction and was meant to add 370,000 b/d of pipeline takeaway capacity in late 2019. However, on March 1, 2019 it was announced startup would be delayed up to a year.

The Justin Trudeau government appears to be making steps to get the Trans Mountain expansion to Burnaby back on track after the August 2018 decision by the Federal Court of Appeal to stop it. Whether it will succeed is unknown. However, if all goes well, it seems impossible this pipeline would be operational before 2020 at the earliest. With a federal election slated for fall 2019 it will be interesting to see how hard the Trudeau administration tries to force the Trans Mountain pipeline forwards prior to the election.

Keystone XL, the pipeline south to the US Gulf of Mexico Coast (USGC) cancelled by the Barack Obama government in 2015, may proceed thanks in part to President Donald Trump. Operator TransCanada is working through approval issues in Nebraska and Montana and appears committed to construction if all the regulatory obstacles can be overcome. It seems unlikely this pipeline would carry any oil before 2021.

Crude-by-rail will move the rest, but this is twice as expensive as pipeline. As producers reach out to contract more rail cars, this cost is rising. There was great enthusiasm for crude-by-rail prior to the 2014 price collapse, but as oil prices declined, shipments dropped off as transportation represented a higher percentage of each barrel sold. As shipments fell, the railways directed their assets elsewhere. Then, when oil prices recovered in markets like USGC, producers phoned back the railway companies. To nobody's surprise, the railroads wanted long-term commitments; they're not unaware that the producers are hoping to replace rail with pipe in two or three years. Meanwhile, new safety regulations requiring replacement of most of the tank-car fleet increased costs further.

Rail capacity is being added but at significant cost. In late October 2018, an increasingly desperate Premier Notley asked the federal government to help pay for more tank cars and locomotives to get more oil moving to ameliorate the price differential.

As for the oil sands, the crusade to keep this product off the market by blocking pipelines is not working. Nor are all future oil sands projects off the table. Teck Resources is pursuing a major new mining operation between Fort McMurray and Fort Chipewyan. Called Frontier, this mine is expected to cost $20.8 billion and will add up to 260,000 b/d for 41 years after production begins in 2026. Imperial Oil is considering another 150,000 b/d project called Aspen.

On October 2, 2018, Shell Canada and its partners announced the approval of the $40-billion Canada LNG export project at Kitimat, BC. This would truly be a game-changer for western Canada's natural gas business by ensuring the removal of 3 bcf/D of natural gas from the western Canadian market, nearly 20% of current production. This will improve prices, cause exploration and development to resume, and may be a key factor in other LNG projects. While construction spending will start soon, it won't carry gas until 2023, and all its production will come from BC. Because the gas market is interconnected, Alberta will see more collateral than direct benefits.

Another factor to help the province's finances will be when many of the new oil sands projects built in the past 10 years come out of their reduced-royalty, capital recovery rate of 5% and go to the full royalty rate of 25%.

For the full year 2017, the Alberta government reported the status of 119 oil sands operations that had gross production of one billion barrels of bitumen for the year. This averages 2.75 million b/d. Of the total, 107 produced oil. The royalty is stated on a per project basis. Only 35 of the 107 on production paid the higher royalty rate, reported in most cases as 27.41%. This means that only 36.7% of the production paid the higher rate in 2017, about one million barrels per day.[338]

In future years when all these projects reach payout, total bitumen royalties payable to the crown could more than double, depending on price, of course. The data reports total royalties for the 2017 year of $2.48 billion. At the full rate the number could be closer to $6 billion, a handsome gain for the provincial treasury.

While lower oil prices will extend the time to capital recovery and full royalties, tough times are helping oil sands operators adjust to market realities. According to CAPP data, the average operating cost per barrel of oil sands produced in 2014 was $31 a barrel. By 2017 this was down to $19 per barrel, $12 or 39% lower. This increase in operating efficiency will contribute significantly to the

---

338 Government of Alberta, Ministry of Energy, "Alberta Oil Sands Royalty Data," *Open Government* (July 25, 2018), https://open.alberta.ca/opendata/alberta-oil-sands-royalty-data1.

long-term viability of this massive resource base and shorten the time until this production makes a significantly larger contribution to Alberta's treasury through production royalties.[339]

Brian Jean is an entrepreneur from Fort McMurray who spent 10 years in Ottawa as a Conservative MP, three years in Edmonton as a Wildrose MLA, and as leader of the Wildrose Party from 2015 until it disappeared through a merger with Alberta's Progressive Conservative party in 2017. An unabashed oil sands supporter, in an interview Jean offered the following comment on "black gold" and Alberta's finances. "I have so much confidence in the oil sands because there's nothing else on God's green earth that will get us out of this financial mess."

If Alberta could figure out how to cut spending by even a modest amount, if world oil prices remain above US$50, if all the new oil pipelines on the table were built so the province's oil sands could fetch something near the world price, and if LNG exports could pull natural gas prices out of the dumpster, then the carbon resources that have paid the freight for Alberta since the province was settled could work their magic again.

CAPP has forecast that based on the projects known to the association in mid-2018, oil sands production could rise by another 1.4 million b/d by 2035. This could make the numbers even better yet. If all the new pipelines are eventually completed, this might even happen.

Of course, this would be in direct conflict with the problems and challenges stated on the Alberta Government's Climate Change website, which predicts a future of weather-driven chaos and calamity unless the climate challenge is addressed. That's why Alberta is taxing carbon and shutting down coal. It is also in direct conflict with the October 4, 2018, IPCC report that the world must reduce carbon emissions by 45% by 2030, not the forecast according to CAPP.

The other major event which could forge a different direction for Alberta is the success of the United Conservative Party (UCP) in the 2019 provincial election. After the NDP won the election in 2015, it was widely believed among non-NDP voters that the NDP only formed its first government because the right-of-centre vote was split between the Wildrose and PC parties. The merger was ratified by the members of both parties in 2017 and a leadership race was held thereafter.

---

339 Canadian Association of Petroleum Producers, *Statistical Handbook for Canada's Upstream Oil Producers* (Calgary, AB: CAPP, 2018). https://www.capp.ca/publications-and-statistics/publications/316778.

Former federal CPC cabinet minister Jason Kenney won. The UCP has already declared it will end the provincial carbon tax if elected and will undoubtedly take Alberta in a different direction than the NDP.

But even if Alberta goes in the opposite direction of the climate change movement in the short term and is able to inject some financial life into its moribund major industry, what about the long haul?

Alberta's massive carbon resources would still remain regrettably landlocked, with geographical and political obstacles to overcome in every direction.

Bill C-69, the new federal legislation designed to revamp the National Energy Board and the pipeline project approval process, looks daunting in its present form. This bill proposes more intervenors, more environmentalists, greater social-impact analysis and—to demonstrate there's a new federal sheriff in town—a per-project review of gender parity.

Bill C-69 makes the final decision political, not regulatory. This means there will never be a clear and predictable path from conception to construction. Critics of the bill are concerned this will send entirely the wrong message to a domestic and international business community that is already convinced that Canada is no longer a safe and transparent place to conduct business.

Even if pipelines help alleviate Alberta's problems of geography, the province will never truly be a master of its own destiny so long as its hydrocarbon-delivery arteries pass through other political jurisdictions, or so long as other levels of government can interfere with its resources and economic future. Further, the concern that Alberta's bread-and-butter carbon-resource industries are a menace to the world's climate will not go away.

But until influencing factors outside the province's control are resolved, there are significant opportunities to do more with less.

Alberta has developed a system of resource regulation that is unrivalled in the major oil and gas producing regions of the world in its complexity. As long as it takes much longer to get anything done in Alberta, investors can put their money in places like Saskatchewan or the US.

Surely there must be some opportunity to bring Alberta's per-capita spending in line with the other larger provinces without—as so many of the vested interests claim—firing doctors or nurses. Cutting provincial spending need not be viewed as a personal health risk. Surely once can check the life spans and mortality rates in BC, Ontario and Quebec to see if this is indeed correct.

There will be an election in Alberta in 2019. In January CAPP and the Calgary Chamber of Commerce both took the position that the province should cut the corporate tax rate back to where it was before the last election. They also called for the government to streamline the oil and gas industry regulatory processes to enhance business competitiveness, and get public spending under control to limit future tax increases.

When the global business environment which has so much impact on Alberta has changed to this degree, how much longer can the province and its governments afford to continue down the current path?

7
POLARIZED POSITIONS PARALYZE PROGRESS

"People demand that their governments solve the problem and the same governments refuse to admit they can't. How is anybody going to fix this mess if we can't work from an agreed-upon set of facts and bridge the gap between polarized factions?"

By now you should be thoroughly confused. Not your fault. This is one of the most complex subjects facing modern society. Because it involves all the people on Earth and the global environment, anybody is qualified to comment. And because it involves everyone, a solution will require unprecedented levels of agreement and cooperation.

The best way for Alberta to get out of the carbon business is to grow it. Yup, that's what we're told. Whatever problems carbon taxes will solve in the long run, getting people to pay them is a huge problem in the short term.

The media loves this story. First, it involves the weather which is, by default, interesting. This is everyone's favourite subject.

Second, it involves bad weather. That storm could have hit me. OMG, it did hit me! I was there, or my friend, or my family were there. Fabulous broadcast images that anybody can relate to.

Thirdly, it involves impending disaster, which is always newsworthy. Think about it. Things are bad but they could be worse. Away we go. People are always worried about the future and because of climate change, now they know why.

Lastly, climate change has become a political battleground. This is a battle of wills between the many concerned individuals who know what mankind must do to preserve its future and everybody else who is being forced to pay for it. Climate change politics are a blood sport. And with fewer wars to cover, the event is an attractive substitute to the media, which wants to fill space and attract readers. Riots in Paris. Burning cars. Growing demonstrations by disenfranchised Alberta oil workers.

Perfect. Alberta, of course, is the subtitle of this book.

It is interesting how people look at information; how they process a multitude of facts and come up with differing truths from the same set of facts. People are biased towards selecting what they want to read or hear and focus on that which confirms their existing beliefs.

For example, if you only read this section's Chapter 2 highlighting all the good news you've probably never heard, you could conclude the following. The world's population is 7.4 billion and growing. The average person is living longer than ever. Poverty is in steep decline, particularly in the last three decades, as are child mortality rates and starvation.

This would suggest that the more people there are on planet Earth the longer and better they live. The best way to improve things would be to have more children. This is, of course, the exact opposite of centuries and decades of postulations, dating back to the Malthusian theory of the late 1700s and the alarming book *The Population Bomb* in 1968.

On the other hand, population growth and per-capita lifestyle improvements have been fuelled by cheap carbon energy, resulting in growing carbon dioxide and other GHG emissions. The world's average temperature is 1°C warmer than it was at the beginning of the carbon-powered Industrial Revolution. Based on these two data sets, you would come to the same conclusion as the IPCC; fossil fuels are the main cause of global warming. Therefore, we must stop using them—the sooner the better.

The climate is changing. It always has. But we're assured this is "accelerated" climate change and the results will be significant with unpleasant consequences, including receding glaciers, rising oceans, and changing weather patterns, including severe storms.

These are two powerful and utterly contradictory conclusions resulting in what are clearly two diametrically opposed views of the future of the world. Carbon has made things better but is now making things worse. Therefore, what got us this far now must go.

Wow.

As we look at the future of Alberta and the planet, there are two solitudes—two distinct opposites—that hopefully readers can agree upon. They hog the headlines but are clear and hopefully not insurmountable obstacles to progress.

On one side we have the extreme fringe of the environmental movement. Notwithstanding the nutbars in the environmental movement who are still thumping the decades-old population control tub, there are a vast number of allegations by climate change alarmists that are simply false. The most egregious is that if only the governments of the world could muster sufficient political will, mankind could make a swift transformation away from fossil fuels to renewables like wind and solar power with minimal disruption to the world's 7.4 billion inhabitants.

Scaring the crap out of everybody and offering the simplest of solutions is the stock-in-trade of climate change alarmists. Even Al Gore admits some of this is over the top. In a recent interview with *PBS Newshour* following the latest IPCC prediction of climate disaster if carbon emissions weren't reduced by 45% by 2030, Gore admitted, "The language that the IPCC used in presenting it was torqued up a little, appropriately – how (else) do they get the attention of policy makers around the world?"[340]

In 2009 Gore told author Hans Rosling, "We need to create fear" (Rosling 2018). This was after broad acclaim for his Gore's 2006 movie *An Inconvenient Truth.*

On October 19, 2018 *CBC News*, in an interview in its new portal and newsletter "What on Earth," climate scientist Michael Mann discussed his infamous "hockey-stick" model that tracked and extrapolated global warming and CO_2 emissions.

Mann talked about how much progress is being made. "We are seeing a dramatic worldwide move away from fossil fuels towards renewable energy." This is true, sort of. That's the conclusion of five forecasts. Except it will take decades to occur, probably longer.

But old habits die hard for professional climate warriors. When asked "What are our challenges?" Mann responded, "The main challenge is defeating the juggernaut that is the fossil fuel industry. They have used their tremendous wealth and influence to block all meaningful efforts to limit carbon emissions and accelerate the transition underway from fossil fuels to renewable energy. The only way that we will defeat them is by turning out to vote and electing politicians who will act in our interest over the special interests."[341]

This is a common refrain. In July 2017, *cdp.net*, originally called the Climate Disclosure Project, disclosed its Carbon Majors Report, titled "The Carbon Majors Database." It lists the 100 companies or entities that have been responsible for

---

340 "Al Gore Calls Trump's Deregulation Proposals 'Literally Insane,'" *PBS News Hour* (October 12, 2018). https://www.pbs.org/newshour/show/al-gore-calls-trumps-deregulation-proposals-literally-insane.

341 "What on Earth?" *CBC News* (October 19, 2018).

71% of global emissions since 1988. The report reads, "Since that time the fossil fuel industry has doubled its contribution to global warming by emitting as much greenhouse gas in 28 years as in the 237 years between 1988 and the birth of the Industrial Revolution."[342]

But it wasn't just the companies. It was also the customers. The report reads, "... emissions *deriving from* [italics added] fossil fuel producers ..." and "Since 1988, more than half of global industrial GHGs can be traced back to just 25 corporate and state producers." Number one? Electricity generators burning coal in China.[343]

Simple problem, simple solution. Get rid of the problem at the source. Consumers get a hall pass.

In fact, oil companies respond to a market by selling people the energy they need. Oil companies are the *only* source of feedstock or fuel for plastics and other petrochemical products, airplanes, ships, big trucks, railways, or light vehicle anywhere but urban centres. China's electricity generators help modernize that country by supplying electricity. These suppliers don't exist just to screw up the planet.

Much is written about how the carbon-resource industries are simply peddlers of addictive drugs, getting unwitting consumers hooked on cheap energy that will ultimately kill them and take the world along for the ride. It's a trick to get people to buy fossil fuel instead of something else.

Except there is no "something else".

Blaming carbon-producing companies for climate change is like holding the world's auto manufacturers responsible for 100% of the vehicular accidents and fatalities. There are no international campaigns to ban cars, although they provide the means by which 1.3 million people die each year. Nor is there a divestment campaign to punish the auto manufacturers.

Love your car. Hate the fuel.

People apparently bear no responsibility. Big Oil is the problem, not you and me. Only a few in the green movement tell the truth. Whatever you think of Bill McKibben and 350.*org*, at least he's honest. He proposes North Americans revise

---

342 "New report shows just 100 companies are source of over 70% of emissions," *CDP: Disclosure, Insight, Action* (July 10, 2017). https://www.cdp.net/en/articles/media/new-report-shows-just-100-companies-are-source-of-over-70-of-emissions.

343 Ibid.

our expectations, stay home, forget international travel, and resume growing our own food. He doesn't address if this model is transferable to the world.

People believe this because today we can sort and stream our information to only hear what we want. In imagining the fix is as easy as outlawing fossil fuels, people have convinced themselves this huge problem will just go away. If the notion that killing climate change by this single action were a disease, it would be called Silver Bullet Syndrome. This disorder is spread by the climate change alarmist industry, which disingenuously spreads its message without question or restraint.

Also unhelpful in this polarization is perpetuating the myth that local governments can solve global problems as long as they have the political will. This is reinforced by the media, even in western Canada. In a column in the *Regina Leader-Post* on October 15, 2018, columnist Murray Mandryk wrote under the headline, "Politicians Need to Understand Climate Change has Big Local Costs."

In Saskatchewan, fighting the federal carbon tax is good politics. But Mandryk calls this short-sighted, writing, "… worries from Ontario Premier Doug Ford or even Saskatchewan Premier Scott Moe that a carbon tax will cost us jobs do seem mockable in the context of the recent IPCC report that the earth only has about 12 years left to avert catastrophe …" He quotes an IPCC co-chair who wrote, "It's a line in the sand and what it says to our species is that this is the moment and we must act now."[344]

While Mandryk is undoubtedly a fine journalist and, like most, well-intentioned, what is obvious is the climate-fear business is doing extremely well; things are going just as planned.

What is important about Mandryk's article is not what the IPCC reported. Predicting doom has been the IPCC's reason for existence since its inception 30 years ago. The problem is the writer's conclusion that if the premiers of Ontario and Saskatchewan did something differently, the future weather of those provinces could, or would, be different. Fear has overwhelmed common sense.

William Nordhaus, the famous climate economist who recently won a Nobel Prize in Economics for his work on the cost and benefits of carbon taxes to reduce emissions, has admitted that getting all the countries of the world to cooperate is highly unlikely, probably impossible. Nordhaus developed a method by which,

---

344 Murray Mandryk, "Mandryk: Politicians Need to Understand Climate Change has Big Local Costs," *Regina Leader-Post* (October 15, 2018). https://leaderpost.com/opinion/columnists/politicians-need-to-understand-climate-change-has-big-local-costs.

if enough wealthy countries bought into carbon taxes as a solution, they could concurrently introduce a form of economic tariffs and sanctions to force non-participants to cooperate.

If it ever gets that far, Doug Ford and Scott Moe won't be invited to the meeting.

Taking an opposing view are the fearless and committed voices who look objectively at the entire climate change disaster phenomenon and refuse to accept that the intentions of the loudest voices are entirely altruist. As a group, those who question any or all of the climate change mantra are commonly referred to and vilified as "deniers". They question that Maurice Strong was noble; that the United Nations does not have a secret agenda for global government or at least expanded influence; that the depopulation movement conceived by Malthus is not best way to save the world; that essential the ingredient for plant growth called carbon dioxide is actually a pollutant; or that those who stand to gain financially from the greatest economic dislocation in history are objective in their actions and public disclosures.

If you read enough about climate change and climate science, there are good reasons to be skeptical. There are certainly other forces affecting the climate, ranging from the behaviour of the sun to ocean currents to, according to one theory, intergalactic radiation bands. Look beyond your favourite and self-satisfying sources of information on the internet and you will learn devastating weather events like hurricanes and droughts were occurring decades ago, when the CO2 content of the atmosphere was much lower. They have forever.

According to *MSN*, three of the five deadliest hurricanes in US history as measured by fatalities took place in 1903, 1900, and 1928. Website *geology.com* reports three of the five strongest hurricanes measured by wind speed to hit the US occurred in 1886, 1935 and 1969.

Hearing the doomsayers claim every forest fire, hurricane, drought, temperature extreme, or heavy snow or rainfall is caused by fossil fuels and man-made climate change could make anyone with an open mind and an interest in history a skeptic.

There are also people who may not question the science, but certainly question the motives of the people who cry louder and more frequently that the sky is falling and certain disaster awaits all who don't take heed and cooperate.

That said, the climate change genie is out of the bottle. People are scared. They want solutions. They want somebody to tell them things will be okay.

While climate-science skeptics who claim fossil fuel emissions are not the main problem may in fact be correct and one day proven so, in today's world millions or even billions of people are never wrong. As the Public Choice Theory has taught us, this will drive politicians and politics to where the votes are. Scared people vote.

Proponents of carbon taxes like the governments of Canada, Alberta, and BC say Canadians must pay carbon taxes to prevent future climate disasters. Except it won't help if the rest of the world doesn't do the same thing or something like it.

Some believe carbon taxes are the morally right thing to do, even if they aren't effective. While this may help people sleep better at night, it deflects everyone's attention from the only true solution, which is reducing carbon emissions.

A regrettably typical response to carbon taxes was contained in a *National Post* column on October 30, 2018, titled, "Heated political rhetoric on carbon tax disservice to Canadian voters." Writing about the political debate, not the contents of the atmosphere, columnist John Ivison reviewed the polar opposite positions of the federal Liberals and carbon tax adversaries Ontario Premier Doug Ford and federal Conservative Party leader Andrew Sheer.

The Liberals have aimed their new carbon tax plan at provinces that don't have one yet. The federal plan is to create a tax/rebate program where the refund may exceed the levies for some households. Ivison wrote, "Both federal and provincial conservatives expressed skepticism about consumers emerging better off from a carbon tax. But voters may be less incredulous when they see the benefit in their tax returns next spring. The Liberals claim 70 per cent of families will emerge better off."[345]

Demanding the Conservatives come clean and admit they don't believe in climate change, Ivison wrote, "... and [if they] consequently intend to do nothing, they should say so and let the electoral chips fall where they may. If they do believe in it, they should produce a plan on how they will combat it." Ivison concluded, "Fighting climate change by raising the price of gas 11 cents over the next five years should be something all but the most ardent skeptics could get behind."[346]

Ivison's article perfect exemplifies the shallowness of political polarization.

---

345 John Ivison, "John Ivison: Heated Political Rhetoric on Carbon Tax Does Disservice to Canadian Voters," *National Post* (October 30, 2018). https://nationalpost.com/opinion/john-ivison-heated-rhetoric-on-carbon-tax-does-disservice-to-canadian-voters Material republished with the express permission of: National Post, a division of Postmedia Network Inc.

346 Ibid.

The only activity that will truly "fight climate change" is an actual reduction in the amount of gasoline consumed, not what it costs. Canadians already pay $0.30 to $0.50 a litre in taxes on gasoline. Canada proved long ago that, for most, demand is impervious to price unless the cost goes up much more than another $0.11 a litre. We must buy fuel for their vehicles in this large, cold country.

On October 30, 2018 federal Environment Minister Catherine McKenna announced New Brunswick Power would be granted an exemption on emission standards for its coal-fired electricity power generation. The same day online network *cbc.ca* reported, "In a climate-policy retreat over the treatment of coal, federal Liberals are proposing to loosen emission standards for power plants that burn the fuel, effectively lowering carbon taxes on each tonne of greenhouse gas released from coal-burning stations, like NB Power's Bellendune, next year (2019) to less than $1."[347]

This is the same environment minister who only a year earlier told the COP 23 Climate Change conference in Bonn, Germany, that Canada was a leader in reducing then eliminating coal for electricity generation and urged the rest of the world to follow Canada's lead. A *Toronto Star* article November 12, 2017, reported, "McKenna said Canada will highlight the 'concrete' action it has taken to reduce emissions, including Ottawa's plan to impose carbon pricing on provinces to reduce the greenhouse gas emissions that cause climate change to keep global warming under 2 degrees Celsius by 2100."[348] That would, of course, include New Brunswick where all 10 MPs are Liberals.

About 12 months later McKenna's press secretary told *cbc.ca*, "A price on pollution is one way to reduce pollution from the electricity sector – but not the only way. One of the most important measures to reduce pollution from electricity is our commitment to phase out traditional coal power by 2030, all while ensuring a just transition for coal workers and communities ..." The article reported the

347 Robert Jones, "NB Power to Dodge Major Carbon Taxes After Ottawa Proposes Looser Rules on Coal Plants," *CBC News* (October 30, 2018). https://www.cbc.ca/news/canada/new-brunswick/coal-fired-power-plants-carbon-tax-1.4882669

348 Alex Ballingall, "Environment Minister touts coal phase-out ahead of climate talks in Germany," *The Star* (November 12, 2017). https://www.thestar.com/news/canada/2017/11/12/canada-uk-team-up-at-climate-conference-in-push-to-eliminate-coal-power.html

about-face "… could benefit New Brunswick consumers by eliminating the need for power rate increases to pay for carbon taxes."[349]

On the other side of the discussion, critics of carbon taxes claim they are just a revenue grab. This is equally untrue. In the marketplace some carbon pricing signals, however distorted or clumsy, will eventually change buying and consumption patterns. It really depends on the amount. If gasoline prices went up another $1 a litre because of huge carbon taxes, the prevailing argument that the people who need fuel would buy it anyway is true. But many would drive smaller vehicles or EVs.

The tragedy of the Canadian carbon tax debate is that it confuses activity with success. "We're on it," claim the politicians. But until these actions result in material and measurable reductions in emissions, the battle is only being fought in the ballot box, not the atmosphere.

The polarization of politics goes beyond opposing views of political parties on whether or not to introduce carbon taxes. The near-universal demands that only governments can and must solve this problem ignores the reality that essential cooperation among governments is difficult if not impossible. If we can't even get governments within Canada to agree, how will this ever work globally?

Further there is a continuous effort by many in the climate change debate to marginalize the private sector; to claim the corporate world is the problem and not the solution. This sentiment excludes an immense pool of capital and ingenuity that could, if properly harnessed, contribute mightily to technology developments and advancements essential to find better ways to use carbon energy today while searching for the replacements of tomorrow.

If any conclusion can be drawn it is surely that governments cannot solve every problem in the world. They don't even know where to start. Political and policy tools that work locally will not work globally. And political success is certainly not the same thing as emission-reduction success.

People demand that their governments solve the problem and the same governments refuse to admit they can't. How is anybody going to fix this mess if we can't work from an agreed-upon set of facts and bridge the gap between polarized factions?

---

349 Robert Jones, "NB Power to Dodge Major Carbon Taxes After Ottawa Proposes Looser Rules on Coal Plants," *CBC News* (October 30, 2018). https://www.cbc.ca/news/canada/new-brunswick/coal-fired-power-plants-carbon-tax-1.4882669

8

IF YOU REALLY WANT TO SOLVE THE PROBLEM...

"What Alberta can contribute to the climate change challenge is to use our knowledge and expertise of energy and technology to create dozens, if not hundreds, of micro-solutions for emission reductions; practical ideas that, once commercialized, would have markets all over the world..."

"**O**ne hundred years from now people will be looking at this carbon warehouse in Alberta and will thank the world that we have it."[350] So spoke Alan Johnson in an interview for this book in the summer of 2018. Johnson is a former Calgary Progressive Conservative MP, coal executive and president of the Coal Association of Canada. A Calgarian, Johnson is the consummate Alberta carbon man, a living dinosaur by the definition of today's anti-fossil fuel movement. He has spent most of his life working with coal and will go the grave talking about all the incredible things that can be done with carbon.

His current project is *aerio*, an H Quest company. H Quest is based in Pittsburgh, Pennsylvania, an American coal and steel town if there ever was one. The website reads, "H Quest Vanguard, Inc. is an early-stage technology company based in Pittsburgh, PA. We are developing a transformational process to derive advanced carbon materials, including graphene and carbon fiber, from low-cost, abundant resources: natural gas and coal."

"Our proprietary low-temperature plasma conversion process can transform these materials in a fraction of a second. This process has no CO_2 emissions and can be deployed at orders of magnitude lower costs than conventional chemical plants employing legacy technologies."[351]

Johnson is partnered with Allen Wright, another long-time Calgarian, former oil company public relations professional and Coal Association president. His

---

350 David Yager, "Oilpatch Innovation Continues Despite Shortage of Capital," *EnergyNow* (November 19, 2018). https://energynow.ca/2018/11/oilpatch-innovation-continues-despite-shortage-of-capital-david-yager/.

351 *H-Quest Vanguard, Inc.* (2018). http://www.h-quest.com/.

company is called CarbonERA, because, as Wright explains, "We're entering into a new carbon era." Over lunch the pair tells your writer that carbon has a brilliant future as a building material; carbon fibre for airplanes, wind turbine vanes, and graphene, a carbon-derivative that is currently the strongest and lightest material in the world.

Johnson expands about where scientists believe the world could go with carbon. Graphene, a new compound made of pure carbon, is proving to be ten times (some say 200 times) stronger than steel and only 5% as dense. Scientific dreamers have long theorized about a "space elevator," a man-made structure that could convey objects beyond the atmosphere without rockets. Graphene may take this one step closer.

Graphene is amazing stuff. Website *graphine-info.com* states, "Graphene is the thinnest material known to man at one atom thick, and also incredibly strong - about 200 times stronger than steel. On top of that, graphene is an excellent conductor of heat and electricity and has interesting light absorption abilities. It is truly a material that could change the world, with unlimited potential for integration in almost any industry."[352]

As it becomes more common and industry figures out how to handle it, graphene will increasingly find its way into products such as jackets that are warmer, golf balls that fly farther, running shoe soles that last longer, sound-blocking materials for vehicle engines, tennis rackets that offer better ball control, smart phones, skis and fly fishing rods.

The pair is seeking early-stage financing of US$10 million to take the laboratory process to a "commercial demonstration plant" to prove their claims. They want to work where there is lots of cheap carbon, like Alberta. The dream is going into mass-scale production of pure carbon as an industrial input for a wide range of products. They are also interested in producing pure hydrogen to power cars or heat houses, noting that if you combust hydrogen and all you get is water. The promoters claim that the process does not use much energy. To prove it, they require what seems like a modest amount of money to save Alberta and the planet.

But raising risk capital in Calgary for anything but marijuana cultivation was all impossible in the fall of 2018. There was a time, not that long ago, when an

---

352 Metalgrass, "What is Graphene?" *Graphene-Info.com* (December 24, 2018). https://www.graphene-info.com/graphene-introduction.

entrepreneur with a good idea could pass a hat around at the right cocktail party in downtown Calgary and raise that kind of money in an hour.

Unfortunately, the private-sector risk capital so essential to fund the transformation of Alberta's economy from what it was to what it should be has been decimated. The combination of reduced commodity prices, mountains of new government regulations and taxes, as well as the lack of market access via pipelines has reduced the valuations of all but the largest companies in oil and gas. Some have restructured, sold, or merged. The rest went broke.

While oil prices and company valuations have recovered in the rest of the world, in Alberta they have not. This has cost the province and Albertans tens, if not hundreds, of billions of dollars.

The enemies of carbon—professional environmentalists, climate change alarmists, vote-seeking politicians, and a generation of people who have no idea how the world got this far before they showed up—have won the battle in Alberta.

But they've lost the war. If the climate change challenge is real—if the world must make a rapid transformation to new power sources free of carbon emissions—it cannot be done with slogans, fear, boycotts, lawsuits or the existing inventory of zero-carbon energy sources. We cannot depend upon miraculous climate-policy unanimity among the world's politically and ideologically diverse countries. And there is no point in perpetuating the lie that the world could go carbon-free tomorrow if Big Oil would just quit financing the climate change-denier movement.

The only solution is technology and innovation and unleashing the capital and creative genius of the private sector, something Alberta's carbon-resource industries have proven they have in abundance. Or did have.

A very informative article to frame this discussion came from Microsoft founder and billionaire turned philanthropist Bill Gates. Released on October 17, 2018 and titled "Climate Change and the 75% Problem," it was published on website *gatesnotes.com*, "The Blog of Bill Gates." Gates in recent years has become a crusader for multiple global causes, using his fortune and reputation to help wherever he can. One of the main projects of the Bill and Melinda Gates Foundation has been helping Africa develop and modernize.[353]

353 Bill Gates, "Climate Change and the 75% Problem," *gatesnotes: The Blog of Bill Gates* (October 17, 2018). https://www.gatesnotes.com/Energy/My-plan-for-fighting-climate-change.

Gates has recently embraced the climate change challenge as another major project and offers followers the opportunity to subscribe to his ideas under the tab, "Get Updates on The Efforts to Stop Climate Change."

To frame the challenge and the solutions, Gates had his researchers develop five questions that illustrate the enormity of the problem. He writes how he was surprised by the answers.[354]

> Question 1. "If cattle were a country, where would they rank on emissions?" The options are from 3rd to 44th. The correct answer is cows are the third-largest emitter.

> Question 2. "How many drivers would have to switch to electric cars to equal the emission saved by replacing one coal power plant with a nuclear plant?" The answers range from 250 to 2.5 million. The correct answer is 2.5 million.

> Question 3. "How has the share of energy produced by fossil fuels changed in the last 20 years"? The responses available are down 25%, down 50%, double, or zero. The answer is zero. Despite all the hype and promises surrounding renewables, they have yet to move the needle according to Gates.

> Question 4. "Which sector of the global economy produces as many GHG emissions as Germany?" Choose between making cars, shipping goods, powering computer data centres or forestry. The answer is heavy transportation which includes cargo planes, ships, trucks and trains.

> Question 5. "Which of these activities will reduce your personal GHG emissions the most over the course of a year?" The options offered are upgrading light bulbs, quit eating meat, avoid one roundtrip trans-Atlantic airplane flight, or replace your car with a hybrid. The answer is quit travelling long distances.[355]

---

354 Ibid.
355 Ibid.

What is fascinating about this quiz and the article accompanying it is they illustrate the enormity of the problem and the impracticality of the most popularly proposed solutions. Gates explains how most folks equate solar panels and wind turbines as the best way to reduce greenhouse gas emissions. Not that simple. The problem with electricity is in the absence of steady sunshine and wind, high-capacity battery storage is required. He writes, "... wind and solar need zero-carbon backup sources for windless days, long periods of cloudy weather, and nighttime."[356]

Gates writes that agriculture is the source of 24% of GHGs. As the quiz suggests, this includes cattle. But it also includes deforestation, which removes CO_2-reducing plants. Burning the cut-down trees to plant crops puts the CO_2 back in the atmosphere.

Another 21% of emissions comes from manufacturing, which includes plastic, steel and cements. Gates observes that cement takes a lot of fossil fuel energy to create and results in a chemical reaction that releases more CO_2.[357]

The next big block is transportation, but that's only 14%. Electric vehicles may help. But Gates observes there are no practical low-carbon options to move airplanes, ships and big trucks.[358]

Buildings contribute 6%. They require energy for heating, air conditioning, lights and all the appliances and other electrical devices inside. Insulation and energy-efficient windows will help, but urbanization will continue to add more buildings consuming more energy.[359]

The other 10% is falls under "miscellaneous" and includes the industries that produce fossil fuels. This validates the carbon industry's assertion that the greatest source of emissions is not producing energy but consuming energy.[360]

It also means that the climate change alarmist tactic of attacking oil and coal companies—while absolving the people who depend upon their products of any responsibility—may be compelling to attract attention, donors, and supporters, but utterly disingenuous in solving the problem. Almost nobody in the ENGO

---

356 Ibid.
357 Ibid.
358 Ibid.
359 Ibid.
360 Ibid.

movement talks about a world without airplanes, air conditioning, home heating, fresh fruit and vegetables in the winter, cement, steel, or plastic.

Bill Gates' entry into the climate change debate is encouraging. Not only does he reframe the challenge, he also reframes the solution. With a growing middle class wanting more energy, the world requires technology, not platitudes, fear mongering, and virtue signalling.

Gates explains, "… we need to invest in lots of research and development, across all five areas, now." He explains several initiatives underway, including his own Breakthrough Energy Ventures, which is investing over US$1 billion, "… into helping promising companies take great ideas from the lab to market to scale. We're using the five grand challenges I mentioned…as the framework for our investments. Every idea we're supporting is designed to solve one of them – and our mission is about to get a big boost from a new partnership in Europe."[361]

Where will the money come from?

Gates states, "We're creating a new way of putting that money to work. Because energy research can take years – even decades – to come to fruition, companies need patient investors who are willing to work with them for the long term. Governments could in theory provide that kind of investment, but in reality, they aren't great at identifying promising companies and staying nimble to help those companies grow."[362]

Gates sees the ultimate solution in the very private sector that so many climate change alarmists prefer to blame. In as polite a way as possible and on behalf of private sector investors like himself who need government support and funds, he has identified the problems without looking to government for all the solutions.

As importantly, and as Nobel Prize winner and carbon tax proponent William Nordhaus has pointed out many times, how do you get global cooperation among governments with such different agendas? How do you deal with the legacy issues surrounding the current distribution of the world's wealth? How do you effectively reduce global emissions without everyone participating? How do you persuade the diverse and competing interests of 7.4 billion people to agree on the same solution?

You don't. Bill Gates, a true globalist, writes, "The world's middle class has been growing at an unprecedented rate, and as you move up the income ladder,

---

361 Ibid.
362 Ibid.

your carbon footprint expands. Instead of walking everywhere, you can afford a bicycle (which doesn't use gas but is likely made with energy-intensive metal and gets to you via cargo ships and trucks that run on fossil fuels). Eventually if you get a motorbike you can travel farther from home to work a better job and afford to send your kids to school. Your family eats more eggs, meat and dairy, so they get better nutrition. You're in the market for a refrigerator, electric lights so your kinds can study at night, and a sturdy home built with metal and concrete."[363]

But Gates is also a humanist who doesn't believe people who live in the underdeveloped world should be denied progress. "All of that new consumption translates into tangible improvements in people's lives. It is good for the world overall – but it will be bad for the climate, unless we find ways to do it without adding more greenhouse gases in the atmosphere. This is undoubtedly a tough problem. It is not obvious what the big breakthroughs will look like. Most likely we will need several solutions to each challenge. Which is why we need to invest in lots of research and development right … now."[364]

A Canadian project in which Gates and Calgary oil tycoon N. Murray Edwards are co-investors is Carbon Engineering Ltd, a Squamish, BC-based company whose website claims is a "clean energy company leading the commercialization of groundbreaking technology that captures CO_2 directly from the atmosphere, and synthesizes it into clean, affordable transportation fuels." In operation since 2009, the company seeks to extract carbon dioxide from the atmosphere and turn it into fuel for vehicles and airplanes. The company hopes to offer "Direct Air Capture" extraction/conversion units that will one day remove 1 million tonnes of CO_2 from the atmosphere annually."[365]

Dave Collyer is a former Shell executive and CEO of CAPP from 2008 to 2014. He was in the middle of the debate between oil sands developers and the climate change movement. At the beginning of this period the oil sands were seen as an engineering miracle and the basis for Prime Minister Stephen Harper's boast of Canada as an "energy superpower." By the end of Collyer's tenure the oil sands were deemed the greatest menace to the world's climate in history. He was interviewed for this book in August 2018.

363 Ibid.
364 Ibid.
365 *Carbon Engineering* (n.d.). http://carbonengineering.com/.

Looking at where to go from here, Collyer says, "It starts with the global picture and how it positions Alberta and Canada. There is an incongruence between the Paris 2°C objective and the other view of the growing middle class. They are incompatible." He doesn't argue with the IPCC's findings but also has a big worldview, the kind you get when you work for Shell. He says without the introduction of disruptive technology that doesn't yet exist, "Adaptation will be a more important consideration." He acknowledges Canadian governments can't say no to signing international emission reduction agreements, but problems emerge when governments try to "overlay these policies on Canada."

Collyer figures Canada should be in the middle, where it has traditionally been on international matters. "Canada should be less dogmatic on targets but should be seen as doing reasonable things. I think we should be with the rest of the world on carbon pricing." But Canada should not position itself as an emissions-target leader.

He believes Canada's most constructive role would be as a "leader in technology," and in doing so "redefine what leadership means in the climate space." It should be a "portfolio" approach, a combination of technology, energy storage, energy efficiency, biofuels and alternative energy. However, Collyer concludes, "These are issues that politicians and political parties have a great deal of difficulty dealing with."

There are lots of good ideas. But many are big government ideas. They are not breakthrough technologies, and they are not necessarily exportable on a micro level to the underdeveloped world, the future carbon emitters Bill Gates understands so well.

Regina is the home of the International CCS Knowledge Centre, established in 2016 by global mining giant BHP and SaskPower. SaskPower was an early adopter of carbon capture and storage (CCS) at its Boundary Dam 3 coal-fired generating plant at Estevan in 2014. BHP is large global resource developer, including coal. Using the installed CCS recovery unit at Boundary as a working laboratory, the objective is to refine CCS technology. The current operation removes about 1 million tonnes per year from the atmosphere and sells it to a nearby oil producer for enhanced recovery. Its information material says this is equivalent to taking 250,000 cars off the road.

CCS Knowledge Centre figures they are ready to move to the next phase with a new a large CCS installation at the nearby Shand generating facility, which is three times as large. Mike Monea, President and CEO, says in a news release, "We are excited about it because many of the common hurdles for large-scale CCS are being addressed and results show that next generation CCS technology will be significantly cheaper, more efficient, and integrate well for renewable energy."[366]

Alberta went down the CCS path in 2007 with great fanfare and expense at the Quest pilot project north of Fort Saskatchewan. Operated by Shell Canada but built primarily with federal and provincial funds, Quest has been operating for three years. Over 3 million tonnes of CO2 are gone, injected into a subsurface reservoir with tremendous CO2 storage capacity. The CO2 is captured at the nearby Shell oil sands upgrader. What used to go into the air is now disposed of underground.

Another CCS initiative with significant government backing because of the expense is the Sturgeon Refinery/North West Upgrader, nearing completion in Redwater, Alberta. When completed it will cost $9.7 billion, nearly double original estimates. It is government supported through Bitumen Royalty In Kind (BRIK) oil sands barrels as feedstock and a commitment by the Alberta Petroleum Marketing Commission (APMC) to pay a set fee for turning BRIK barrels into market products like diesel fuel. The plan is for the market value of the finished products to be greater than the market value of a barrel of raw bitumen. If the spread is positive, it's a stroke of genius. If the spread is negative, the Alberta government—owner of APMC—could be on the hook for billions.

This was related to a grand plan in 2007 to capture CO2 from Sturgeon, ship it south via the Carbon Trunk Line, then use the gas to increase oil recovery from mature oilfields in central Alberta, the same model as Saskatchewan. After years of false starts, the project got off the ground in August of 2018 when it was announced a 240-kilometre pipeline would be built to Lacombe to ship CO2 from the Sturgeon Refinery and the Redwater fertilizer facility.

These are classic political-action projects that politicians love, and the public believes are helpful. But this type of capital-intensive solution overlooks dozens if not hundreds of smaller ideas which, if broadly adopted, could lead to meaningful long-term carbon reductions in global markets because of their broader application and lower cost.

---

366 "Large Scale CCS Ready for 2nd Generation," *Carbon Capture Journal* (September/October 2018). https://ccsknowledge.com/news/large-scale-ccs-ready-for-2nd-generation.

Horizon Oilfield Solutions Inc. was founded by Josh Curlett and his brothers. Interviewed for this book, Curlett said the company supplied two of their original technologies to drilling operations before the downturn. However, in 2015 they undertook what Curlett calls a "post-correction diversification strategy." They rebranded as CLEANTEK to focus less on the drill bit and more on investing in new technologies to diversify their revenue base. With government policy shifting towards minimizing industry's carbon footprint, CLEANTEK developed and launched new technology-based equipment to make clients more money and help reduce their environmental impact.

A main CLEANTEK product is remote location lighting systems. Using a combination of solar panels, batteries, high-output custom-designed LED lights, and computerized systems management, their products can reduce emissions by 80% from conventional and more popular diesel generator systems. The updated version results in 95% lower emissions than conventional systems using advanced batteries. The website says, "This innovative design was built for use anywhere on the planet."[367]

Working with recently-acquired Apollo Energy Services, CLEANTEK retooled an existing patented lease lighting system called "Halo" for drilling rigs. Because of its reduced weight, existing drilling rig masts could be retrofitted without upgrades to mast hoisting rams. Halo lights an entire worksite from a single location at the top of the mast plus cuts fuel consumption and maintenance.

Another CLEANTEK's invention turns wastewater into steam by dehydrating clean water from concentrated contaminants. Industrial wastewater—which is generated in billions of barrels in North America alone—has for years been trucked away for subsurface disposal. CLEANTEK separates the water from contaminants by vapourizing the water, thereby leaving a low volume of solid for disposal. Less trucking and handling plus fewer emissions.

One of the company's most promising technologies uses exhaust heat from a processing facility, such as a gas plant or drilling rig, as the sole source of energy. This reduces trucking required for water disposal by up to 90% in some cases, thus cutting carbon emissions from diesel fuel. Curlett estimates that the first gas-plant deployment will result in approximately 9,000 less truckloads of water in the initial 10-year operating span.

---

367 "SolarHybridLi™," *CleanTEK* (2018). https://www.cleantekinc.com/.

Curlett concluded, "Every technology we develop is a step change in its respective industry."

Another creative technology is aimed specifically at overcome pipeline opposition and oil-spill concerns. It's a means to solidify oil sands into puck-sized pieces and ship it in open rail hopper cars in the same way as gravel or elemental sulphur is currently carried. CN Rail has been investing in the *CanaPux* process for several years. It's a method of mixing oil sands with waste plastic into pieces about the same size as a bar of soap. Thanks to the plastic, the final product actually floats, making it easy to recover if spilled into water.

In late 2018 CN, along with a First Nations investment partner, Wapahki Energy Ltd., revealed a plan to build a $50-million pilot plant on the Heart Lake First Nation reserve in the of oil sands country in northeastern Alberta. The plant would convert up to 10,000 barrels of oil per day into CanaPux. A separate facility would process up to 300 tonnes a day of waste plastic to add to the oil sands.[368]

Hydraulic fracturing has revolutionized the oil and gas in Alberta and North America. It uses large volumes of water. Shipping the water to the wellsite generates emissions, then more emissions are generated in disposing of about half of the returned water. Trace Water Solutions founder Pat Carswell, who was interviewed for this book in summer 2018, explained he wants to clean and recycle water, not dispose of it. Trace's brochure reads, "Our revolutionary technology takes frac flow-back water and turns it into 100% fully-reusable, alternative water."

According to climate watchers, water is going to become precious. Since we're not going out of the oil business anytime soon, Carswell figures his company can help the industry do the old job better, one of many micro-steps to green what we've got. His brochure reads, "… 90% of water is not re-used during fracking, making it one of the most unsustainable processes in oil and gas."

Carswell's solution is not another regulation but innovation. But after seven years on this project, he's frustrated at the rate of adoption. Money is very tight, and oil company clients are afraid to try something new if they don't have to. The current solution is to do it the old way—the wrong way—at a reduced cost, not change gears and rethink the entire process. Carswell laments, "Industry wants to change without risk, they want to be the 3rd or 4th adopter, not the first."

---

368 Canadian Press, "Indigenous Energy Company Touts Job Creation in Joint Venture to Make Bitumen Pucks," *Global News* (January 17, 2019). https://globalnews.ca/news/4860615/bitumen-pucks-indigenous-energy-firm/.

As Alberta's economy adapts to new market realities, policy-makers must be cognizant of how legacy assets and processes can become obstacles to progress. Nobody knows this better than Advantage Cogen Energy Ltd., a Calgary-based company run by conventional oil and gas veteran Earl Hickok. His company is partnered with M-Trigen Inc. of Houston which specializes in "micro" natural gas solutions for heat, cooling and electricity. Hickok prefers to use the popular jargon "cogen": cogeneration of combined heat and power.

In an interview Hickok explained how he and his partners became attracted to finding more ways to use Alberta's bountiful supplies of cheap natural gas more effectively. Cogen has been around for years. It is common in the oil sands where process steam is generated using the waste heat from on-site natural gas electricity generators. Smaller-scale units have immediate applications in small commercial buildings, medical facilities, recreation centres, and even high-density residential projects.

The trade term is "distributed power": the so-called smart grids that allow people to generate their own electricity in the same way renewables like solar panels and wind turbines are used. All the power is generated on site as required. With cogen the heat is also harnessed. Where Advantage provides value over centrally generated electricity—coal, gas or nuclear—is in the re-capture and deployment of the waste heat conventional systems release through the exhaust. Further, transmission lines lose 7 to 8% of the power generated due to resistance in the electrical cables, something called "line loss."

Regarding Alberta, Hickok asks the question, "If we could wipe the board clean and start all over and had natural gas in all the places we need it, which we do, could you not generate electricity more efficiently and effectively? The answer is yes."

While gas and the generating technologies are cheap enough, there are government-imposed market forces that are impairing customer acceptance and growth. One is that Advantage and its clients must pay carbon taxes on natural gas while their competitors in the electricity and distribution business are, in many cases, exempt during this period of transition from coal. The other is that the province has put a cap on electricity prices.

That said, these challenges are temporary. While not as clean as renewables, cogen is a 24/7/365 power supply. They provide the backup renewables require when it is dark or calm. Further, every Alberta consumer's power bill contains expenses for transmission infrastructure whether or not the lines are in use, as is the case with wind power in southern Alberta. Advantage's solution makes a lot of sense for gas-rich province like Alberta.

Disrupting existing technology is a challenge. After 50 years of processing and production, the oil sands are legacy operations, defined as using recovery technologies that are proven and are coming down in cost. Why try something new?

Because regardless of the technical and commercial successes, the environmental challenges are legion. Therefore, the innovation process continues as people come up with new ways of extracting bitumen from reservoir materials and turning it into products and profits using less energy, less water and with a more benign environmental footprint.

Such is the case with *NUWORP*, an acronym for Novel Ultra-Low Water Oil-Sands Recovery Process. NUWORP is the patented brainchild of oil sands veteran Maurice Dusseault and partners Jesse Thé and Roydon Fraser. They have come up with a new way of replacing Karl Clark's legendary hot water extraction process that in only three years will date back a century.

NUWORP's presentation describes how the team worked backwards from the solution to devise a means to get there. The solution eliminates tailings ponds, minimizes heat loss, reduces the need to mix then separate ingredients (water in; water out), reduces alternate heat sources, gets rid of coke and elemental sulphur, returns clean sand back to the mine, involves oxygen-free pyrolysis and gasification, and adopts the Fischer-Tropsch process to create diesel fuel, gasoline, petrochemical feedstock and other fuels considerably cleaner than raw bitumen.

Unfortunately, that's the end of the NUWORP story at the present time. Like many new oil sands and hydrocarbon enhancement processes, it is only at the design and engineering stage. But the point is that the economic prize of figuring out a cleaner and more effective way to exploit the massive oil sands resource remains huge, and human ingenuity remains infinite.

As with Advantage's cogen solution, if one were starting fresh with oil sands exploitation and recovery today, there are a lot of different approaches than those created nearly 100 years ago using a far different science, technology, and oil sands experience knowledge base.

One of the main GHGs emitted through oil and gas industry operations is methane. A primary component of natural gas, methane is a valuable commercial product; unintended releases are referred to in the industry as "fugitive emissions."

Governments have vowed to legislate and enforce reductions in industrial methane emissions from various sources.

One of the main GHGs emitted through oil and gas industry operations is methane. A primary component of natural gas, methane is a valuable commercial product; unintended releases are referred to in the industry as "fugitive emissions." Governments have vowed to legislate and enforce reductions in industrial methane emissions from various sources.

In the oil and gas industry, a culprit in the emissions of methane and other GHGs is the hole in the ground created when wells are drilled. When a well has reached the end of its productive life, it is sealed with cement ("plugged and abandoned"). Unfortunately, it has been discovered that over time the cement seal can shrink and GHGs may leak to the surface. This is referred to in the industry as "surface casing vent flow." As more existing producing assets mature, developing new well-abandonment and plugging technology has been a labour of love for the industry during the prolonged downturn since 2014.

This is a global problem that could be solved by the Alberta carbon industry.

Dale Kunz of Winterhawk Well Abandonment Ltd. of Calgary has developed a novel suite of tools that can find and repair the source of surface casing vent flow. Kunz is a petroleum technologist with decades of experience in all aspects up of upstream oil and gas operations.

Winterhawk has several technologies. The Casing Reshaping Tool (CRT) swells the production casing to the point where it mechanically shuts off methane flows up the outside of the casing. This technology will cost a fraction of previous methods. Field tests have demonstrated the effectiveness of this application.

In cases where the methane flow is low volume, it is often difficult to pinpoint the source of the leak. In such instances Winterhawk's Mechanical Diagnostic Tool (MDT) can be employed. The tool is set in a shallow point in the wellbore then moved downward. At various intervals it elastically expands the casing while methane flow from surface is simultaneously monitored. If there is no change after a specific expansion, the leak is determined to be higher up in the wellbore. If expansion reduces the flow, it's evident the source is below. Once the location is determined, the CRT is called into action to seal the leak.

Change and technological adaptation can only occur if there is capital to fund it. Nobody knows this better than Jim Ross, a former investment banker who, with partner engineers Steve Price and Steve Kresnyak, established Expander

Energy Inc. For 15 years the company has been working to commercialize new hydrocarbon-processing technologies to create what their website calls, "the next generation of synthetic fuels. ... Synthetic fuels technology has become simpler and more cost effective and is poised to play an integral role in supplying alternative, environmentally friendly fuels for decades to come."[369]

But until recently, the only thing Expander expanded was its inventory of patents and list of potential financial backers that said no. But the company finally found enough money to build a demonstration plant under the name Rocky Mountain GTL Inc. near Carseland, east of Calgary. GTL stands for "gas-to-liquids." The plant will convert 5.5 MMcf/D feet per day of natural gas into 500 barrels of high-performance diesel fuel.

Ross is now Rocky Mountain GTL's CEO. In an interview in September 2018, he explained that in the life cycle of the natural gas from reservoir to diesel, there is a 15% reduction in GHG emissions from refined diesel fuel. The diesel fuel the company produces is chemically pure, with no sulphur and no aromatics. This means a fuel with improved combustion performance, long catalytic-converter life, and no sulphur dioxide or particulate emissions. Ross says once proven, the technology is scalable and, as importantly, exportable. After years of trying, Expander is also working on another project to convert biomass to diesel. The plant for this pilot project is located near Edmonton.

R&D (research and development) is most successful in a business environment where risk capital is plentiful and the rewards are great. That description applied to Alberta at one time, but not today. The prize for new and innovative ways to cut emissions and improve efficiency are high now that carbon is taxed. But the funds to pursue them are scarce.

But this is not the first time Alberta's great oil and gas gravy train has been off the tracks, when capital had little interest in investing in the province. The 1980s were similar in many ways. Back then the Getty administration had introduced the Junior Capital Pool Program, or JCP, and the Alberta Stock Savings Plan (ASSP). The latter offered a 30% provincial tax credit for investing in certain listed Alberta equity offerings.

The JCP was the 1980s downturn version of corporate crowdfunding, except investors got a piece of the action through equity ownership. It allowed

---

369 "About Expander Energy Inc.," *Expander Energy Inc.—Innovative Energy Solutions* (n.d.). http://www.expanderenergy.com/about.html.

entrepreneurs to raise some seed capital from friends and family—all equity, not debt—and get on with creating a new business. Some of the companies that were founded as JCPs had spectacular commercial success. While not all were successful, investors lost their own money, not the government's. All the government did was give investors a reason to put their capital at risk.

No macroeconomic analysis of Alberta's equity investment programs of the 1980s has been done. But if it were, it would undoubtedly prove the results were spectacular. Your writer was struggling to make a living and raise a family in the battered oilpatch of the 1980s. Through the combination of the JCP and ASSP, he, partners and investors went on to help create a new, exportable drilling technology that at one point had a market value of over a billion dollars. The Alberta treasury recovered its tax deduction for that company's seed capital many times over.

Alberta's NDP government introduced the Alberta Investor Tax Credit (AITC) in early 2017 to provide a 30% provincial tax credit for investors prepared to put capital into specific sectors such as IT, cleantech, health technologies, tourism, and digital technologies such as animation and video games. The government had to sort through the applications of various companies to determine if they fell into one of the acceptable categories. This resulted in initial delays but the program is working. Rocky Mountain's Jim Ross credits this program for helping raise the money for his company's GTL project. Dale Kunz of Winterhawk says his company also was also able to secure a much-needed capital infusion with the help of AITC. At the end of 2018 the province had approved about $28 million in tax credits, which helped unlock about $100 million in investment.

However, R&D remains the wrong place for government direct investment. Not only can bureaucrats often not pick winners, but government must be transparent in telling taxpayers where their money is going and forthcoming about when the investments have failed.

When faced with a dead-end investment, private investors can cut and run before they lose more money or accept the loss and use it as a tax write-off against investment gains in other areas. All venture-capital models show that of 10 investments, most fail, a few break-even, and one is a spectacular success, which covers the other nine.

Because governments aren't using their own money, they're inclined to prop up losers long after they should. Regrettable but anticipated losses in the normal course of R&D can become huge losses for political reasons. Alberta did this is a big way with failed attempts at industrial diversification in the 1980s.

Ensuring the private sector has enough disposable income to fund R&D on climate change solutions is not the direction Alberta is currently going. Since 2015, Alberta and Ottawa have both introduced policies that reduce disposable income and discretionary investment (including R&D) capital through higher taxes on almost everything. The governments then trickle taxpayer money back to the private sector on certain projects with investment decisions often made by people with no experience or knowledge in the field in which they are investing.

What Alberta needs is a reverse carbon tax: financial incentives for those with a stake in Alberta's future because: 1) they live here; 2) they are prepared to open their wallets; 3) they will provide financial support for somebody they know and trust that has new technology and vision. What Alberta can contribute to the climate change challenge is to use our knowledge and expertise of energy and technology to create dozens, if not hundreds, of micro-solutions for emission reductions; practical ideas that, once commercialized, would have markets all over the world.

This is what Canadian carbon-resource companies are good at. But they need carrots, not sticks, to create a new generation of coal, oil, and gas production that meets the world's needs with significantly reduced emissions until entirely new energy platforms are developed that meet the zero-carbon challenge.

There are four realities facing Alberta. Understanding and acting appropriately to all three will fundamentally shape our future.

The first is that our carbon-resource industries are still valuable and always will be. The world is not going out of the coal, oil, and natural gas business anytime soon, and neither is Alberta. Alberta is an ethical, honest, clean, safe, transparent, regulated, and advanced producer of coal, oil, and gas. We have nothing to apologize for and no one to answer to but ourselves.

The world's energy industry is not leaving or refusing to invest in Alberta because of our social or environmental record. Capital is avoiding Alberta because it is not welcome where it matters the most: the income statement.

The second is that current Canadian policies of appeasement of climate change activists have so far failed and will never succeed. Progressive politicians can state, publicly and ad nauseam, how much they care. They can confirm their deep concern with a patchwork of ill-conceived policies that other Canadians can hardly wait to eliminate. But such actions are not going to change anybody's mind.

The so-called "social license" is a fraud. Taxing, regulating and capping oil sands crude production will not miraculously persuade opponents to stop hating the product, particularly those opponents who have constructed an economic livelihood from crusading to stop fossil fuels by every means possible.

The third reality is that while Alberta must continue on the path of contributing towards reducing emissions to combat climate change, the province must do it differently. This is what ethical global citizens do. Canada's carbon-resource industries should not be penalized, They should be encouraged to find, foster, develop, and commercialize products, processes, and technologies that can make the production, transportation, and consumption of fossil fuels as efficient as possible and with increasingly lower emission levels.

Lastly, Albertans must accept then support the provincial government reviewing its size, structure and cost. The bureaucracy as it exists today was built on revenues from resource industries that are, to quote OPEC at the 2019 World Economic Forum in Davos, "under siege". While it is unlikely Alberta will be exiting the coal, oil and natural gas businesses anytime soon, economic reality dictates a thorough review of expectations and provincial spending so the economy doesn't go in one direction while the government goes in another, as has been the case since late 2014.

Albertan and Canadian entrepreneurs, scientists, and companies would enthusiastically support using our knowledge and ingenuity to help change the world in the right business environment. From the fundamental understanding that the private sector is better equipped to figure out what solutions are required and where, the focus switches to the business environment most likely to foster technology and innovation.

This environment is not the one we have.

Although NDP Premier Rachel Notley's overall fiscal regime has clearly not been supportive of the province's carbon-resource industries, she is correct when she states Alberta needs new pipelines to ensure the economy stays strong enough to fund an orderly transition. Selling oil cheap doesn't reduce emissions. It only impoverishes its owners and makes and denies funds required to do the job better or do something else entirely.

The climate change challenge is real. But if we really want to solve the problem, we'll all have to get involved in several ways.

First, acknowledge it is everybody's problem. That includes you. Get involved. Read. Research. Vote.

Second, concede real emission reductions cannot be achieved by virtue signalling. Understand the problem and issues enough to realize what works and what doesn't.

Third, recognize that the private sector and free markets are the solution, not the problem. None of the major technological advancements in our history came out of any government's Department of Saving the World, run by selfless scientific wizards under the wise direction of whoever you voted for in the last election.

Last, governments can and must help by directing and leading the charge. But they can accomplish far more by unleashing market forces and human ingenuity than by cooking up a mish-mash of wealth-destroying programs and subsidies accompanied by a seemingly limitless supply of moral admonishments and soothing words, or by delivering a false message that we don't have to worry because they've solved the problem.

Trust me. Do this and we'll make it.

Epilogue

"There are very few citizens of poorer countries leading the charge on climate change. They're too busy trying to eke out an existence. Fortunately, the wealthy will speak on their behalf. Because for them the necessities of life have been replaced by the anxieties of life."

Except for the Preface, I have avoided writing this book in the first person. What I prepared was a comprehensive set of facts and statistics from which readers could form their own views and opinions.

When I stared this book, I didn't know what the conclusion would be, only that the current state of the Canadian climate change debate, climate policy, and energy policy had deteriorated to the point of absurdity.

There were three major triggers that inspired me to write this. The first was when Prime Minister Justin Trudeau promised to ban oil tankers on BC's northern coast during the 2015 federal election campaign. This move effectively killed the Northern Gateway pipeline. As Alberta's economy stumbled again in 2018 because of the collapse of domestic oil prices, the federal government contracted amnesia about how that action might have in any way affected Alberta's current economic predicament or international investor confidence.

The second was the fierce to-the-point-of-irrational opposition to Energy East in Quebec, particularly by Denis Coderre, then the mayor of Montreal. That western Canadian oil could replace imports from questionable international producers and aid the province of New Brunswick became secondary to chasing votes from anti-oil populists in Quebec.

The third was BC NDP Premier John Horgan and his partner in power—at least at the time—BC Green Party Leader Andrew Weaver and their utter intransigence to the economic implications of blocking the Trans Mountain pipeline and using whatever powers the province could muster to ensure it was never built.

What any of these three, partisan, shameless, and economically damaging interventions into the economic future of Alberta might cost the province, its people, its industry, and the country was less important than rewarding voting supporters. The most disingenuous aspect is the opponents of oil and Alberta have wrapped themselves in the shroud of moral purity by claiming their actions will save the world from climate change. If this were true, I'd join the crusade, as would hundreds of thousands of my fellow Albertans.

All three have contributed to Alberta enduring the lowest market-set oil prices in the world in the fall of 2018, resulting in massive economic damage and disruption to industry, the province, and Albertans. Without achieving any reduction in global carbon emissions. When and how does this end?

The unprecedented mass protests by Alberta oil workers was promoted by the hypocrisy and unfairness of those who believe they are saving the world by inflicting economic misery on Alberta.

For decades we Canadians have been told our country is all about fairness. Income. Gender. Sexual orientation. Race. Employment. Opportunity. Economically persecuting Albertans because they produce and export oil flunks the fairness test, hands down.

Many Canadians believe they can change the future climate by opposing the oil sands and pipelines. They routinely advise Albertans that if we get with the program we can miraculously transition our economy from fossil fuels to renewables.

The hypocrisy comes from how few, if any, are willing to make any changes to their own lifestyles or energy-consumption patterns. Albertans and the province's economy are the sacrificial lambs on the altar of climate change mitigation. This is wrong and Alberta oil workers know it.

The ultimate indignity is that Alberta's critics are happy to buy their oil from any other country in the world—regardless of its human rights and environmental record—before they will park their car or forfeit their next airplane flights.

No matter how the future unfolds, what Alberta should not be is what it has become; the only major producer of still-important and still-essential carbon resources that is being forced to commit economic suicide with the assistance and support of many of fellow Canadians.

After 10 months of studying the politics of climate change, I can no longer be persuaded the crusade is exclusively populated by altruistic people with honorable intentions.

Because the deeper I dug and the more alarming rhetoric I discovered, the more I realized none of the people leading the debate from any side have put forward a workable plan to actually fix the problem in a way whereby the cure isn't worse than the disease. The only thing I found that made sense was an article by Bill Gates, which I quoted.

Let me define making sense. A program or policy to replace fossil fuels that reverses the remarkable and near-continuous progress of the human condition in the last 200 years doesn't make sense. Never has the world been better equipped to solve this problem than it is today. But the solution proposed by many climate change alarmists is to dismantle the only sector that might actually succeed: private industry.

Global warming came onto my radar screen over 30 years ago when I was the editor of an oilfield trade magazine. The linkage between fossil fuels, carbon dioxide, and the temperature was in the news. To find out what this meant, in 1988 I asked the Canadian Petroleum Association (now CAPP), the lobby group for the exploration and production companies, to explain it for me.

Technical Vice-President Hans Maciej talked about climate change in geological terms, saying, "We produce oil from the Devonian formation at Leduc and we ski on the Devonian at Lake Louise. Of course the climate is changing. It always has."

CEO Ian Smyth was more circumspect, clearly concerned about what the future would hold for member companies and the world. But in 1988 the primary concern of Alberta's oilpatch was economic survival following the oil price collapse only two years earlier. That Alberta would experience another oil boom was unknown, as was the fact that the oil sands and pipelines to tidewater would become battlegrounds for the future direction of the country and, in some quarters, the world.

There are repeated allegations today that Big Oil intentionally concealed from its customers the fact that it knew its products negatively impact the atmosphere and ultimately the climate. This is a staggeringly shallow accusation. The Rio Earth Summit to tackle the global warming problem was in 1992, and the Kyoto Accord to establish a price on carbon and global emissions trading market was negotiated in 1997. For those who struggle with math that is 27 and 22 years ago respectively.

If no one ever reads any history, then the climate change disaster industry can continue with its misleading narrative that Big Oil knew all along, tricked hapless consumers into buying its products, and therefore should be forced through litigation into paying for damages past, present, and future.

Imagine if, in 1988, global oil producers had collectively agreed with emerging climate concerns and announced to their customers they would be withdrawing products like gasoline, diesel, and jet fuel from the market. Sorry about that, but this is for your own good. What would the public and political outcry have been? Oil executives would have all been arrested for intentionally destroying the economy!

Shake your head, dear reader, and please think this through.

As noted in the Preface, the state of Alberta's glaciers concerns me. I have been thinking for years about how Albertans should have a plan for when the glaciers are gone. Maybe we can install cisterns in homes and buildings to capture the spring run-off as winter snow melts. Why just look at it? Why not save some of that water? We can do that. Climate change in Alberta has always been a problem, in my view.

But the warmer it gets, the sooner the day will come when the glaciers of the Rocky Mountains are gone. And the Andes. And the Alps. And the Himalayas. And all the people who depend on melting snow and ice for their water supply will be in trouble.

Alberta has been identified and vilified as the worst of the worst in the global climate change crisis. The oil sands—a technological marvel for the scientists and engineers who spent 300 years trying to turn the one of the world's largest oil deposits into money and fuel—has been branded public enemy number one in the battle to control global warming.

It's just oil. The idea that a coordinated attack on 3% of the world's oil production is going to change anything is so fundamentally disingenuous it defies logic. But if there is anything this book has hopefully demonstrated, it is that logic is in short supply when it comes to climate change.

Climate alarmists try to brand the problem as so awful and so urgent the impossible becomes possible and governments of the world will cooperate with coordinated policies and action. This will not work. It will never work. The sooner people quit preaching and start planning—deal with what can be done and quit telling everyone what should be done—the sooner we may actually confront then solve the problem.

Until I researched this book, I never really understood one of the most fascinating forces at play in this debate: the power of the internet. As more people (including me) are learning, the internet is not entirely benign. Through social media sites like Facebook, millions of people can now live in their own separate universes, reading, viewing, and believing only what they want and sharing information and experiences only with the people who think and act like they do. People can reinforce the views they already hold, and avoid, block, or ignore contrary points of view.

This is the only possible explanation I can derive for the repeated assertion that fossil fuels can be replaced immediately if only the awful companies that produce their products can be brought to heel. Or regional governments are justified in suing large oil companies for damages that haven't yet occurred, on behalf of people unwilling to make material changes to personal consumption or behaviour.

In my research I found personal responsibility for fossil fuel consumption was sorely lacking. I encountered little understanding that a rapid phase-out of fossil fuels without an economically viable replacement would have a huge negative impact on billions of people. When I started this book, I had no idea of the vastness of the chasm that exists between what people want the world to be and what the world actually is.

Whatever the science, or the allegations, or the cause, climate change is real if for no other reason than millions of people in Canada and hundreds of millions more around the world believe it is. Which is why the politics, governments, the industry, and people's idiosyncrasies have been discussed. The climate change issue, and possibly the direction of the whole world, is now increasingly and ultimately driven by what people want to hear or believe, or think they want to hear and believe. I'm not sure we can run the whole world only by reinforcing everyone's preconceived convictions.

I read all I could find about the climate change movement and the people behind it. It was disturbing at times. This nugget is worthy of repeating, an article from the July 16, 2018 edition of *The New York Times* titled, "Raising My Child in a Doomed World – Some would say the mistake was having our daughter in the first place." Author Roy Scranton is an English professor. He lamented how guilty he felt he had created human life in a world that was, in his view, doomed by the activities of his fellow 7.4 billion earthlings.

God help us all. This little girl needs protection from her father, not mankind. The poor child had not yet soiled her first diaper and her father had preordained her fate. This is the colossal arrogance of the "only I know best" attitude of so many nowadays. Scranton's view of the world is in the mirror: just himself. It's all about me. What world?

Another aspect of people when they're worried is that they gravitate towards comforting words. They want to believe somebody is doing something about what's troubling them and, by extension, things will be okay. Two politicians who gained lasting fame for their steady hand during times of crisis were Calgary mayor Naheed Nenshi during the devastating floods in southern Alberta in 2013 and New York mayor Rudy Giuliani after the 9/11 terrorist attacks in New York City in 2001 that killed thousands.

Both provided near-continuous public support via every manner of electronic and print media, saying something like, "We'll be okay. If you're seeing or reading this, you'll be safe." Not one single basement in Calgary was cleaned any sooner, nor was one grieving family member of the deceased in New York relieved of any misery.

But they provided what people wanted to hear. Whether or not they were good mayors up to that point, or in Nenshi's case whether or not he is still a good mayor as this book is written, is irrelevant. Comforting words. Wanting to hear it is human nature. Think of your mother.

Canada is surely the king of political comforting words. Climate change? We've got your back on this one, folks. The energy-efficient light bulb will arrive at your house shortly. Here's $14,000 to buy a new Tesla. Yes, we're going to levy carbon taxes to combat climate change and prevent terrible weather, but most of you won't pay it because of the generosity of our rebate programs.

The latest Canadian federal government message is not only are we solving the problem, but it won't cost you a dime. No guilt, no expense. Now that, ladies and gentlemen, is progressive twenty-first century government! We'll pay you to let us solve the problem so you can hop in the old motor home this summer and hit the open road, just like you've always done. Take a remorse-free vacation and fly somewhere warm next winter. If your conscience needs further assuaging, buy a carbon-offset credit for your flight.

Reducing carbon emissions to save the climate doesn't come from paying more for gasoline because of a carbon tax only to collect a rebate or buying an offset because you took an airplane flight. It comes from burning less gasoline and jet fuel.

The real solution to the climate change challenge lies with research, development, technology and the private sector. There is an enormous economic opportunity for the dreamers of the world today. The prize for inventing and commercializing a replacement for fossil fuels with a reliable, affordable, and benign energy source could be the greatest financial home run in history. The flaw is so many people remain convinced government knows best, and so many of the thought leaders on the subject either blame or don't trust the private sector.

If climate change is such a huge challenge, you'd think more people would gravitate towards solutions that work and be considerably less doctrinaire. But that's the problem. Climate change isn't just about the climate. It is also about human nature, and the ways by which some of our character flaws can be exploited.

To use a currently popular term, climate change is a first-world problem. Only the relatively wealthy (on a worldwide scale) have the time to dedicate to undermining the economic system and energy sources that have provided them their comfortable lifestyle. There are very few citizens of poorer countries leading the charge on climate change. They're too busy trying to eke out an existence. Fortunately, the wealthy will speak on their behalf. Because for them the necessities of life have been replaced by the anxieties of life.

Wow.

As Bill Gates so excellently articulated, solutions won't come from big taxes and central planning. They will come from patient and sophisticated private investors carefully using their own money with a long-term commitment to the project. They will all be private. If they think it will work private investors won't change course or withdraw funding after the next election as governments might do.

After all, how do you change the world with a political system where the government of the day only has a four-year shelf life? The priority of a sitting administration is to ensure its own re-election by catering to the wishes of the voters who put it in power, even if those people don't represent the majority of the citizens it governs.

I am not an anarchist. Government policy is essential. Controlling negative externalities like pollution, exhaust emissions and crime is what governments are for. That's what police departments do. Governments and their legal authority to penalize and punish are the only sure ways to force people and businesses not to do things in their own interest that negatively impact others.

But the only solutions that will work globally are the ones that can be transferred to and exploited by the developing world—the billions who want just a slice of the health and prosperity Canadians take for granted. Wind and solar power generated in Alberta isn't going to solve a pollution problem created in China. Nor will more electric cars or low-energy light bulbs in Toronto or Edmonton.

Quebec can't export its morally advanced green hydroelectric power to India to replace coal-fired electricity.

What conservative politicians must do—for the benefit of the country and the world—is admit climate change is a problem, not a socialist plot. Having conceded there is cause for concern, they must further concede the solution lies in government policy and provide the appropriate direction. People want to hear that. They want leadership. They want peace of mind. They also want to help. If they believe it will work, they will do their part. If conservative governments don't provide leadership voters will migrate to those who will.

The next step is for conservative politicians to reinforce they can accomplish more by doing less. Then carefully explain how. The government must remove itself from the business of picking winners and let the market decide.

This must also be accompanied by reducing the size and cost of government. As Albertans must surely be figuring out, it is unsustainable to simultaneously shrink the economy and expand the government. That's what will happen as Canada exits the carbon-resource business. It's a gargantuan economic hole to fill. Until now, all the solutions to the climate challenge have been higher costs and larger government—for your own good of course. But the so-called "populist" movements around the world, highlighted by higher carbon taxes and the "Yellow Vest" protest riots in France, show that ordinary people are at the financial breaking point.

On January 21, 2019 the insolvency and bankruptcy protection unit of national accounting, tax, and consulting firm MNP LLP revealed in its regular "MNP Consumer Debt Index" that since September of 2018 the number of Canadians who could become insolvent if their monthly bill increased by just $200 increased to 46%. MNP further stated that "number of Canadians who say they don't make enough to cover their bills and debt obligations increased … to 31%."[370]

More taxes, more programs, and more government may indeed help protect the climate. If you don't have a job, then you can't afford a car or take a vacation and must live in a smaller house. That will certainly reduce emissions. But if people can't make ends meet, which disaster they face first—financial crisis or awful weather—becomes irrelevant. Opposition to carbon taxes is not fomented only

---

370 Grant Bazian, "Canadians significantly more worried about debt, interest rates and personal finances compared to September," *MNP* (January 21, 2019). https://mnpdebt.ca/en/blog/canadians-significantly-more-worried-about-debt-interest-rates-and-personal-finances-compared-to-september.

by climate change deniers. It is increasingly supported by ordinary people simply trying to make ends meet.

The last step will be a negative carbon tax. Through the tax system, governments will replace sticks with carrots and provide incentives for people and companies prepared to invest in: lower-carbon fossil fuel production and extraction processes; zero-carbon energy; substitutes for steel and cement; less energy-intensive steel- and cement-manufacturing processes. The decisions on the best paths forward will be set by the market, not the government.

Alberta's carbon-resource industries can and should be rebranded from the world's enemy to its salvation. Alberta companies, educational institutions, entrepreneurs, inventors, and scientists must be challenged, encouraged, and supported to be recognized as world leaders in low-emission hydrocarbon exploration, production, transportation, and by-product refining and manufacturing processes.

This will result in hundreds, perhaps thousands, of tiny economic decisions that will collectively have an enormous impact on the atmosphere and the economy, not the photo ops and press releases elected politicians prefer as they use the public's money at least in part for their own benefit.

The government's obligation will also be to ensure there is enough liquidity in the system so that when entrepreneurs go looking for financial support, somebody besides the government actually ponies up some risk capital. This means governments must understand and accept how continued growth in taxes and spending reduces the amount of money available for other activities.

Large-emitter carbon taxes should be kept in place to continually remind industry what it is supposed to be doing. Maintain carbon taxes on the business supply chain so companies are cognizant that much of their carbon footprint is through industrial inputs like materials, transportation, and services. Consumer carbon taxes should stay in place so nobody forgets the task at hand.

However other taxes—like the GST, PST or HST, depending on the province—should be cut so carbon levies are truly revenue-neutral. Corporate taxes must be reduced to ensure businesses can deliver goods and services at the same price and therefore not place a further economic burden on customers and the economy. A company that produces and uses less carbon will be more profitable because the carbon taxes paid will be immediately offset by other taxes payable. Cut carbon, count the cash, and watch the market work miracles. If carbon taxes are offset by other tax cuts, behaviour can be changed and emissions can be reduced without inflicting financial penalties on the greater economy.

With other consumption taxes reduced, individuals who can figure out how to use less carbon will benefit economically without the need for government departments to manage and distribute the rebate cheques current carbon tax proponents are so proud of. And prevent politicians from earmarking precious tax dollars for feel-good, photo-op, climate-battling initiatives. Such boondoggles are extremely expensive, usually of little or no use in international markets, and will barely register in the reduction of global GHG emissions.

Imagine saving the world by shrinking the government, not expanding it. What a thought. Maybe I'm the dreamer. Will it work? It has never been tried, so it is impossible to say no.

An ideal embraced by conservatives is the power, freedom, and responsibility of the individual. This means the intellectually honest position is that climate change is your problem, not the government's.

Sorry folks, but the climate change solution is you. And me. Each and every one of us. In Alberta, in Canada, and around the world. This is pure conservative politics, if such a thing exists any longer. We should be aware of the challenge, be confronted with pricing decisions that change our behaviour, and have confidence our governments have properly directed the private sector.

The private sector has done this many times before. It was industry, not government, that created the food, medicines, shelter, and clothing that has enabled the world to support more many people in more safety, security, and comfort than ever before at a population level predicted to be impossible.

Individual and group resistance to the cost of compliance with the prevailing climate manta is well demonstrated. One need only look at the "Yellow Vest" protests in France over more increases in the price of fuel. So it's reasonable to assume the reaction to current strategies requiring bigger costs and bigger government will be negative. It is clearly much easier to tell people what to do than force them to do it.

As climate change moves from a public-policy and media debate to legislation and government decisions with real economic disruption and losers, it is no longer a test of political will. The shift from talking to doing is forcing a realistic assessment and review of whether this plan can or will ever achieve its stated objectives.

Ultimately, it is a moral decision. For each of us. Me? I've made mine. I am a passenger on the great train of life, not the engineer or even the conductor. Mankind is going to do what it does. But we humans have a proven instinct for self-preservation no one seems to believe or remember. Making sure what we do doesn't hurt others is the right thing. Which is why we should address climate change. I also remain a big fan of democracy, even with its warts. Like Winston Churchill said, "… it has been said that democracy is the worst form of Government except all those other forms that have been tried from time to time."[371]

Unlike so many, I have never felt qualified to decide what the maximum population of the world must be, when we have reached that level, and what the world should look like when we're done.

I also believe the billions in the world who don't have electricity should be able to get it if they want it. I believe more people should be able to move from poverty to the middle class even if they need carbon-based energy to do it. If that causes more global warming until we come up with workable and global solutions, I accept that as well. If the solution is adaptation because we can't stop global warming fast enough, then make sure the most vulnerable have the financial capacity and technological resources to adapt.

All of this will be much easier to accept if I knew we were collectively doing something that might actually solve the problem at some point in the future. If Alberta's carbon industries were encouraged, not vilified, I am confident they could do more to reduce emissions than all the tax/rebate/renewable-subsidy programs Canadian governments have concocted and will continue to concoct until we have an intelligent debate on the subject that leads to doing some else. Technology driven by free market forces is a powerful tool. Given the right incentives, change will happen faster and sooner than anyone can imagine today.

As I have written, I am not a climate change denier. The climate has always changed, and always will change. It has been proven over the centuries that the composition of the atmosphere can and will affect the weather. Controlling emissions make sense. It always has and will continue to do so.

But I am most certainly a denier that Canadian governments can and must save the world no matter what the cost to the economy, and their current policies will make a meaningful contribution. If I have accomplished anything in this book it

---

371 Winston Churchill, *PARLIAMENT BILL, HC Deb 11, 444* (November 1947): 207. https://api.parliament.uk/historic-hansard/commons/1947/nov/11/parliament-bill.

is hopefully that people understand how many ways the climate change issue has been distorted, hijacked and misdirected, and how little progress can and will be made until more people acknowledge the complexity of the challenge.

I have great confidence in the future of mankind and, with a few adjustments, my country and home province Alberta.

You should too.

References

Dabbs, Frank. 1993. *Ralph Klein, A Maverick Life*. Vancouver, BC: Greystone Books.

de Mille, George. 1969. *Oil In Canada West, The Early Years*. Calgary, AB: George de Mille, printed by Northwest Printing & Lithography Ltd.

Doern, G. Bruce, and Glen Toner. 1985. *The Politics of Energy*. Agincourt, ON: Methuen Publications.

Easterbrook, Gregg. 2003. *The Progress Paradox: How Life Gets Better While People Feel Worse*. New York, NY: Random House.

Ehrlich, Paul. 1968. *The Population Bomb—Population Control or Race To Oblivion*. New York, NY: Ballantine.

—. 1968. *The Population Bomb—Population Control or Race To Oblivion*. New York, NY: Ballantine.

Epstein, Alex. 2014. *The Moral Case for Fossil Fuels*. New York, NY: Penguin Group.

Ferguson, Barry Glen. 1985. *Athabasca Oil Sands: Northern Exploration 1875 – 1951*. Edmonton, AB: Alberta Culture/Canadian Plains Research Centre.

Ferguson, Niall. 2017. *The Square and the Tower—Networks and Power from the Freemasons to Facebook*. New York, NY: Penguin Random House.

Fitzgerald, J. Joseph. 1978. *Black Gold With Grit: The Alberta Oil Sands*. Sidney, BC: Gray's Publishing Ltd.

Foster, Peter. 1983. *Other People's Money: The Banks, the Government and Dome*. Toronto, ON: Collins Publishers.

—. 1979. *The Blue-Eyed-Sheiks: The Canadian Oil Establishment*. Toronto, ON: Collins Publishers.

—. 1982. *The Sorcerer's Apprentices—Canada's Super-Bureaucrats and the Energy Mess*. Don Mills, ON: Collins Publishers.

—. 2014. *Why We Bite the Invisible Hand*. Toronto, ON: Pleasaunce Press.

Gray, Earle. 1982. *Wildcatters—The Story of Pacific Petroleums and Westcoast Transmission*. Toronto, ON: McClelland & Stewart Ltd.

Hanson, Eric. 1958. *Dynamic Decade*. Toronto, ON: McClelland & Stewart Ltd.

Hofmeister, John. 2010. *Why We Hate The Oil Companies: Straight Talk From An Energy Insider*. New York, NY: Palgrave MacMillan.

Klassen, Henry C. 1999. *A Business History of Alberta*. Calgary, AB: University of Calgary Press.

Klein, Naomi. 2015. *This Changes Everything: Capitalism Vs. The Climate*. Toronto, ON: Vintage Canada.

Lomborg, Bjorn. 2007. *Cool It: The Skeptical Environmentalist's Guide to Global Warming*. New York, NY: Vintage Canada.

—. 2001. *The Skeptical Environmentalist—Measuring the Real State of the World*. Cambridge, UK: Press Syndicate of the University of Cambridge.

Lonn, George. 1968. *Dusters and Gushers*. Toronto, ON: Pitt. Pub. Co.

MacGregor, James G. 1981. *A History of Alberta*. Edmonton, AB: Hurtig Publishers Ltd.

Manning, Preston. 2002. *Think Big*. Toronto, ON: McClelland & Stewart Ltd.

Marsh, John. 2011. "Untitled essay about the Lougheed years."

McConaghy, Dennis. 2017. *Dysfunction: Canada after Keystone XL*. Toronto, ON: Dundern.

McInnis, Edgar. 1969. *Canada*. Toronto, ON: Holt, Rinehart & Winston of Canada, Limited.

McKibben, Bill. 2011. *eearth: Making a Life on a Tough New Planet*. Toronto, ON: Vintage Canada.

Morton, Desmond. 2001. *A Short History of Canada*. Toronto, ON: McClelland & Stewart Ltd.

Morton, Ted, and Meredith McDonald. 2015. *The Siren Song of Economic Diversification: Alberta's Legacy Loss*. Calgary, AB: University of Calgary.

Newman, Peter C. 1993. *Promise of the Pipeline.* Calgary, AB: TransCanada Corporation.

Nikiforuk, Andrew. 2009. *Tar Sands: Dirty Oil and the Future of Continent.* Vancouver, BC: Greystone Books.

Nikiforuk, Andrew, Sheila Pratt, and Don Wanagas. 1987. *Running On Empty: Alberta After The Boom.* Edmonton, AB: NeWest Publishers Ltd.

Pinker, Steven. 2018. *Enlightenment Now: The Case For Reason, Science, Humanism and Progress.* New York, NY: Penguin Random House.

Pratt, Larry. 1976. *The Tar Sands, Syncrude and the Politics of Oil.* Edmonton, AB: Hurtig Publishers Ltd.

Richards, John, and Pratt, Larry. 1979. *Prairie Capitalism.* Toronto, ON: McClelland and Steward Limited.

Richardson, Lee. 2012. "Lougheed: Building a Dynasty and Modern Alberta From the Ground Up." *Policy Options.* June 1. Accessed December 20, 2018. http://policyoptions.irpp.org/fr/magazines/the-best-premier-of-the-last-40-years/lougheed-building-a-dynasty-and-a-modern-alberta-from-the-ground-up/.

Rosling, Hans. 2018. *Factfulness: Ten Reasons We're Wrong About the World and Why Things Are Better Than You Think.* New York, NY: Flatiron Books.

Simon, Julian L. 1981. *The Ultimate Resource.* Princeton, NJ: Princeton University Press.

—. 1996. *The Ultimate Resource 2.* Princeton, NJ: Princeton University Press.

Yergin, Daniel. 1991. *The Prize.* New York, NY: Touchstone.

Index

Dictionary of Acronyms and Abbreviations

AER	Alberta Energy Regulator
AGM	Annual General Meeting
AGTL	Alberta Gas Trunk Line
AGT	Alberta Government Telephones
AIP	American Institute of Physics
AITC	Alberta investor Tax Credit
AOGCB	Alberta Oil and Gas Conservation Board
APGA	American Public Gas Association
APMC	Alberta Petroleum Marketing Commission
APNGCB	Alberta Petroleum Natural Gas Conservation Board
ASSP	Alberta Stock Savings Plan
BC	British Columbia, province of
bcf	billions of cubic feet
bcf/D	billions of cubic feet per day
b/d	barrels (of oil) per day
BNEF	Bloomberg New Energy Finance
b/o	barrels of oil
boe	barrels of oil equivalent
BRIK	Bitumen Royalty in Kind
BTU/BTUs	British Thermal Units
CAPP	Canadian Association of Petroleum Producers
CCS	carbon capture and storage
CDE	Canadian Development Expense
CEE	Canadian Exploration Expense
CERI	Canadian Energy Research Institute
CFC/CFCs	chlorofluorocarbon(s)
CIA	Central Intelligence Agency

CNR	Canadian National Railroad
CO2	carbon dioxide
COP	Conference of the Parties
CPC	Conservative Party of Canada
CPI	Consumer Price Index
CPP	Canada Pension Plan
CPR	Canadian Pacific Railroad
CWNG	Canadian Western Natural Gas
dilbit	diluted bitumen
E&P	exploration and production
EIA	Energy Information Administration (American)
ENGO	Environmental Non-Government Organization
EPA	Environmental Protection Agency (American)
ERCB	Energy Resources Conservation Board
ESG	environment, social and governance
EV	electric vehicle
FEA	Frontier Exploration Allowance
FIRA	Foreign Investment Review Agency
GCOS	Great Canadian Oil Sands
GDP	Gross Domestic Product
GHG/GHSs	greenhouse gas(es)
GSC	Geological Survey of Canada
GTL	gas to liquids
IBC	Insurance Bureau of Canada
IEA	International Energy Agency
IISD	International Institute for Sustainable Development
IMF	International Monetary Fund
IPCC	Intergovernmental Panel on Climate Change
IPL	Interprovincial Pipeline Company
IRAP	Industrial Research Assistance Program
JCP	Junior Capital Pool

KXL	Keystone XL pipeline
LNG	liquified natural gas
LSD	legal Subdivision
LTO	light tight oil
mcf	million cubic feet
MIT	Massachusetts Institute of Technology
MMBtu	million British Thermal Units
MMcf	million cubic feet
MMcf/D	millions of cubic feet per day
NAICS	North American Industry Classification System
NASA	North American Space Administration
NDP	New Democratic Party
NEB	National Energy Board
NEP	National Energy Program
NGGKT	Natural Gas and Gas Liquids Tax
NGL/NGLs	natural gas liquids
NGO	Non-Government Organization
NOP	National Oil Policy
NRC	Natural Resources Canada
NRF	New Royalty Framework
NWT	Northwest Territories
NYC	New York City
ODI	Overseas Development Institute (British)
OECD	Organization for Economic Cooperation a Development
OFS	oilfield services
OMERS	Ontario Municipal Employees Retirement System
OPEC	Organization of Petroleum Exporting Countries
PC	Progressive Conservative Party
PgC	1 trillion tonnes of carbon
PGRT	Petroleum and Gas Revenue Tax
PIP	Petroleum Incentive Program

PPM	parts per million
PSAC	Petroleum Services Association of Canada
R&D	research and development
SAGD	steam assisted gravity drainage
SEC	Securities and Exchange Commission
Socred	Social Credit Party
tcf	trillion cubic feet
TCFD	Task Force on Climate-Related Disclosures
TCPL	TransCanada Pipe Lines Inc.
TQM	Trans Quebec and Maritimes Pipeline
UK	United Kingdom
UN	United Nations
UNFCCC	United Nations Framework Convention on Climate Change
US	United States, United States of America
USGC	United States Gulf (of Mexico) coast
WCEL	West Coast Environmental Law
WCS	Western Canada Select benchmark crude oil price
WCSB	Western Canadian Sedimentary Basin
WTI	West Texas Intermediate benchmark crude oil price
WWF	World Wildlife Fund

Acknowledgments

"Banking on Climate Change: Fossil Fuel Finance Report Card 2018" by Lorne Stockman, Rainforest Action Network, BankTrack, Indigenous Environmental Network, Sierra Club, Oil Change International, and Honor the Earth (March 28, 2018). https://www.ran.org/bankingonclimatechange2018/

Black Gold With Grit – The Alberta Oil Sands by J. Joseph Fitzgerald (1978).

The Blue-Eyed Sheiks – The Canadian Oil Establishment by Peter Foster (1979).

Dysfunction: Canada after Keystone XL by Dennis McConaghy (2017).

Eaarth: Making a Life on a Tough New Planet by Bill McKibben (2011).

Excerpts from ENLIGHTENMENT NOW: THE CASE FOR REASON, SCIENCE, HUMANISM, AND

PROGRESS by Steven Pinker, copyright © 2018 by Steven Pinker. Used by permission of Viking

Books, an imprint of Penguin Publishing Group, a division of Penguin Random House LLC. All

rights reserved.

"A Global Review of Insurance Industry Responses to Climate Change" from *The Geneva Papers on Risk and Insurance – Issues and Practice* by Evan Mills (July 2009).

"Heated political rhetoric on carbon tax does disservice to Canadian voters" from *National Post* by John Ivison. Material republished with the express permission of: National Post, a division of Postmedia Network Inc.

"Jack Mintz: End the denial and admit it: Canada has a competitiveness problem," from *Financial Post* by Jack Mintz (May 9, 2018).

"Jack Mintz: Carbon Pricing has been Fully Exposed as Just Another Tax Grab," from *Financial Post* by Jack Mintz (April 10, 2018).

"Jack Mintz: The Liberals ignored my carbon tax plan. Theirs is much worse," from *Financial Post* by Jack Mintz (October 25, 2018).

Every reasonable effort has been made to contact copyright holders; any errors or omissions in the above list are entirely unintentional. If notified, the author will be pleased to make any necessary corrections at the earliest opportunity.

Printed in Canada